Also in the Variorum Collected Studies Series:

NATHAN SIVIN
Medicine, Philosophy and Religion and Ancient China: Researches and Reflections

JOHN HEDLEY BROOKE
Thinking about Matter: Studies in the History of Chemical Philosophy

MAURICE CROSLAND
Studies in the Culture of Science in France and Britain since the Enlightenment

ROBERT FOX
Science, Industry and the Social Order in Post-Revolutionary France

ROY M. MacLEOD
Public Science and Public Policy in Victorian Science

A. RUPERT HALL
Science and Society: Historical Essays on the Relations of Science, Technology and Medicine

TREVOR H. LEVERE
Chemists and Chemistry in Nature and Society, 1770–1878

A.I. SABRA
Optics, Astronomy and Logic: Studies in Arabic Science and Philosophy

JULIO SAMSÓ
Islamic Astronomy and Medieval Spain

SILVIO A. BEDINI
Science and Instruments in 17th-Century Italy

DAVID A. KING
Astronomy in the Service of Islam

ALLEN G. DEBUS
Chemistry, Alchemy and the New Philosophy, 1550–1700

R.W. HOME
Electricity and Experimental Physics in 18th-Century Europe

WALTER PAGEL
From Paracelsus to Van Helmont: Studies in Renaissance Medicine and Science

VARIORUM COLLECTED STUDIES SERIES

Science in Ancient China

Nathan Sivin

Science in Ancient China

Researches and reflections

Ashgate
VARIORUM

Aldershot · Burlington USA · Singapore · Sydney

Published in the Variorum Collected Studies Series by

Ashgate Publishing Limited
Gower House, Croft Road,
Aldershot, Hampshire GU11 3HR
Great Britain

Ashgate Publishing Company
131 Main Street
Burlington, Vermont 05401–5600
USA

Ashgate website: http://www.ashgate.com

Reprinted 2000

ISBN 0-86078-492-4

British Library Cataloguing-in-Publication Data
Sivin, Nathan
Science in Ancient China: Researches and reflections
(Variorum Collected Studies Series; CS 506)
I. Title II. Series
509.31

US Library of Congress Cataloging-in-Publication Data
Sivin, Nathan
Science in Ancient China: Researches and Reflections / Nathan Sivin.
p. cm. - (Variorum Collected Studies Series: CS506)
Includes index. (cloth; alk. paper)
ISBN 0-86078-492-4
1. Science–China–History. 2. Science, Ancient. I. Title. II. Series: Collected Studies: CS506.
Q127.C5S58 1995 95-30588
509.51–dc20 CIP

The paper used in this publication meets the minimum requirements of the American National Standard for Information Sciences – Permanence of Paper for Printed Library Materials, ANSI Z39.48–1984. ∞ ™

Printed by Biddles Limited, Guildford and King's Lynn

VARIORUM COLLECTED STUDIES SERIES CS506

CONTENTS

This book contains xii + 314 pages

INTRODUCTION

This volume and its companion, *Medicine, Philosophy and Religion in Ancient China*, contain a selection of essays published from 1966 (the year I received my doctorate) to the present. My main response to being reminded of that long expanse of time is surprise. It has passed like a moment, with no sense of changing preoccupations, of staleness, of old issues done with. This is, I suppose, a variant on the old but true idea that, no matter what the course is called, every teacher has one thing to teach, namely his own curiosity and what it has turned up.

What motivated all these studies is precisely that: an intense curiosity about how people in a civilization very different from that of Europe have gone about understanding Nature (whether external or represented in their own bodies) and defining their relation to it. The study of East Asian science has played more than a small role in the growing awareness that Europe only realized a few of the possibilities of science and technology. Studying other cultures historically is at least as serviceable a way as philosophic speculation to learn about ways of thinking that Western traditions never evolved.

If there is anything unconventional about my work, it is due to a dilettante's equal familiarity with the advancing frontiers of the history of science and medicine on one hand and Chinese studies on the other. Most people on one side think of the other as it was a generation or two ago. Fence-straddlers are no more objective than anyone else, but at least we are seldom tempted to treat the conventional wisdom on either side reverently. That is especially true of these two disciplines, because their styles are so different.

The main difference between technological history in 1966 and 1995 is the realization that Nature and the body are not objective entities, but were differently imagined, invented and reinvented in diverse cultures. These years have spanned two major fads—the study of ideas while ignoring their settings, and the study of social circumstances while ignoring what scientists thought about—as well as several minor ones. All of them have left me uneasy, but they have frequently yielded useful tools and insights.

It has long since ceased to be news that the shared activity of human beings, not some force inherent in the character of physical reality, evolves science. When I hear a colleague in the course of a lecture breaking the news about its "social construction" for the thousandth time, I feel irresistibly impelled to leave the room. The point is well taken (at least when taken sparingly), but the usual implication that one need not know how scientists individually act, reason, and imagine robs it of most of its value. I realize that either/or stances are essential to the success of fads, but it also makes them ephemeral. Dealing with the complexity of events in the real world calls for all the subtlety and interconnection that we can grasp. That complexity is precisely what makes the practice of history zestful.

Classical sinology, still obsessed with texts and still tending to treat them as sacred objects, has barely noticed the fruitfulness of new methodologies, and has rolled unseeing like a juggernaut over wave after wave of intrepid social constructionists, literary deconstructionists, discourse analysts, and other representatives of post-modernism. That sublime oblivion has left me equally uneasy. Nevertheless, scholarship is very much a common enterprise, and new light often shines from old lamps. I greatly admire, and eagerly draw on, the quite fresh understanding of early Chinese thought that a number of young scholars have created since ca. 1990 using philological methods that have not changed a great deal in the past thousand years. I have probably profited as much from those rolling along on various bandwagons as from those who are not aware that they exist. I have set down some reflections on such disparate styles of scholarship in Chapter VIII.

As in most research that aims at more than cleaning up details of what is already understood, the key is finding the questions that are likely to be fruitful. No system, no readymade method, generates such questions. They are as likely to jump out of a chat with someone as puzzled as myself, or to simply pop up in the shower, as to crystallize from rumination.

Led by serendipitous questions, I have immersed myself at one time or another in all of the main dynasties and most of the traditional sciences. A few of the essays gathered in this first volume, *Science in Ancient China*, are records of such focussed explorations (Chapters II, IV, and V), and they are the majority in its companion, *Medicine, Philosophy and Religion in Ancient China*. Other essays in the present volume set down occasional attempts to step back and see what my investigations have revealed about the larger objects of curiosity (in particular, Chapters I, III, VII, and VIII). A couple orient readers who are encountering the Chinese technical traditions for the first time (Chapters VI and IX).

Do not approach these chapters as attempts at definitive statements. Laymen may speak of this or that book as definitive, but scholars know that the best we can do is strive tentatively to understand. My own strivings, because I have chosen breadth at the inevitable expense of depth, are particularly tentative. A great cosmologist once said to me that he would know his field was seriously deficient in vitality if his own work were not obsolescent within five years of the time it was published. I would be delighted if the velocity of the fields that interest me were as great.

It will be obvious to any reader that my quest for Chinese visions of Nature has led me consistently to the boundaries of science, or rather the boundaries between the rational, abstract, taxonomic, and sometimes quantitative modes of the Chinese sciences, and other kinds of knowledge. Although those borders are clear-cut in modern China (which acknowledges no contact between science and "superstition"), in early China, as in medieval Europe and recent Islam, knowledge was meant to be unitary. Intellectuals defined categories to facilitate the flow of thought, not merely to keep it from being contaminated. This continuity across the spectrum of experience was crucial in making sense of the polymathic career of Shen Kua (1031-1095), which encompassed almost every conceivable challenge to historical understanding (Chapter III). I later discovered that a strong sense of the limitations of empirical knowledge, and the complementarity of other kinds of knowledge, were as engrained in mathematical astronomy as they were in the qualitative sciences (see Chapter IV of the second volume). This relationship renders widespread platitudes about the ties between science and institutionalized religions (especially Taoism) highly problematic. I groused at such clichés in a 1978 essay, and I hope to have set them to rest in an essay published for the first time in *Medicine, Philosophy and Religion in Ancient China.*

In the course of this exploration, it has become ever clearer that in trying to make sense of ancient science in any culture, the old model of the march of progress is a potent distraction. Shen Kua was not a modern scientist who just happened to be born at the wrong time; but then neither was Nicole Oresme or John Dee. None of them was trying to do modern science. Where positivism still reigns, who the precursors of modern knowledge were and how early they discovered this or that are still the central questions. The gradual realization that Chinese science was not stumbling toward today's knowledge has led to vast confusion on the parts of Sinologists unaware of the much broader issues that have engaged technological historians for the past generation. It is not rare to find, as a reaction, various "proofs" that Chinese were not "really" doing science, were not capable of "real" theory or "real" induction, or only imposed irrelevant abstractions on top of technological problem-solving. Such proofs require a certain innocence of what ancient treatises on quadratics, the equatorial-ecliptic transform, the fine working of human metabolism, and so on, say. In every instance, "real" turns out to be defined by today's European standards. Scholars of medieval Europe are not so dismissive about the mathematical ventures of the theologians they study.

The distraction has turned out to be most stultifying in connection with an odd but fashionable historical exercise. Cocktail-party chit-chat aside, people who take history seriously normally become extremely evasive when asked why something didn't happen in a certain time or place. They answer that the only way to explain why event A didn't happen at time B is to give a historical account of why C happened at D instead.

By far the most frequent exception to this evasiveness is the question of why, since the Chinese have always been so clever, they did not invent modern science before Europeans did. For well over a century scholars, rather than answering that the real issue is why Westerners invented it before Chinese did, have hazarded an assortment of guesses that, as they accumulate, have left no one satisfied. Nevertheless they continue to accumulate. Adding hopeful conjectures to the pile remains a fashionable exercise to this very day in China. This is such an unlikely phenomenon that, in the 1981 lecture that appears here as Chapter VII, I tried to shift the focus to a more interesting issue, namely why, despite no progress generation after generation in answering it, despite the crude positivism of the underlying assumptions, and despite the obvious fallacies in the multitudinous answers, scholars continue to pose the "Scientific Revolution Problem." The essay has not succeeded in ending such distractions. The fact that no one has proved it wrong-headed is no consolation. But the questions it raises are now discussed more often than they used to be. That is not bad.

This opportunity to revise a few of my old writings has let me incorporate more recent insights and to state the old ones more clearly. I have also corrected many small errors, and one large one (see the Retrospect of Chapter IV). Most unsettling, this retrospect has shown me how badly I wrote in the late 1960's. After great efforts to convince my students that the fogginess and verbosity of what passes for academic writing have nothing to recommend them, I find too much fustiness in my own. I am glad to have had the chance to crop some of it out.

When citing Chinese sources, I prefer to give Chinese characters in the same form as in the original publication data, whether traditional or simplified. *Meng ch'i pi t'an* 夢溪筆談 refers to the original Brush Talks from Dream Brook, completed before 1095. *Yuan k'an Meng ch'i pi t'an* 元刊梦溪笔谈, although a photographic reproduction of the oldest extant edition (1305), was first published under that title in 1975, and gives its publication information in simplified characters. When I refer to the period between 1635 and 1644, I use "1635–1644"; "1635/1644" refers to some unknown moment within that interval.

Since several of these essays have carried their own dedications, it would be excessive to give this volume a formal dedication. But the associations of each chapter are too rich to pass over without acknowledgement. Consider this book and its companion an offering to my teachers, colleagues, and students, past, present, and future.

N. S.
Chestnut Hill
3 March 1995

***PUBLISHER'S NOTE**: The articles in this volume, as in all others in the Collected Studies Series, have not been given a new, continuous pagination. In order to avoid confusion, and to facilitate their use where these same studies have been referred to elsewhere, the original pagination has been maintained wherever possible.*

Each article has been given a Roman number in order of appearance, as listed in the Contents. This number is repeated on each page and quoted in the index entries.

ACKNOWLEDGEMENTS

I am grateful to the following publishers and organizations for permission to publish original or revised versions of papers included in this volume: E. J. Brill (for Chapter II); the American Council of Learned Societies (III, V); the Polish Academy of Sciences (IV); University of California Press (VI); the MIT Press (IX).

In addition to my thanks for contributions to the individual chapters noted in them, I would like to thank the Diet Library, Tokyo, and the Harvard-Yenching Library for permission to reproduce illustrations from books in their collections. The studies on which these essays are based have been supported by the National Science Foundation, the John Simon Guggenheim Foundation, the Japan Foundation for the Promotion of Science, the Department of Humanities, MIT, and the School of Arts and Sciences at the University of Pennsylvania. The School's Educational Technology Services solved a number of difficult problems involved in scanning the illustrations in Chapters III and V. I reiterate my gratitude to the Institute for Research in the Humanities, Kyoto University, the Chinese Academy of Sciences, the China Academy for Traditional Chinese Medicine (both of Beijing), the Needham Research Institute, and St. John's College, Cambridge, for hospitality over many years. Among my most concrete debts are those to Mary Beyer, who has served resourcefully as my secretary for many years, and to my wife Carole, who is always broadening my horizons.

I

Chinese Conceptions of Time*

LET ME BEGIN by not feeling free to ignore the question "Why should the way the ancient Chinese conceived time be worth reading about?" I cannot claim to come bearing occult secrets of the mysterious East. I suspect that the greatest mystery of Cathay is that there is so much less mystery than we like to imagine, and that in fact what so outrages our sense of reality about Chinese thought is that it can be so remarkably matter-of-fact. Since the Scientific Revolution—or since the Renaissance, or since the Middle Ages, depending upon whose view you favor—our perceptions have moved so far away from the ordinary experience of mankind that we find it very difficult to recognize common sense when we hear about it. The Chinese conception of time is based on the cycle of natural change through the seasons, and on the regular motions of the celestial bodies, both of which keep going on without the world's changing to speak of. It is based, in a word, on the time-bound change that most of mankind sees and feels. Our own everyday time metric is largely derived from the gradual unwinding of coiled springs in old-fashioned timepieces, and the rapid reversals of polarity of a current in electric clocks. Our overall frame of time is long and narrow and pointed upward, rising from the squalor of primitive savagery to the blessed state in which every family in the world will have two television sets and will replace them both every year. Since the perfectly simple and straightforward are so far removed from us, it is only natural to consider them exotic.

If we look at it from that point of view, this exploration of another culture's ideas can perhaps provide a somewhat longer perspective from which to view several of our assumptions about

---

*This article is respectfully and affectionately dedicated to Joseph Needham, F.R.S., Master of Gonville and Caius College, Cambridge University, who is making the Chinese scientific tradition accessible to the West, as a felicitation on his sixty-fifth birthday.

ourselves, and to see more clearly what makes us often feel uneasy about them. I do not mean to announce as a new discovery the idea that modern science is simply not commonsensical, and that there is no point in wishing, as those of us who are not scientists or philosophers often do, that it made more comfortable sense in everyday language. But because we've grown up with the fruits of modern science all around us, we have learned to consider obvious many things that would have been—or, as a matter of record, were—rejected as utterly ridiculous in the ancient world. I am thinking, for instance, of how little disturbed we are by the fact that our surest and most powerful knowledge of the physical world comes from applying mathematics, a set of ideal relations, to the phenomena on the earth around us, which aren't ideal at all. Aristotle, who has never been rivalled as a master of horse sense, was perfectly convinced that there could be no mathematical physics. How could you talk about a cartwheel resting on the ground as though it were a circle on its tangent, he thought, because circles are perfect figures which touch their tangents at only one point, but no cartwheel anywhere is a perfect enough circle to touch the earth only in a single point. But *we* don't mind a bit, because we have learned what abstraction can do.

Perhaps the best remedy for this ignorance about our loss of innocence—I suppose that the process of maturing always involves a loss of innocence—is to take a fairly close look at how nature, or time, or space, was viewed back in the days when common sense counted for something. That was when the earth was at the center of the universe, and the celestial bodies moved symmetrically about it for our convenience and edification, and because we played our part in the universal harmony by living decently in society and by performing certain ritual acts from time to time. Of course, now that we have given up on symmetry after another, and settled for being residents of a middling planet of a middling sun located about two-thirds of the way out in an undistinguished galaxy, the whole issue of common sense is philosophically much less compelling, and we become more hardened to living with some of the paradoxical notions about physical reality that our modern theories entail.

We are used to thinking of the Scientific Revolution as a more or less well-defined series of events which began somewhat over four hundred years ago. It is obviously not over, but at least we are learning to live with it, and to think in its terms, and to make decisions which take its consequences into account. But in Japan the Scientific Revolution began only a century ago; in China it began only about fifty years ago. All I mean by this is that science began to figure in the thinking of a few people who were making decisions which affected national development. It would not be too much to say that the minds of the majority of the population of Asia remain almost completely untouched by the fundamental new scientific ideas of the last half-millennium. I do not mean to deny that their lives have been altered more or less by the material impact of modern technology, which is another matter. I just want to point out that a great many Eastern thinkers still hold very different ideas from ours, about man and his place in the universe and what the universe is and how it works. This is not an occasion for exasperation or a feeling of superiority on our parts, but simply one more reason to stop basing our decisions and policies on the conviction that all decent people see the world pretty much as we do, and on the corollary that, so long as our intentions are good, anyone who doesn't want what we want is wicked.

Now let us look directly at this Chinese philosophy of time. It is dead now, in the sense that it is too late for it to develop any further as philosophy, but it is still alive in people's minds. While now it has to compete with, or be absorbed into, the Chinese brand of Marxism if it is to survive, there was a time when most people in China thought of time in very much the same way—although some, who were preoccupied with the political and moral problems of building a stable society, thought about it much less than others.

Perhaps the greatest contrast with our view lies in the fact that, like most of the ancient world, ancient China very largely lacked a conception of cosmic progress, at least in the early stages.[1] There was no feeling (outside of myths) that the cosmos had evolved from some primitive beginning, and was headed in some definite direction —that it had been worse, or that it would become better. Until Buddhism arrived, with very different conceptions, people did not believe that the material world had been created, nor that it had been designed or controlled. No one could put a definite limit on the past, just as there was no reason to limit the world's extension into the future. It is not surprising that so much of Chinese argument proceeds by finding analogies in the past.

Once the Chinese began by thinking of nature as steady and constant, they had to do something about the inconvenient fact that every individual thing in nature is always changing, however quickly or slowly. Their approach to science began to crystallize at the point when they became conscious that this change was not random at all, but that it was rhythmic. Every organism passes through a cycle of birth, growth, maturity, decay, and death, and it does so at its own speed, in its own good time. Chinese thinkers called this aspect of the being of a thing its *tao*, a word which literally means "Way."[2] They were aware that each individual's *tao* was unique and non-interchangeable, and that any attempt to rush it or slow it down was bound to be futile. They embodied this insight in parables like the one about the farmer who was in a hurry to see his crops come up, so every day he went out and gave each sprout a good hard tug. By the time they all died, he had come a long way toward an appreciation of the *tao* of plants.

The next step was to realize that these impermanent and individual life rhythms were tied together by, and in fact depended upon, certain basic overall rhythms which repeated themselves again and again. All of the smaller cycles—of light and darkness, of warmth and cold, of wet and dry, of activity and rest—seemed to resolve themselves in the alteration of the seasons, the great rhythm which also controlled all of the time-bound human activities of agriculture. The flux and eternal change which surrounded man began to take on order when it was seen as a fluctuation back and forth between two seasonal poles. The cycle of light and dark and back to light in the course of twenty-four hours was, in its outline, parallel to the cycle of hot and cold and back to hot in the course of the year. We can think of it as a movement between polar opposites as the Chinese did, and make a general equation by calling the opposites *yin* and *yang*, feminine and masculine, dark and light, passive and active, and so on. Or, what would doubtless be more comfortable to us, we can think of it as maximal and minimal states of a variable (day length, temperature, etc.) and graph it as a cosine curve.

Take the daily cycle of light. It is not a very profound observation that noon, the point when light is brightest and dark is least, is the instant when darkness begins to increase again at the expense of light. Or, in a man, the prime of life is just the moment when the long slide into old age and death becomes inexorable. In other words, the extremes are the points in a cycle where a change of direction takes place, so it is in their nature to be unstable. If we wanted to be pretentious about it, we could phrase this law of Chinese science as "Maximal and minimal states of any variable are unstable." In human affairs just as in nature,

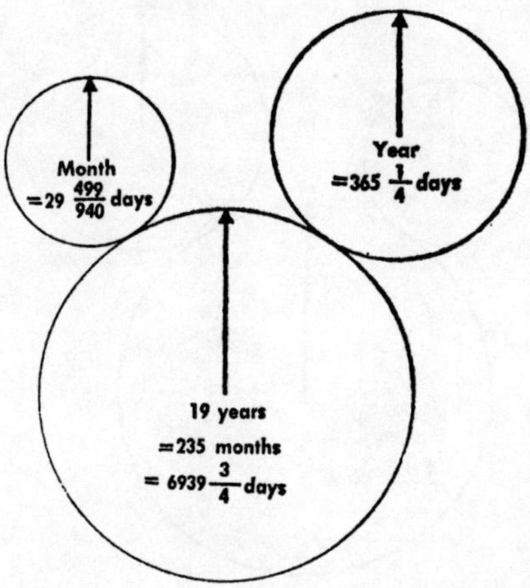

FIG. I. THE BEGINNING OF A LARGE CYCLE: WINTER SOLSTICE AND NEW MOON.

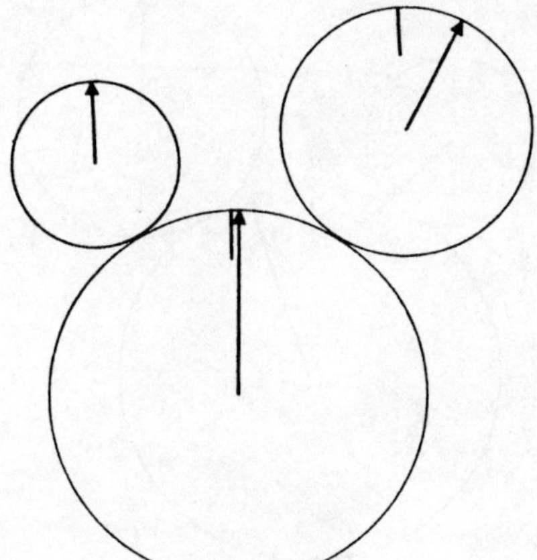

FIG. II. ONE MONTH LATER

the tricky times came at the upper and lower points of the cycle. To be too successful, too rich, ar too happy, was dangerous, for then things had to change for the worse. Generally the Chinese cultivated the area in the middle where the opposites were better balanced.

So there seemed to be *a* Way of Nature, an underlying pulse which went on and on, harmonizing all the little pulses of natural phenomena. Insofar as people in an agricultural society were unprepared to ignore the seasonal rhythms, they came to think of nature as a great organism itself, with a *total* life rhythm generated out of the harmony of all its parts, including man.

By this time they were aware that the key to this overall *tao* of nature lay in the sky. The cycle of light and darkness was the cycle of the sun and moon. The cycle of the seasons was tied directly to the yearly motion of the sun among the stars. The regularity which made the world go on while everything in it changed could be referred to the utterly reliable motions of the celestial bodies, and so the problem was in essence solved. That is why one of the Emperor's main functions was providing an accurate calendar which showed the motions of the sun, moon, and planets. If it was the Emperor's responsibility to see that social activity did not violate the *tao* of nature, it was naturally a matter of state security to understand just what the rhythms were.[3] This wasn't a matter of burning incense and mumbling things. It was done by professional

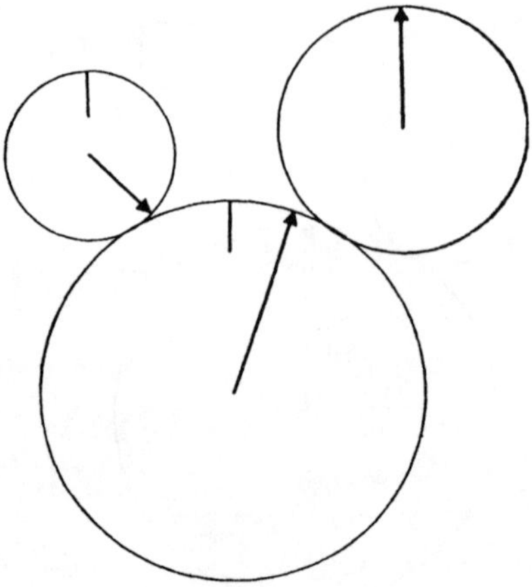

Fig. III. One Year Later

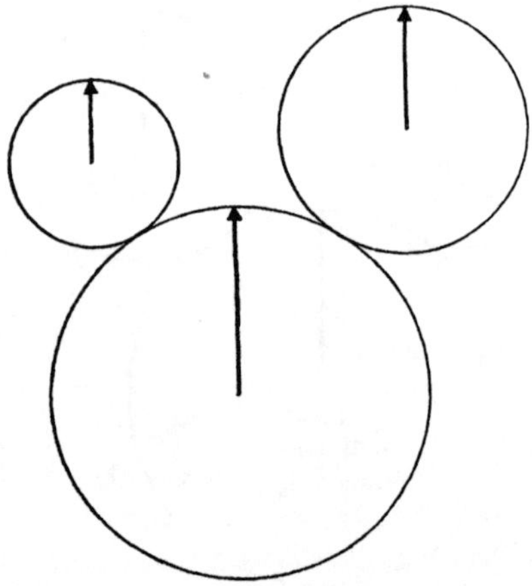

Fig. IV. The End of a Large Cycle: Nineteen Years Later

astronomers, who were very decent mathematicians and respectable bureaucrats at the same time.

The astronomers began, much as astronomers did anywhere in the ancient world, by marking off the basic cycles of day, month, and year. They did this by determining when certain basic celestial events recurred. When the sun went from its highest point in the sky all the way round to return to its highest point, that was the day. When the moon went from invisibility back to invisibility, that was the month. When the sun came back to a certain position among the stars, that was the year.[4] A calendar that really uses the moon—ours is very abstract and no longer does—has to deal with a simple mathematical problem: twelve months do not make a year. If real months, real events in the sky, are counted as months, then a month lasts either twenty-nine or thirty days, and twelve months, on the average, add up to only 354 days. But still the sun takes 365 days to go around, so an extra eleven days have to be accounted for one way or another. In their earliest mathematical astronomy, the Chinese realized that, as the excess accumulated, in nineteen years it added up to almost exactly seven extra months, so that nineteen years was equal to 235 ($[12\times 19]+7$) months. This was the smallest number of real years which equalled a whole number of real months. The year and the month could start *together* —that is, a new moon could fall on the winter solstice, which theoretically began the astronomical year—only once every nineteen years. We can imagine this situation as that of two wheels (one for the month and one for the year) both driven by a larger wheel (for 235 months = 19 years.) If both begin with their pointers set at zero, only when the large wheel has turned once will the pointers on both smaller wheels read zero again. If we number each day by a combination of both pointer readings, each day will have a unique number until the nineteen years have passed and the readings repeat themselves.

Now each pointer reading actually stands for a different position of the sun and moon, and thus for a different day of the year and day of the month. To make an ordinary calendar which tells us when years and months start and when to insert the seven extra months, we simply count off these days. The point is that we need count them off for only nineteen years, because we know that the combinations of sun and moon positions will repeat themselves forever in nineteen-year stretches.

This problem was solved by the elementary mathematical process of finding the lowest common multiple, the smallest whole number which is a multiple of the number of days in a month and days in a year. Nineteen years is approximately 6940 days, and the average number of days in a month (29.53) divides into it to give 235.

Chinese astronomers went on in the same way to build longer cycles, longer rhythms which would harmonize more and more of the celestial motions, so that their calendar could predict the positions not only of the sun and moon but those of the planets too. Since Jupiter, for instance, took roughly twelve years to make a circuit of the firmament, and the basic sun-moon cycle was nineteen years, it would take all three luminaries 228 (12×19) years to return to their starting positions together. There is no point in carrying the mathematics of this sort of problem any further. It is obvious that by the time they had worked in all five of the planets, eclipses, and so on, the lowest common multiple of all their periods was bound to be an enormous number.

In fact, when the first complete system of these cycles was worked out, about two thousand years ago, it turned out that the overall cycle was 23,639,040 years long. It would take that long for the big wheel which drove all of the little wheels to revolve once. You could define an initial state of the universe, when all the cycles began to go—when, on a new year's day which was also the first day of the month, the sun, moon, and all the planets were lined up next to each other like a string of pearls. That was in a very fundamental sense the moment when time began. It would take more than 23 million years for the same lineup to recur. Then, for all practical purposes, time would be beginning all over again. A new world age would be starting.

Regardless of the fact that time flowed on, and could be measured as objectively as you like with sundials or water clocks, every moment was unique, every moment had its *tao*, because it marked a unique combination of the positions of the heavenly bodies. Moments exactly a world age apart, on the other hand, were fundamentally the same. Their *tao* was the same because they represented the same beat in the rhythm, the same event in the cycles of the natural world.

This was what the Chinese originally meant by a world age. They could think of time starting all over again every twenty-three million years, without in any way compromising their conviction that the physical world was eternal. Only later on did some thinkers, adopting the Buddhist idea of world ages beginning with a creation of the world and ending with its destruction, envision cycles of material existence rather than just cycles of time. But this twenty-three-million-year cycle was only the beginning. Some of the world ages worked out by later astronomers were so immense that questions about what happens when one ends became quite academic.

Now we are left with a final question: What happens to human history when it is measured against such stupendous stretches of time? The Chinese were well aware that human society—by which they meant Chinese society, naturally enough—was changing. Their constant use of the past to comprehend the present and shape the future—their historical sense, in other words—may have emphasized the elements whose relevance or whose connection was not wiped out as time passed, but it also meant that anyone willing to think in terms of historical change had plenty of data available.

When the Chinese thought about history, that is to say about human activity as conditioned by time, they were simply dealing with their old problem in a new context: In the world of affairs everything is changing, and every man seems to be seeking his own individual good—which they also called his *tao*—but despite this flux, this never-ending flurry of taking and doing and making and wanting in every direction, the overall profile of society changed very slowly and in a more or less measured way. Again the laws of constancy, the overall, governing rhythms of this change, were sought, and we should not be surprised that the answer which made sense to most Chinese thinkers was cast in the form of cycles. This seems all the more inevitable when we remember that the earliest Chinese historians were by occupation court astronomers.

The Chinese sense of history differed radically from ours in that it proceeded by discrete dynasties, which lasted as long as the descendants of a ruler could continue to inherit his throne. By the time not too many dynasties had flourished and gone their way, people were aware that they very easily could be thought of as a kind of organism, which passed through periods of growth, maturity, and decay like any other life cycle. Dynasties ordinarily began when someone who was dissatisfied with the old distribution of power took over. He tended to be a vigorous man, well able to cope with the necessity for balancing political change and stability. Because he was bringing in a new order, he was often able to ignore vested interests an established dynasty would not be able to ignore—at the price, to be sure, of coping with the interests which were vested in him. But this flowering in the early years of a dynasty would gradually lose its bloom, for any number of reasons. The tremendous tasks that had to be done on a regular basis just to keep such an enormous empire in one piece tended not to get done. Gradually the factors which led to dissatisfaction snowballed, and the decline, like that of a dying man, accelerated and became irreversible. This was usually looked at by Chinese historians as a decay in the virtue of the emperor. Emperors who came at the very end of a dynasty, those who got beaten in other words, were almost always portrayed as corrupt. It is striking that some of them, in accounts written *before* they got beaten, appear to be decent enough people.

Educated Chinese, then, tended to think of the past as a series of cycles sharply demarked by changes of dynasty, and to place people by which dynasty or which reign-period they lived in. But how does the dynastic cycle fit into the overall scheme of time? The answer seems obvious enough: just like any other cycle. In fact there was even a "long cycle" of five successive dynasties, which came of applying to history the set of five categories which were widely used to characterize the segments of any cycle of change. This answer is only partly true, though, for it would be a gross distortion to picture the succession of dynasties mechanically as revolutions of a single rigid wheel or wheel system. Man's moral potential, his option to choose a harmonious way of life, could not be so freely compromised.

The explanations which were given for what determines the longevity of a dynasty are too various even to sum up here, but what matters is that they do not depend on a preset span which no human will or effort can affect. For Confucius, who formed so much of Chinese thought, man's social concerns came first, and were to be dealt with by social means. Among Chinese philosophers, he is one of those least aware of the cosmos.

Now we begin to glimpse the intricacy that was required to do justice to so packed an aspect of human experience. There was the world age, compounded of familiar celestial cycles but itself so immense that (if we imagine it as a circle again) a piece of its edge the length of a human life would be indistinguishable from a straight line. There was the linear time of the water clock, of hours, days and years simply counted off, which could be as abstract and utilitarian as our own, but was still checked against the revolutions of the sun and moon. This ambivalence is equally apparent in a history which has room for both a consciousness of change and progress, and an awareness of the recurrence of situations and relationships that makes the past vitally relevant to the present. There are yet other complexities, as must be expected of so comprehensive a set of conceptions, but they do not violate its harmonious wholeness in the minds of those who lived by it.

One has to regret its passing away, for it preserved, in a beautifully symmetrical, balanced structure, many profound insights about how our experience is bounded by time. It could satisfy the necessities of everyday experience, of natural philosophy, of poetry and other forms of magic, in a unity which for our purposes we have had to rend apart. It represents one of man's great successes in his attempt to make sense of time, to impose order on the seemingly chaotic succession of individual moments by tying them, with unvarying measures, to eternity.

NOTES

1. It would be very easy to find exceptions to many of my general statements, for my aim is to present some new reflections as simply as possible. Readers who wish to go on to a systematic and detailed description will find a brilliant one, full of references to other important studies, in Joseph Needham's *Time and Eastern Man. The Henry Myers Lecture 1964* (London: Royal Anthropological Institute of Great Britain and Ireland, 1965).

2. Many aspects of *tao* do not fall within the dimension of time. Perhaps I overemphasize its temporal aspect here, but I do so in order to demonstrate how much has been missed by underemphasis in the past.

3. The purpose of the ephemerides was usually described as regulation of agriculture, whose rhythms were fundamental. It is a mistake to consider that the whole story, because the tasks of farming were timed by the sun, not by the moon or planets. For that matter, it is difficult to imagine a Chinese peasant's really needing to consult a calendar to know when to plant his crops.

4. To be a bit more technical, the day was defined from midnight to midnight and the month from one conjunction to another. The astronomical year was determined by winter solstice, the southmost point of the sun's annual path.

II

COSMOS AND COMPUTATION IN EARLY CHINESE MATHEMATICAL ASTRONOMY

This paper is dedicated as an anticipatory "萬歲" for his sixty-fifth birthday to Willy Hartner, who gave me my first glimpse of two grand traditions—ancient astronomy and its historiography.

Abstract

The usual description of Chinese mathematical astronomy as a fundamentally practical, empirical collection of techniques reflects not national character but a conscious choice at a certain point in history. The two great systems of the first century A.D. were necessarily founded upon and greatly conditioned by philosophical assumptions about the simple cyclical character of the celestial motions. Astronomers were forced to incorporate mediocre prediction methods for lunar eclipses and planetary motions, which their postulates were too crude to fit. Techniques of very high accuracy could have been discovered and used with no more sophisticated mathematics, but they could not have been assimilated to the formal character of Chinese astronomy as a whole. The dilemma was resolved over the next few centuries, not by astronomers' substituting new assumptions more conformable to the complexity of the phenomena, but by their becoming indifferent toward cosmology.

Contents

Tables

Figures

When we think about Chinese astronomy at all, we tend to think of it as a fundamentally practical, empirical art, a collection of mathematical techniques toward a more or less purely political end. We lack the documents to support even a tentative evaluation of the connections between theory and practice in the germinal — or pre-bureaucratic—phase [1]). For later ages sources abound; their study in the light of modern astronomy has given rise to the commonplace that the computations at the basis of the Chinese calendar were as independent of any physical model of the world as those of ancient Babylonian astronomy, on the basis of still very incomplete evidence, appear to have been. To be sure, the Chinese science in its maturity operates without assumptions about the real motions of the physical luminaries. When Chinese astronomers speak explicitly about the structure of the world, they use the common-sense geocentric language which satisfied our ancestors too. But cosmological speculations are justly characterized as intermezzos in Chinese astronomy, which could have got along very well without them.

Between the origin of Chinese astronomy and its full flowering as a mathematical science in the Sui and T'ang, the sense of cosmos almost completely dropped out. It is impossible to imagine astronomy beginning, those immense labors of recording and analysis first being brought to yield laws, had there been no sense of the universe as a system, a dynamically balanced model of abiding reality to be contrasted with the phenomenal flux of terrestrial experience. In the Han dynasty sources of the first century A.D. this consciousness can still be discerned, transformed almost out of recognition. What we find is a section of the calendrical treatises devoted to deriving the fundamental astronomical constants from a *yin-yang* and five-elements analysis of cycles of change, patterned on the metaphysics of the Appendixes to the Book of Changes (see below, pp. 8-9). But the attempt to construct a deductive foundation for astronomy, an application of the basic conceptions of natural philosophy parallel to their applications in medicine, alchemy, and geomancy, was patently arbitrary—which is not to

---

[1]) Early astronomy has been surveyed with great authority in Henri Maspero, "L'astronomie chinoise avant les Han," *T'oung Pao*, 1929, 26: 267-356. Homer H. Dubs, "The Beginnings of Chinese Astronomy," *Journal of the American Oriental Society*, 1958, 78: 295-300, although not consistently reliable for astronomical interpretations, cites some important additional sources.

claim that theoretical rigor was highly prized elsewhere in Chinese science—and remained irrelevant to the work of prediction. The deductive element appears in later calendrical treatises as occasional antiquarian exercises rather than as sustained and serious attempts to account for what is.

But what can be said about the structure of the predictive techniques themselves? If we examine the formal relations of the Han constants in the light of their astronomical applications, it becomes clear that we are contemplating a great system of cycles, a mathematical cosmos far abstracted from, but in theory capable of generating, the successive configurations of the physical sky. In another two centuries this was no longer true; the calendrical treatises had become collections of astronomical techniques, much more sophisticated, to be sure, but whose interrelations were largely vestigial.

The sense of cosmos, if it had indeed existed, and had survived in China long enough to play a fruitful part in the formation of mathematical astronomy, is safely buried in the very professional systems of the Chinese art's last millennium. But what killed the conviction that astronomy could be physics as well as mathematics? Was it merely the blighting hand of bureaucracy, offering technicians security in a hierarchy which had nothing to gain from theory? Was it, as in Europe between Plato and the Copernican Revolution, the vested interests of philosophers? In China too did they reserve to themselves the right to reason out what the universe was like, and leave to astronomers only the job of supporting them mathematically, not of contributing to the improvement of philosophy? Both of these factors are part of what must assuredly be a very complex answer, but neither alone would have sufficed. It is a matter of historical record that cosmology had much to offer both political theory and administrative practice—which after all is why room was found for treatises on judicial astrology and mathematical astronomy in the Standard Histories. Chinese astronomers, far from being subservient to the a priori vagaries of philosophers, paid them remarkably little attention.

I should like, in what follows, to examine a third possibility. My point is that a careful consideration of the Han astronomers' mathematical procedures can indicate the presence of grave contradictions between their assumptions about the necessary character of the celestial motions on the one hand, and the necessity for accurate predictions on the other. This internal crisis, which reveals itself in a variety of ways, is so serious that a radical realignment is hardly to be marvelled at.

Form and Content in the Early Calendrical Treatises

Tension arose in the first place because eclipse prediction was only one aspect of a highly integral and stereotyped system of mathematical astronomy. The form in which the calendrical art was transmitted was decisively conditioned, as in ancient Mesopotamia, by the importance of astrology to the security of the state. [1])

Celestial phenomena which could not be predicted were ominous in the fullest sense of the word: they were omens. Every solution to a problem of astronomical prediction meant removal of one more source of political anxiety. It is well known, for instance, that in

---

[1]) I have not devoted much space to characterizing Chinese astronomy, since much of the ground has been covered adequately in Western languages. The reader will find an excellent general and bibliographical introduction to the concepts, methods, and tools of Chinese astronomy in Joseph Needham, *Science and Civilisation in China*, vol. III (Cambridge, England, 1959); see also his *Time and Eastern Man* (London, 1965), p. 9, note 2. Needham's references to earlier work on many astronomical problems are so complete and conveniently set out that in general I do not duplicate them below. It is necessary, however, to supplement Needham's Bibliography B with the anonymous "Minkoku irai no Chūgoku tenmonkai kōsaku gaikyō 民國以來の中國天文界工作概況" (A survey of the work of the astronomical profession in Republican China) in the intelligence periodical *Chūgoku bunka jōhō* 中國文化情報, June 1941, no. 28, pp. 1-29 *passim*.

Detailed descriptions of the early calendrical treatises are accessible in the well-known articles of Wolfram Eberhard and his collaborators, to whom I must acknowledge a debt whose magnitude will be obvious to anyone familiar with their work. For references, see Eberhard, "Index zu den Arbeiten über Astronomie, Astrologie, und Elementenlehre," *Monumenta Serica*, 1942, 7: 242-266. The most important technical analyses of traditional Chinese astronomical systems are found in the many works of Yabuuti Kiyosi and his collaborators, the most germane being Nōda Chūryō 能田忠亮 and Yabuuti [Yabuuchi Kiyoshi 藪內清], *Kansho ritsurekishi no kenkyū* 漢書律曆志の研究 (Researches in the Treatise on Harmonics and Calendrical Astronomy of the Han history; Kyoto, 1947). A valuable discussion of Han chronology, based on archeological as well as literary documents, is provided in Ch'en Meng-chia 陳夢家, "Han chien nien-li-piao hsü 汉簡年曆表叙" (Prolegomena to chronological tables based on the Han wooden tablets), *K'ao-ku hsüeh-pao* 考古學報, 1965, no. 36, pp. 103-149. Of the voluminous literature on astrology, Shigeru Nakayama, "Characteristics of Chinese Astrology," *Isis*, 1966, 57: 442-454, is most compendious and least blemished by reductionism.

the early Standard Histories observations of solar eclipses were recorded and interpreted in the Imperial Annals or the Treatise on Five-Elements Phenomena—and planetary phenomena in the Annals or the Treatise on Astrology—while for lunar eclipses, which could be predicted, it was sufficient to publish the method of their calculation in the Treatise on Harmonics and Calendrical Astronomy 律暦志. The eclipse of the sun never lost its astrological significance; in the absence of spherical geometry, very few successful predictions were possible until a time when the solar eclipse's ritual significance had been rendered immutable by centuries of precedent. But in the Han, before the institutional aspects of astrology had jelled, it is not impossible to find cases in which astrological relevance was removed by rational explanation. In the Treatise on Five-Elements Phenomena of Pan Ku's 班固 (A. D. 32-92) History of the Former Han 漢書, the new moon visible on the last calendar day of the month 朓, and the old moon visible on the first day 仄慝, were omens of the ruler's laxity or overstrictness. But reporting of these events was soon dropped, and they were rationally explained in the calendrical treatise of Ssu-ma Piao's 司馬彪 (240-306) Continuation of the Han History 續漢書: "From its adoption at the beginning of the Grand Inception period [24 December 105 B. C.], the Triple Concordance 三統 astronomical system was used for over a hundred years. The calendar ran slightly behind the phenomena, so that the new moon occurred earlier than the calendar predicted. The [true] conjunction would take place on the last day of the month in some cases, and the moon would appear on the first" [1]).

---

[1]) The system which Ssu-ma Piao calls Triple Concordance is usually called Grand Inception (cf. p. 11 below).

Han shu (*Han shu pu chu* 補注, Basic Sinological Series ed., 1959 reprint), 27: 2451-2452; *Hsu Han shu* (*Hou Han shu chi chieh* 後漢書集解, Basic Sinological Series ed., *chih* 志 2), p. 3389, emending "月見" to read "月或朔見" as in *T'ai-p'ing yü lan* 太平御覽 (Chung Hwa Book Co. reprint of 1960), 16: 10a; Ch'ien Pao-tsung 錢寶琮, "漢人月行研究" ("The motion of the moon as understood by the people of the Han dynasty"), *Ch'ing-hua hsueh pao* 清華學報, 1935, 17: 47-48. Wang Hsien-ch'ien's 王先謙 (1842-1918) compilations of annotations to the Han histories, cited above, are indispensable for the study of early astronomy.

For the events leading up to the printing of Ssu-ma Piao's treatises together with the *Hou Han shu* in the T'ang, see Hans Bielenstein, "The Restoration of the Han Dynasty. With Prolegomena on the Historiography of the *Hou Han Shu*," *Bulletin of the Museum of Far Eastern Antiquities*, Stockholm, 1954, 26: 16-17.

An astronomical system was a complete set of mathematical techniques for calculating an ephemerides which provides both positions and dates of characteristic phenomena for the sun, moon, and planets. Once a system was officially adopted, it became part of the Emperor's ritual paraphernalia—not simply because a calendar was needed for agriculture, to which the motions of the moon and planets were entirely irrelevant, but even more fundamentally because the ability to predict moved celestial events from the realm of the ominous to that of the rhythmic and intelligible. The Emperor was thus enabled to know Nature's Tao so that his social order might be kept concordant with it. Failure of the official system to predict was necessarily a sign of moral imperfection, a warning that the monarch's virtue was not adequate to keep him in touch with the celestial rhythms. The Chinese theory of the natural order and the political order as resonating systems, with the ruler as a sort of vibrating dipole between them, imposed on the history of astronomy an insatiable demand for increased precision—far exceeding, in the area of the calendar, any conceivable agricultural, bureaucratic, or economic necessity [1]).

---

[1]) I adduce the progressive general improvement from one major astronomical system to another, and above all the remarkable fact that apparent solar motion was successfully substituted for mean solar motion (after many tries) in official calendars based on the Great Expansion 大衍 (727) and later calendrical treatises. Repeated attempts were made to incorporate the apparent motion of the moon, equally irrelevant and unjustified in terms of mundane applications of the calendar. For authoritative evaluations of technical developments, see Yabuuti, "Astronomical Tables in China, from the Han to the T'ang Dynasties," in *Chūgoku chūsei kagaku gijutsushi no kenkyū* 中國中世科學技術史の研究 (Tokyo, 1963), pp. 445-492, continued in "Astronomical Tables in China from the Wutai to the Ch'ing Dynasties," *Japanese Studies in the History of Science*, 1963, 2: 94-100. "Astronomical table" is Yabuuchi's translation for "*li* 曆," which I prefer to render "astronomical system" or "calendrical treatise" as the sense demands.

The reader will note that I have adopted a policy of translating reign titles. Although this is far from established practice, I find that consistency in discussing ancient astronomy demands it. "*T'ai ch'u*," which I render "Grand Inception," was both the name of a calendar reform which took effect at a moment in which, as we shall see, all the major calendrical cycles began together, and the name of a regnal period under the aegis of that reform. The semantic significance of "*t'ai ch'u*" as applied to the calendar reform is patent; I do not feel free, therefore, to deny or ignore the sense of the same words in connection with the reign. The spirited argument against translation in Mary C. Wright's "What's in a Reign Name: The Uses of History and Philology," *Journal of Asian Studies*, 1958, 18: 103-106,

An astronomical system, if given official status, became inviolable, not to be tinkered with or dismembered by technicians [1]). Of the half a hundred systems which saw service in the last two millennia (as many again were proposed but not adopted), some were superseded merely as one more sign of the new order which a change of ruler or reign period was supposed to bring, with a very minimum of real change in computational techniques. But most systems were discarded for precisely the reason that they could not adequately predict eclipses, or because someone presented a better scheme (or at least a new one) for computing the ephemerides.

The calendrical treatises which have been preserved intact in the two Standard Histories of the Han are handbooks of mathematical astronomy, but they are much more—in fact they are meant to be systems which include the totality of cosmological knowledge, incorporating the several numerological traditions popular in the Han. This is patent in the section of the Triple Concordance treatise in which the numerical values of several of the fundamental cycles are derived from the *yin-yang* duality (represented in this context as earth and sky) and the five elements.

> . . . The Book of Changes says: "The celestial 1, the earthly 2, the celestial 3 . . . The celestial numbers are five, and the earthly numbers are five. When the numbers are properly distributed [among the five elements], each plays a complementary part in the whole. Then the celestial numbers are [i.e., total] 25, the earthly numbers are 30; the numbers of heaven and earth together are 55. By this number [natural] change is brought to completion and the spiritual beings set in motion."
>
> Further, adding the final [*yin* and *yang*] numbers gives 19; permutation has gone as far as it can and so there is a transformation [which begins the cycle again]. Thus [19] is the Intercalation Divisor. Triple the celestial 9, double the earthly 10; this is the Coincidence Number [47]. Triple the

---

reduces to the point of view that, despite our willingness to expend time and thought on finding true equivalents in another language, in many cases we may never know all of the manifold and complex meanings invested in a reign name, or that if we did we often would be unable to condense our understanding into two English [or even German ?] words. This is a salutary caution, but its ring of authenticity has more to do with the finitude of human endeavor than with the problem at hand. The same argument may be applied in principle to Chinese poetry, which sinologists (among others) feel justified in translating despite general agreement that it is untranslatable.

[1]) The Astronomical Bureau could violate the spirit of this precept in emergencies by adopting a new technique "on a provisional basis". We shall see below that in the Later Han a lunar eclipse technique was so used for fifty-six years (p. 59).

celestial number 25, double the earthly number 30; this is the Phase Coincidence Cycle [135]. Multiply it by the Coincidence Number; this is the cyclic return of solstice and new moon, the Coincidence Month [513 years]. After nine [Coincidence Months] the Epoch Cycle [4617 years] begins again [1]).

The method of the astronomical handbook proper was equally formal: cycles were determined for the phenomena to be represented, and, by a process which amounts to finding lowest common multiples, larger cycles were constructed to contain and subsume series of smaller ones. The system was made integral, when this process was done, by a "great year" cycle, like an immense wheel driving a congeries of graduated smaller wheels arranged in subsystems. It was then necessary to find the epoch, to determine just how long ago the largest cycle had begun. Then the state of any of the smaller cycles, which by definition began at the same time, could be determined by a counting process no more complicated than computing the positions of the hands of a watch when one knows the time elapsed since midnight. The technical presentation in the Han treatises is much simpler and more schematic than in Ptolemy's Almagest, for the magnitudes of the various Chinese cycles have already been justified metaphysically, as we have seen, by relating them to the fundamental numbers; it is not considered necessary to record the observations by which they must have been derived originally. There remains only to list the numerical values in an order which makes their hierarchic subordination apparent, and to the counting-off process by which any celestical phenomenon may be predicted [2]).

---

[1]) *Han shu, 21A*: 1684. The constants which occur in this excerpt will be explained anon.

The first paragraph is quoted from *I ching*, "Hsi tz'u," I, 9. See *The I Ching or Book of Changes*, tr. Richard Wilhelm, 2 vols. (New York, 1950), I, 331-333.

[2]) The points made in this paragraph will be developed in greater detail later. Readers who want a more elementary introduction to the role of concordance cycles in Chinese thought are referred to my "Chinese Conceptions of Time," *The Earlham Review*, 1966, 1: 82-92.

I use the word "metaphysics" neither in the original meaning (as a special designation for the books which follow the Physics in the Aristotelian corpus) nor in the modern meaning (if the word "meaning" applies), a straw man for empiricists who have been misled by a wishful interpretation of Newton *et al.* to believe it feasible and desirable that science be free of ontological associations, and to think that the sole function of such associations in the past has been to mire the inexorable march of positive knowledge.

TABLE 1

Numerological correspondences in the Triple Concordance Treatise

Element	*Directional Correlate*	*Heaven (yang, odd)*	*Earth (yin, even)*
WATER	north	1	6
FIRE	south	7	2
WOOD	east	3	8
METAL	west	9	4
EARTH	center	5	10

In the earliest astronomical system of which we have adequate records, the Grand Inception 太初 system of the Han Martial Emperor, the epoch was the Astronomical New Year 天正 of the Grand Inception reign period (24 December 105 B. C.).This particular moment of time was simultaneously the winter solstice, the first day of the Astronomical First Month, and the first day in the sixty-day cycle by which days were recorded [1]). If no more than

---

I take "metaphysics" in the much more modest classical sense of "thought about how the world must actually be constituted so that known physical laws apply." While there is indeed no *a priori* reason why science should involve such theoretical speculation, practicing scientists from the beginning to the present, lacking the superior detachment of the academic positivist philosopher, have perversely insisted on referring their discoveries to an intelligible world. In doing so they have consistently posited entities, connections, and models which defy empirical verification or operational definition, and which are in the final analysis justified largely by esthetic criteria.

[1]) The Astronomical First Month of the year was defined as the month in which the winter solstice occurred. The civil first month differed from it by a number of months which varied from state to state and period to period. One of the key features of the Grand Inception reform (Teng P'ing 鄧平, Lo-hsia Hung 落下閎 and others, 104 B. C.) was shifting the civil first month from the month before the Astronomical First Month to the second month after, where it remained until modern times. In the Han the same term is used to denote the first day of the Astronomical First Month, distinguished by later astronomers as 天正朔. I render the latter sense as "Astronomical New Year." See Liu T'an 劉坦, *Chung-kuo ku-tai chih hsing sui chi nien* 中國古代之星歲紀年 (Recording of years according to Jupiter and the Year Star in Chinese antiquity; Peking, 1957), pp. 173-188, Ch'en Chin-sien [Chen-hsien 陳振先], "The Anomalous Calendars of the Ch'in and Han Dynasties," *Chinese Social and Political Science Review*, 1934, 18: 157-176,

a calendar for days, months, and years was needed, their cycles could be counted off, and the successive locations of the sun and moon predicted as a consequence, from the celestial configuration at the epoch.

So far as we know, a new method of constructing solar and lunar tables, which perform the functions we conventionally associate with the idea of a calendar, was as far as the Grand Inception reform went. Another century passed before the schema was extended to provide a universal system of astronomy. In the words of the Continuation of the Han History, "By the Epochal Erection 元封 period of the Martial Emperor [calculations based on the current system] no longer accorded with the celestial phenomena; the Emperor called together specialists 術士 who produced the Grand Inception system, with epoch in year 14 of the sexagenary cycle. In the time of Wang Mang, Liu Hsin 劉歆 made the Triple Concordance system, with the Superior Epoch set at a Great Planetary Conjunction in sexagenary year 47, thirty-one Epoch Cycles before the Grand Inception epoch [1])."

---

and "秦末漢初之正朔閏法及其意義" (Late Ch'in and early Han calendrical techniques and their significance), *Kuo wen chou-pao* 國聞週報, 1934, vol. 11, nos. 4, 5, 6, 7, 10, 13, 16, 18, 20, 23, and 26 (all installments are paginated separately).

The date provided for Astronomical New Year of 104 (by which I mean the Chinese year which corresponds most closely to 104 B. C. of the Julian calendar) is based on my calculation of the true conjunction (see below, p. 23); it differs from the conjunction date given in Tung Tso-pin 董作賓, *Chung-kuo nien li chien p'u* 中國年曆簡譜 (Taipei, 1960), by one day.

What with the varying lengths of reign periods and calendar years, the completely abstract sexagesimal day and year cycles immensely simplified keeping track of intervals between events, and thus served as a backbone for chronology. I would hesitate, nevertheless, to consider their inclusion in the system of cosmological cycles nothing more than a matter of mathematical convenience. Their utility gave them numerological validity. They were by no means the only cycles whose initial point did not correspond to an observable event.

1) *Hsu Han shu* (*chih* 3), p. 3499; *Shih chi* 史記 (*Shih chi hui chu k'ao cheng* 會注考證, reprint, Taipei, n. d.), *26*: 10-15; Chang Hung-chao 章鴻釗, *Chung-kuo ku li hsi i* 中國古曆析疑 (Resolutions of problems concerning ancient Chinese astronomical systems; Peking, 1958), pp. 81-82. There has been considerable discussion as to whether Liu Hsin's sytem was new or simply copied from that of Teng P'ing and his collaborators; see, for instance, Ch'en Tsun-kuei 陳遵嬀, *Chung-kuo ku-tai t'ien-wen-hsueh chien shih* 中國古代天文學簡史 (A short history of ancient Chinese astronomy; Shanghai, 1955), p. 38, note 8. The position that Liu took over the Grand Inception calendrical methods and constants, but with great originality extended them

The calendrical complex of the Triple Concordance system. In order to see what this new step amounted to, it will be necessary to review several structural features of the Triple Concordance system of constants.

LUNATION 月 $29\frac{43}{81}$ days [1]).
YEAR 歲 $365\frac{385}{1539}$ days.

The precision of this constant is specious. The fraction comes from a trivial modification of the previously current value, $365\frac{1}{4} = 365\frac{385}{1540}$.

CONCORDANCE CYCLE 統 1539 years = 19,035 lunations = 562,120 days.

Since this cycle contains an integral number of days, months, and years ($1539 = 81 \times 19$), it defines simultaneous recurrence of solar and lunar events at the same time of day—e.g. coincidence of new moon and winter solstice at midnight.

EPOCH CYCLE 元 1, 686, 360 days = 4617 years = 3 Concordance Cycles.

Since this cycle contains a number of days exactly divisible by 60, it defines the recurrence of solar and lunar events (spaced by the "official" constants) on a given day in the 60-day cycle—e.g. coincidence of winter solstice and new moon at midnight on sexagenary day #1. The Triple Concordance system was named after the three Concordance Cycles which make up the Epoch Cycle.

---

into a universal system which became the pattern for his successors, is the only one which accounts for all the evidence. This interpretation can be traced back at least as far as Hsu Kan's 徐幹 (170-217) *Chung lun* 中論 (*Han Wei ts'ung-shu* 漢魏叢書, Han-fen-lou 涵芬樓 reprint of 1925), *B*: 12a.

The Epochal Erection period was named for the Martial Emperor's performance of the Altar-erection Sacrifice 封 in 110.

[1]) In order to make the structure of the system as clear as possible I discuss only a few of the most relevant constants, and cite them in highly condensed forms. The lunation, for example ($29\frac{43}{81} = \frac{2392}{81}$ days) is actually expressed as two separate integral constants, the Day Rule 日法 (81) and the Lunation Rule 月法 (2392).

The constants discussed are mean figures, of course. Any particular calendar month was either twenty-nine or thirty days long, and the interval between two true conjunctions (i.e. the length of a true lunation) could vary by as much as a day.

We can look at this set of constants as a complex of circles turning upon each other (Figure 1). The Epoch Cycle simply specifies what

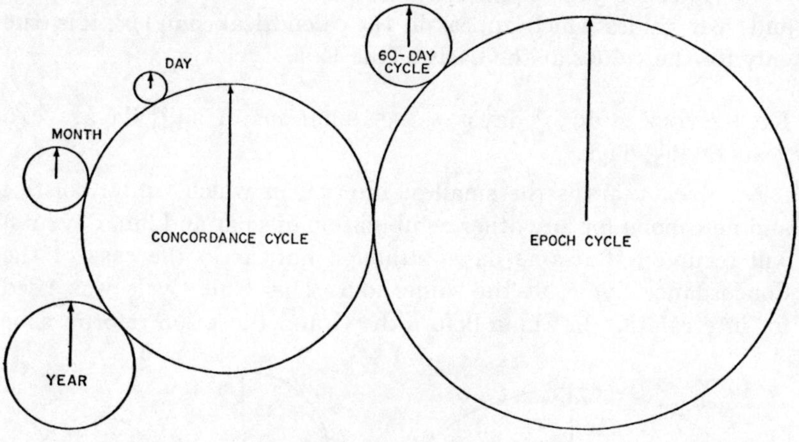

Figure 1. System of calendrical constants in the Triple Concordance treatise In a scale model, circumference would be proportional to length of cycle. The rotating arrows all point upward at the same time only once every 4617 years.

motion of the integral system is needed to return all cycles simultaneously to their original orientations. In such a system, if we know the original orientation and the number of revolutions any one circle has passed through at any given moment, we can predict the orientation of any other circle. In other words, by counting the time elapsed from epoch we can compute the date of any event with respect to the winter solstice (what day of the tropical year the event falls on), conjunction (what day of the month it falls on), sexagenary day cycle, and hour.

Eclipse complex. The small set of constants which performs the ordinary calendrical functions is not the only one driven by the Epoch Cycle. Another set allows prediction of eclipses, incorporating for this special purpose an ancient intercalation cycle simpler than the calendrical complex's Concordance Cycle.

PHASE COINCIDENCE CYCLE 朔望之會 135 lunations = 23 lunar eclipses [1]).

The properties of this eclipse cycle will be examined further on.

[1]) Note that the element "coincidence 會" serves what might be called

Coincidence Month 會月 6345 lunations = 1081 eclipses = 513 years = 27 Rule Cycles.

The lunation-year equation does not hold for the offical month and year values which appear in the calendrical complex; it is true only for the values at the basis of the Rule Cycle.

Rule Cycle 章 $6939\frac{3}{4}$ days = 235 lunations of $29\frac{499}{940}$ days = 19 years of $365\frac{1}{4}$ days.

Nineteen years is the smallest interval in which winter solstice and new moon (or any other combination of solar and lunar events) will recur on the same day—although not, as in the case of the Concordance Cycle, at the same hour. This Rule Cycle was used for intercalation in China before the Grand Inception reform, as is

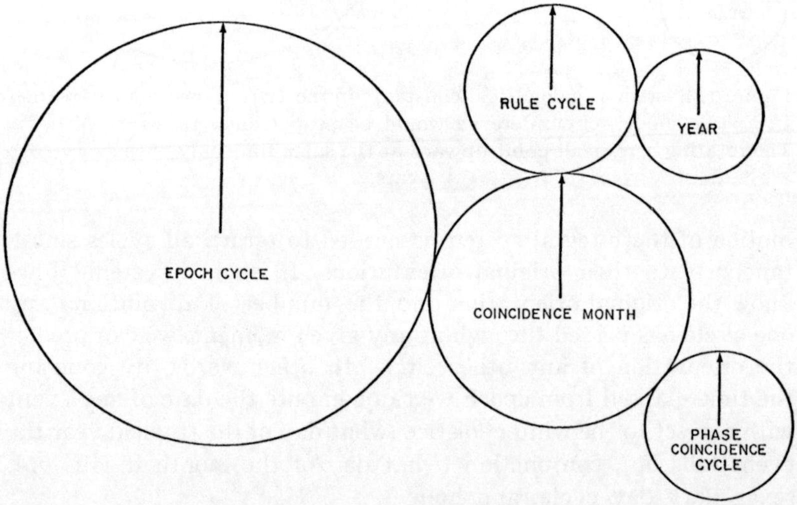

Figure 2. System of eclipse constants in the Triple Concordance treatise. The year cycle—and the month and day cycles, which are also driven by the Rule Cycle, but are omitted in this diagram—represent the "obsolete" values discussed below. The initial positions of both the calendrical complex and the eclipse complex recur at the same moment.

evident from the use of ancient values for the year and lunation. The lunation value is not as precise as it looks; it is simply derived

---

an acronymic function in the names of more complex cycles below. It stands for "phase coincidence," and merely indicates that the 135-month eclipse cycle is a component of the larger cycle in question. The astronomical meaning of "phase coincidence" is explained below (p. 40).

from the practical approximation 235 months = 19 years ($29\frac{499}{940} = \frac{19 \times 365^1/_4}{235}$). Since nineteen years of twelve months each makes a total of only 228 lunations, the Rule Cycle implies adding seven intercalary months every nineteen years [1]). The lack of rigor involved in perpetuating obsolete year and lunation values was overlooked because of the Rule Cycle's simplicity and its adequacy in this special application—converting years into months for purposes of eclipse prediction. The Concordance Cycle is 81 Rule Cycles.

EPOCH CYCLE 元 4617 years = 9 Coincidence Months.

This cycle unites the calendrical and eclipse complexes, and thus serves as the period for recurrence of a lunar eclipse on the first full moon after a new moon which occurs at midnight on the winter solstice, on the first day of a sexagenary cycle. I specify a lunar eclipse at the full moon rather than a solar eclipse at the new moon simply because only lunar eclipses could be predicted at this time. The point is that the relevant eclipse cycle can be counted off from the same epoch as the other cycles.

Planetary complex. The third set of constants, which clearly went beyond the Grand Inception schema, brought the planets together into a single system which was bound into the grand overall structure. First of all, a Synodic Cycle (called Minor Recurrence Cycle 小復 for Mercury and Venus, and Minor Circuit 小周 for the three classical outer planets) was determined for each planet, in which an integral number of Synodic Revolutions (called Recurrence Cycles 復 for the inner planets and Appearance Cycles 見 for the outer) was completed in an integral number of years (see Table II). The Year Number Cycle 歲數 was obtained when the Synodic Cycle was multiplied by a Masculine (216) or Feminine (144) Factor 乾〈〈 策 as specified.

The function of these multipliers was to tie into each subsystem the famous Jupiter Cycle, Liu Hsin's (d. A. D. 23) short-lived innovation which amounted to no less than defining the mean year as $\frac{1}{12}$ of the sidereal period of Jupiter (or its invisible counter-rotating correlate the Year Star 太歲)—that is, the interval required for the

[1]) The Rule Cycle was also used in the West at least as early as the fifth century B. C. It is commonly called the "Cycle of Meton" to commemorate the man who proposed it in Athens. Otto Neugebauer, *The Exact Sciences in Antiquity*, reprint of second ed. (New York, 1962), p. 7; Bernard R. Goldstein, "A Note on the Metonic Cycle," *Isis*, 1966, 57: 115-116.

TABLE II

Constants of the Planets in the Triple Concordance System

Planet	*Synodic Period, Years (modern value)*	*Synodic Cycle*		*Synodic Period, Years (from Synodic Cycle)*	*Factor*	*Great Period, Years*
Jupiter	1.092	12 years	= 11 revolutions	1.091	144	1728
Venus	1.598	16	10	1.600	216	3456
Saturn	1.035	30	29	1.034	144	4320
Mars	2.135	64	30	2.133	216	13824
Mercury	0.317	64	202	0.317	144	9216

planet to pass through one duodecimal Jupiter Station 次 —rather than as the interval between passages of the sun through the winter solstice. According to Liu's "station-skipping rule 超辰法," Jupiter passes through 145 stations in 144 years, so that if the solar year is to be maintained as a convenient approximation for civil purposes, one extra year must be counted for every 144 passages of the sun through the winter solstice. This amounts to a sidereal period of 11.92 years for Jupiter, a fair approximation to the modern value of 11.86 years. In the 1728 solar years of Jupiter's Year Number Cycle the planet passes through 1740 stations, completing exactly 29 sexagenary year cycles. [1]

The lowest common multiple of the Year Number periods governed the repetition of any overall state of the five luminaries. It was designated the

GREAT PLANETARY CONJUNCTION CYCLE 五星會終數 138, 240 years.

If the five planets were in general conjunction—like strung pearls, as the cliché went—at a given time, they could again be in general

[1]) *Han shu*, *21B*: 1707-1717; for exact values of the Synodic Periods see *21B*: 1722-1736.

The Masculine Factor is evidently metaphysical camouflage. We see from Table II that the Synodic Cycles of which this factor is the multiplier are both twice as long as need be. In effect the multiplier is $2 \times 216 = 3 \times 144$, so that the Jupiter Cycle is still included.

The ephemeral centrality of Jupiter in the astronomy of the Hsin interregnum has been much studied in the West, especially by de Saussure, but still awaits definitive treatment. The most important contribution in recent years is the monograph of Liu T'an cited on p. 10 above. On pp. 25-26 Liu provides evidence that the Jupiter/solar year conversion was not yet known when the Grand Inception system was worked out.

conjunction 138, 240 years later, and not before. Since the interval is divisible by 60, successive great conjuctions would always fall in the same year of the sexagenary year count [1]. General conjunctions have been reported consistently in the Chinese histories, although they do not correspond to this very aprioristic cycle, and were not taken into account in constructing it. From the Imperial Annals (not the calendrical treatise) of the History of the Former Han we learn that a general conjunction took place in 205 B.C. According to modern calculations, the planets were actually strung out over 33° of right ascension. The criterion was apparently that they be located (more or less) within one lunar mansion, which in this case happened to be a wide one [2]. The conjunction of 205, although undoubtedly known to Liu Hsin and his colleagues, is ignored in the planetary complex of the Triple Concordance system.

Great Year. From the Great Planetary Conjunction Cycle of 138, 240 years is derived the cycle

2, 626, 560 years =

5120 Coincidence Months = 19 Great Planetary Conjunction Cycles.

In this period, since the Coincidence Month reconciles the Phase Coincidence and Rule Cycles, the "obsolete" periods for months and years are included. Three times this period,

7, 879, 680 years =

5120 Concordance Cycles = 57 Great Planetary Conjunction Cycles, gives the concordance of planetary periods, eclipse cycles, sexagen-

---

[1] *Han shu, 21A*: 1696.

[2] *Han shu, 1A*: 27 and *26*: 2241. J. K. Fotheringham is responsible for correcting the traditional date of this conjunction; see *The History of the Former Han Dynasty*, tr. Homer H. Dubs (Baltimore, 1938), I, 151-153. The date assigned to the conjunction in the *Han shu* is still a matter of contention. Cf. Nōda, *Tōyō tenmongakushi ronsō* 東洋天文学史論叢 (Collected papers on the history of astronomy in East Asia; Tokyo, 1943), pp. 348-349, and Liu T'an, pp. 139-141.

The classic definition of a universal conjunction occurs in the Han apocryphal book *Shang shu wei K'ao ling yao* 尙書緯考靈曜 (cited in *K'ai yuan chan ching* 開元占經 [724], small xylograph in 24 vols., *5*: 3b): "At the beginning of a month [which is at the same time] sexagenary day 1 and winter solstice, the sun, moon, and five planets begin together at the start of the mansion Herdboy 牽牛, the sun and moon like a suspended jade annulus (*pi* 璧) and the five planets like strung pearls." One finds in the Ch'ien-lung Emperor's astrological compendium (*Ch'in ting*) *Hsieh chi pien fang shu* 欽定協紀辨方書 (preface dated 1742), *1*: 24b the notion that at epoch the seven luminaries are lined up one mansion apart.

ary year cycles, and "official" values for months and years (see p. 12). The last step is to multiply by 3 again to derive the

Grand Polarity Superior Epoch 太極上元 23, 639, 040 years = 5120 Epoch Cycles = 171 Great Planetary Conjunction Cycles.

This is the universal concordance, the "Great Year" period. It is counted from the veritable beginning of time, at which at midnight on day #1 of a sexagenary day cycle which begins year #1 of a sexagenary year cycle, at new moon on a winter solstice, the sun, moon, and five planets are lined up in conjunction—and on the next full moon there is a lunar eclipse. This stupendous concatenation of celestial events is repeated at the end of the cycle [1]).

But this set of specifications has nothing to do with the winter solstice which began the astronomical year corresponding to 104 B.C. Then the planets were scattered all over the sky. How far along was the Great Year then? This question could be answered in principle, since the periods of the planets were known, if one were to count by Epoch Cycles from the beginning of time, thus mantaining the year-eclipse-sexagenary concordance which also characterized the winter solstice of Astronomical New Year 104. One would simply compute at each step the positions of the planets (which would be different each Epoch Cycle) until a point was reached when their distribution approximated that seen in the sky at the time of that solstice. The great mathematician Tung Yu-ch'eng 董祐誠 (1791-1823) has shown that the problem could have been solved easily enough by traditional cycle manipulation, beginning with the datum that one planet was so many years along in its Great Period in 104. It is no longer possible to be sure how many planets, and which planets, were used, but in any case the period from Supreme Ultimate Epoch to the beginning of 104 was determined to be 31 Epoch Cycles or 143, 127 years [2]).

---

1) *Han shu, 21A*: 1696-1697.

2) *Han shu, 21B*: 1811.

In the case of Mars, which Tung used, this datum would hold true only five times in each Grand Polarity Superior Epoch Cycle, since the concordance cycle for Mars' Great Period and the Epoch Cycle is one-fifth of the "Great Year." At each of these five moments the distribution of the other planets would be markedly different. See Tung's *San t'ung shu yen pu* 三統術衍補 (reprinted in *Han shu pu chu, ch.* 21B), pp. 1852-1856 (also reprinted from his collected works in *Hsi hsueh fu ch'iang ts'ung-shu* 西學富强叢書). A similar demonstration using Jupiter has been published by Shinjō Shinzō in his *Tung-yang t'ien-wen-hsueh shih yen-chiu* 東洋天文學史研究 (Researches in

The day exactly 143, 127 years before the epochal solstice was, then, the beginning of time. It became the "first day" for every astronomical purpose; all celestial events were counted off either from it or from an intermediate cycle, just as for chronological purposes modern astronomers follow Scaliger in numbering Julian days from 1 January 4713 B.C.

Aside from inexact constants, and the arbitary character (in terms of mathematical astronomy) of the Great Planetary Conjunction Cycle, the most obvious shortcoming of the Triple Concordance system is that it incorporates two different sets of values for the mean year and mean lunation : one, based on an older tradition, which reflects the Rule Cycle, and the other, the "official values," meant to be a closer approximation but actually less accurate.

Modifications in the Later Han. The Quarter Day 四分 system (Li Fan 李梵, Pien Hsin 編訢 *et al.*, ca. A.D. 85) is often described by Chinese historians as regressive, since the lengths of the tropical year and lunation are, as before the Grand Inception reform, directly related to the Rule Cycle. They are the year and month values which we have just seen preserved in the eclipse complex of the Triple Concordance system. As Table III demonstrates, this reversion is an improvement not only in terms of consistency but in point of accuracy [1]).

TABLE III

Comparison of Calendrical Constants

System	*Tropical Year, Days*	*Lunation, Days*
Triple Concordance	365.2502	29.53086
Quarter Day	365.2500	29.53085
Modern	365.2422	29.53059
Accuracy of Quarter Day constants, 1 day per	128 years	310 years

the history of Oriental astronomy; tr. Shen Hsuan 沈璿; Shanghai, 1933), pp. 477-478. Shinjō sets up an indeterminate equation, which the ancients in effect would have solved by counting one Epoch Cycle at a time and trying values, and shows that the first possible solution is the number employed.

[1]) The system is named for the fractional part of the mean year value.

Early in the fifth century the central problem of reconciling the year and month was dealt with in another way, when Chao Fei 趙斐 first abandoned

The following constants of the Quarter Day system are counterparts of those we have examined in the Triple Concordance system, but the scheme is somewhat simpler.

YEAR $365\frac{1}{4}$ days

LUNATION $29\frac{499}{940}$ days, derived from the Rule Cycle.

The Concordance Cycle is no longer needed to deal with the "official" values of year and lunation. Instead, the

OBSCURATION CYCLE 蔀 27, 759 days = 76 years = 4 Rule Cycles, which includes an integral number of days and thus governs repetition of phenomena at midnight or at any other specified time of day, is made to concord with the sexagenary cycle for day count in the

ERA CYCLE 紀 555, 180 days = 9253 day cycles = 1520 years = 20 Obscuration Cycles.

This cycle defines the interval between simultaneous recurrences of new moon and winter solstice at midnight on the first day of a sexagenary day cycle. The sexagenary year count, which appeared only in the planetary complex of the Triple Concordance system, was transferred into this calendrical complex by radically redefining the

EPOCH CYCLE 4560 years = 3 Era Cycles,

which now becomes the interval for recurrence of the Era Cycle phenomena in the first year of a sexagenary year cycle.

It is particularly significant that the eclipse complex is not included; eclipse cycles are no longer computed from the beginning of an Epoch Cycle but, as we shall see, from the beginning of time. They are counted off either by the old Coincidence Month (renamed Year Number Cycle 歲數) of 513 years or, according to another method, by an auxiliary cycle, the

OBSCURATION COINCIDENCE CYCLE 蔀會 2052 years = 27 Obscuration Cycles = 4 Coincidence Months.

---

the Rule and Obscuration Cycles for greater complexity. He substituted the relation 600 years = 7421 lunations, of which 221 must be intercalated. The most advanced calendrical treatises tended to adopt increasingly complex intercalation cycles. Yabuuti, "Astronomical Tables in China, from the Han to the T'ang Dynasties" (see page 7), p. 448.

Twenty Obscuration Coincidence Cycles are required to include the Epoch Cycle. This comprehensive interval is defined in the treatise as the

EPOCH COINCIDENCE CYCLE 元會 41,040 years = 9 Epoch Cycles [1]).

The circumstance that in the Quarter Day system the cycle which culminates the calendrical complex is still called the "Epoch Cycle" merely obscures the fact that fundamental changes have taken place, changes which imply a considerable difference in the texture of cosmic reality. Above the level of the Epoch Cycle, similarity with the Triple Concordance system practically vanishes. The whole process of building up to a Grand Polarity Superior Epoch Cycle and then locating the present moment from observations of planetary positions was simply short-circuited. The length of the grand cycle which drove the system was not even computed. At one point the period of the Great Planetary Conjunction Cycle is given as 2, 999, 162, 158, 026, 300 years [2]). This formidable increase over the previous 138, 240 years is a natural result of improvement in knowledge of the synodic periods of the planets.

TABLE IV

Constants of the Planets in the Quarter Day System

Planet	*Synodic Cycle*		*Synodic Period, Years* (from Synodic Cycle)
Jupiter	4725 years =	4327 revolutions	1.092
Venus	4661	5830	1.599
Saturn	9415	9096	1.035
Mars	1876	879	2.134
Mercury	1889	11908	0.317

As might be expected from the abandonment of Liu Hsin's canonical value for the Synodic Cycle of Jupiter, there is no longer

[1]) *Hsu Han shu* (*chih* 3), pp. 3434-3438.

[2]) Pp. 3455-3456. The figure given in the text, 2, 990, 162, 100, 582, 300 years, is more than usually corrupt, but it is corrected in *Han ssu fen shu* 漢四分術, the great mathematician Li Jui's 李銳 (1765-1814) commentary on the mathematical techniques of the Later Han treatise. This monograph and Li's analogous *Han san t'ung shu* 漢三統術 and *Han ch'ien hsiang shu* 漢乾象術 are printed in his collected works, *Li shih i shu* 李氏遺書, and have been excerpted liberally in modern commentaries upon the early histories. For the reader's convenience I cite the latter.

any role for the Masculine and Feminine Factors to play; the Great Planetary Conjunction Cycle is merely the lowest common multiple of the five synodic periods [1]). Finally, as the text explains, it is "multiplied by the Obscuration Cycle constant in order to bring it into accord with the Epoch Cycle." In other words, multiplication by 76 gives the cycle which combines the calendrical-eclipse and planetary complexes, the "Great Year" of 227, 936, 324, 009, 998, 800 years. This figure so satisfactorily approximates infinity that its precise magnitude is beside the point; the value of the "Great Year" constant is not in fact given in the treatise at all [2]).

As a reflection of concern with what might be called "literary numerology," the age of the world was derived not from computation of planetary positions and their relation to a universal conjunction, but from statements in three apocryphal traditions of interpretation based on the Spring and Autumn Annals to the effect that 2, 760, 000 years had passed from the beginning of time until the capture of the fabulous *ch'i-lin* 麒麟 animal (481 B.C.) which ended the Spring and Autumn Period of the Chou dynasty [3]). To find the calendrical epoch, it was merely neccessary to go sufficient years past 481 B.C. to bring 2, 760, 000 up to an integral number of Epoch Cycles. Since this number did not lie anywhere near the time of Li and Pien—the two nearest Epoch Cycles would begin in 1681 B.C. and A.D. 2879—they settled on the beginning of the nearest Era Cycle, or 161 B.C., as the epoch. But the ephemerides was calculated, and days were numbered, from the beginning of time.

Comparative utility of Triple Concordance and Quarter Day systems

To what degree was the Quarter Day system astronomically superior to its predecessor? We have seen from Table III that the newer system's constants for tropical year and lunation, although

---

[1]) The values derived for Mercury and Venus have been doubled to facilitate comparison with the modern values given in Table II.

[2]) Li Jui calculated the figure; as printed in his commentary (*Li shih i shu, Wen hsuan lou ts'ung-shu* 文選樓叢書 ed., *4B*: 3b) it is unmarred by typographical error. A misprint has crept into the *Hou Han shu chi chieh* version (p. 3456.4).

[3]) *Ming li hsu* 命曆序, *Yuan ming pao* 元命苞, and *Ch'ien tsao tu* 乾鑿度. See *Hsu Han shu* (*chih* 2), pp. 3413-3414, and Wolfram Eberhard and Rolf Mueller, "Contributions to the Astronomy of the Han Period III. Astronomy of the Later Han Period," *Harvard Journal of Asiatic Studies*, 1936, 1: 228-230.

more ancient, are slightly better. The improvement would amount to about one day in eight hundred years for the lunation value, and one day in five thousand years for the year value, and would thus be altogether negligible over the roughly two hundred years during which each system was actually used. Even over a much longer period these improvements would be swamped by the comparatively low accuracy of the constants (for the Quarter Day system, one day in 128 years for the year, and one day in 310 years for the lunation). All that can fruitfully be compared is the accuracy of predictions for the period in which the methods were meant to be applied.

To begin with the ordinary calendrical functions, one could not want a better date with which to test the Triple Concordance system than its calendrical epoch. This day is *defined* by the coincidence of conjunction, winter solstice, and the inception of a sexagenary day cycle, all at midnight. The accuracy of this definition must inevitably affect calendrical phenomena predicted for other dates. As recourse to historical longitude tables proves, the coincidence corresponds to what could have been in the sky. The sexagenary cycle is a pure counting cycle, so its coincidence with the new moon cannot be impugned by any astronomical observation. The winter solstice falls within a day of the true conjunction [1]).

Astronomical First Month, 104 B. C.	Conjunction (Solar longitude = lunar longitude)	Solstice (Solar longitude = 270°)
Modern Computation	10: 30 A. M. 24 December 105	6: 00 P. M., 23 December 105

It would be misleading to take the calendrical epoch (161 B.C.) of the Quarter Day system for comparison. Since its basis was astronomically arbitrary, this initial date is so far removed from the time for which predictions were wanted that substantial differences in accuracy are to be expected (the system was in official use A.D. 85-263). A much fairer test of calendar-making reliability would be

[1]) Bryant Tuckerman, *Planetary, Lunar, and Solar Positions . . . at Five-day and Ten-day Intervals*. Vol. I. 601 B. C. to A. D. 1. Vol. II. A. D. 2 to A. D. 1649 (Philadelphia, 1962-1964). My calculations here and below have been reduced to local time at Yang-ch'eng 陽城 (long. 113° E.), the traditional location of the Imperial Observatory, and rounded to the nearest half-hour. Shigeru Nakayama and Owen Gingerich have kindly run computer checks on my abacus computations.

based on a date within the period of employment. Arbitrarily choosing the Astronomical First Month of A.D. 100, and performing calculations as directed in the treatise, one finds the accuracy of the solstice determination, both absolutely and with respect to lunation, to be inferior [1]).

Astronomical First Month A. D. 100	Conjunction	Solstice
Predicted	1 December 99	25 December 99
Modern Computation	4: 30 A. M., 1 December 99	5: 30 P. M., 23 December 99

How, one is prompted to ask, can a discrepancy of between one and two days not have been revealed by observation? It was at this time a necessity, and later became a matter of ritual, that a gnomon was used to find the day on which the sun's noon shadow was longest. Because the rate of change of shadow length is minimal in the vicinity of the solstices, this method is exceedingly imprecise. Shigeru Nakayama has estimated that an error of 1 centimeter in shadow length gives four to five days' error at winter solstice [2]). The upshot is that the Later Han astronomers had no reason to be dissatisfied with their predictions for December 99.

There was consistent improvement in knowledge of planetary periods, but the constants adopted in the previous system were already remarkably accurate. Whether greater practical ability to predict planetary phenomena was actually a factor in the success of the Quarter Day system is a moot point. Only mean values are given for the various planetary cycles, and there is no indication that variations were accounted for—as they must be if verifiable predictions are to be expected. Throughout the Han many visible planetary phenomena, particularly conjunctions, occultations, and "trespassess" of the planets upon certain constellations, were still in the realm of the ominous [3]).

While there was no change in the fundamental eclipse prediction

---

[1]) *Hsu Han shu* (*chih* 3), pp. 3443-3444. Eberhard and Mueller, pp. 209-212, renders these procedures corectly in every respect.

[2]) "Accuracy of Pre-Modern Determinations of Tropical Year Length," *Japanese Studies in the History of Science*, 1963, 2: 101-102.

[3]) Eberhard and Mueller, p. 209; Nakayama, "Characteristics of Chinese Astrology," p. 446-447.

cycle—that is, throughout the Han lunar eclipses were computed by use of the interval $5\frac{20}{23}$ (or $\frac{135}{23}$) lunations—the great modification in alignment of the Phase Coincidence Cycle with the other cycles (and thus in the epoch from which eclipses were counted off) makes a comparision advisable.

One of the fundamental specifications of the Triple Concordance system was that the lunar eclipse cycle began on 9 January 104 B.C., the first full moon after the Astronomical New Year. If this initial condition was not based on observation, the predictive value of the cycle would appear to be simply tossed away. But that was in fact the situation; according to modern calculations there was no lunar eclipse anywhere in the world between 13 August 105 and 29 December 104. Both of *these* dates can be predicted by the Triple Concordance method. It can be calculated, that is to say, that lunar eclipses will take place in the appropriate months, although only the second eclipse was visible in China [1]). One wonders whether the determination of the Martial Emperor to make the Astronomical New Year of 104 a "grand inception" from the cosmic point of view overruled a natural expectation that the eclipse which begins the cycle be prominent, or at very least observable.

Even if the counting-off technique of the Triple Concordance system were capable of forecasting every eclipse which takes place, only half of the predictions, on the average, would be confirmed in China. In order to reap the fruits of calendrical astronomy at all, the judicial astrology of the time must have incorporated a rule of this sort: An eclipse seen but not predicted is an omen; that an eclipse is predicted but not seen has no astrological significance [2]). Since eclipses were predicted at intervals of five or six months (by adding

[1]) At 5: 45 A. M., Yang-ch'eng time, magnitude 10.4 on a scale of twelve units. Here and below I determine visibility for the maximum phase by the method of Theodor von Oppolzer, *Canon of Eclipses*, tr. Owen Gingerich (New York, 1962), p. xxxiv. For the Chinese prediction method, see *Han shu, 21B*: 1743, translated in Eberhard, Mueller, and Robert Henseling, "Beiträge zur Astronomie der Han-Zeit. II," Preussische Akademie der Wissenschaften, *Sitzungsberichte* (Phil.-Hist. Klasse), 1933, 23: 943-944. For the reader's convenience I retranslate the procedure, using the terminology which appears throughout this paper, in Appendix A below.

[2]) In his *Li-tai jih-shih k'ao* 歷代日食考 (Researches on solar eclipses through history; Shanghai, 1934), pp. 62-71, Chu Wen-hsin had shown that in the T'ang solar eclipses were over-predicted too. In "Characteristics of Chinese Astrology" (p. 446) Nakayama cites Suzuki Takanobu 鈴木敬信 to the effect that in Japan, where records of lunar eclipses were published, a quarter of those noted prior to 1600 could not have been visible.

increments of $5\frac{20}{23}$ months and rounding off to the nearest full moon), a simple criterion would have sufficed to deal with the by no means negligible problem of whether an eclipse is predicted but not seen, or whether instead the closest visible eclipse is seen but not predicted. If the interval between forecast and nearest observation were five or six months (or, conceivably, eleven or twelve months, if the gap happened to be twice as long as usual), the situation would be astrologically insignificant. If, on the other hand, an eclipse were seen one to four (or seven to ten) months earlier or later, it would clearly be unpredicted, and thus ominous. The epochal eclipse of the Grand Inception period, since it precedes the eclipse of 29 December 104 by twelve lunar months, is an astrologically valid interpolation.

With the aid of this criterion we may proceed to evaluate the long-term accuracy of the technique. Table V presents in summary form data on twenty additional predictions, ten for the first eclipse in each of ten successive civil years beginning 104 B.C., and a corresponding series for a period about 150 years later. In the first group, only five of the ten predictions correspond to an eclipse visible at Yang-ch'eng, the site of the Imperial Observatory. The other five are all either six or twelve months from an observable eclipse, and thus must also be reckoned as successful predictions.

This perfect score was not maintained over the period in which the Triple Concordance system was actually used, as the second group, a representative sample, demonstrates. Four of the ten eclipses between 48 and 57 were confirmed by observation. Four more would have been vindicated by interpolation; the Chinese astronomers could not have known, as we do, that the prediction for 49 fell only one month after a lunar eclipse visible outside of China—nor that an eclipse was visible in Western Europe on 11 February 54. The forecast for 53 was an unambiguous failure; that for 57 was an ambiguous failure, since the eclipse of 11 December 56 could not have been viewed less than about two hundred miles east of Yang-ch'eng[1]). Of the ten predictions, in other words, three

[1]) In the Han reports of eclipses were received irregularly from elsewhere in China; the historical importance of this factor seems to have been minor. See, for instance, *The History of the Former Han Dynasty*, tr. Dubs, I, 289 and 338. Even if there had been observatories all over China, as we know there were not, the increase in eclipses observed would not sensibly affect the import of this or subsequent arguments.

TABLE V

Two Series of Lunar Eclipse Predictions by the Triple Concordance Technique

Year	*Calendar month predicted*	*Date of full moon*	*Nearest actual eclipse*	*Nearest visible eclipse*	*Interval, months*	*Evaluation*
104 B.C.	5	4 Jul		29 Dec 104	6	Interpolated
103 B.C.	5	23 Jun		23 Jun 103	–	Confirmed
102 B.C.	5	13 Jun		7 Dec 102	6	Interpolated
101 B.C.	4	2 Jun		7 Dec 102	6	Interpolated
100 B.C.	3	22 Apr		22 Apr 100	–	Confirmed
99 B.C.	3	11 Apr		11 Apr 99	–	Confirmed
98 B.C.	2	1 Apr		1 Apr 98	–	Confirmed
97 B.C.	2	21 Mar	25 Sep 98	1 Apr 98	12	Interpolated
96 B.C.	97/12 intercalary	8 Feb		8 Feb 96	–	Confirmed
95 B.C.	96/12	29 Jan	29 Jan 95	4 Aug 96	6	Interpolated (unknown success)
A.D. 48	5	15 Jun		15 Jun 48	–	Confirmed
A.D. 49	4	4 Jun	6 May 49	29 Oct 49	5	Interpolated (unknown failure)
A.D. 50	3	26 Apr		26 Apr 50	–	Confirmed
A.D. 51	3	15 Apr		15 Apr 51	–	Confirmed
A.D. 52	2	3 Apr		8 Oct 51	6	Interpolated
A.D. 53	2	23 Mar		21 Feb 53	1	Known failure
A.D. 54	1	11 Feb	11 Feb 54	1 Feb 55	12	Interpolated (unknown success)
A.D. 55	54/12	1 Feb		1 Feb 55	–	Confirmed
A.D. 56	55/12	21 Jan	27 Jul 55	1 Feb 55	12	Interpolated
A.D. 57	56/12	9 Jan		11 Dec 56?	1	Known failure?

NOTE: An entry is provided in the "Nearest actual eclipse" column only when the eclipse which occurs closest to the date predicted would not have been visible at Yang-ch'eng.

were failures in an absolute sense, and one or two would have been known as failures.

Was the epoch somehow at fault, or is the basic eclipse cycle bound to lose time eventually like a cheap clock? A close examination of the modifications which appear in the Quarter Day treatise is clearly called for. There are two methods, which count off the interval from the beginning of time by different cycles. Their basic functions are not precisely the same; the first computes the last eclipse preceding, and the second computes the first eclipse following, the Astronomical First Month of the year designated. An auxiliary formula for the first technique, however, makes the two capable of predicting the same event. Since the two have never been critically compared in a Western language, I translate and explain them below, and calculate the month of the first eclipse of A.D. 100 as directed by both techniques.

(1) Quarter Day Technique, First Method (*Hsu Han shu* [*chih* 3], pp. 3439, 3450-3452)

> To find the year in the current Obscuration Coincidence Cycle when calculating lunar eclipses, divide [the number of the year counted from] the Superior Epoch by the Epoch Coincidence Cycle (41040 years), and divide the remainder by the Obscuration Coincidence Cycle (2052 years) . . . The remainder is the number of the year in the current Obscuration Coincidence Cycle.

Years elapsed to 161 B. C.	2, 760, 320
Interval to A. D. 100	260
Years elapsed to A. D. 100	2, 760, 580.

A. D. 100 is no. 2, 760, 581 in the Superior Epoch Cycle.
$\frac{2760581}{41040}$ = 67 Epoch Coincidence Cycles, remainder 10901.
A. D. 100 is no. 10901 in the 68th Epoch Coincidence Cycle.
$\frac{10901}{2052}$ = 5 Obscuration Coincidence Cycles, remainder 641.
A. D. 100 is no. 641 in the sixth Obscuration Coincidence Cycle.

> Method of calculating lunar eclipses: Take the number of the year in the current Obscuration Coincidence Cycle and subtract 1. Multiply by the Eclipse Number (1081). Divide by the Year Number (513). The integral part of the result is called Accumulated Eclipses 積食; the remainder is called Eclipse Remainder 食餘.

Substracting 1 converts from the number of the year back to years elapsed up to Astronomical New Year (abbreviated A. N. Y. below). $640 \times \frac{1081}{513} \frac{\text{eclipses}}{\text{year}} = 1348\frac{316}{513}$ eclipses in current Obscuration Coincidence Cycle up to A. N. Y. 100. Discarding the remainder (which represents eclipse parts accumulated between the last eclipse and A. N. Y.) means that the next cycle will be counted off only up to the eclipse itself.

> Accumulated Eclipses[1] is multiplied by the Month Number (135) and divided by the Eclipse Rule (23). The integral part of the result is called Accumulated Months 積月; the remainder is called Month Remainder 月餘分.

$1348 \times \frac{135}{23} = 7912\frac{4}{23}$ months to eclipse preceding A. N. Y.

> Accumulated Months is divided by months per Rule Cycle (235). The remainder is the number of months in the current Rule Cycle. First casting out 除 intercalations in the current Rule Cycle [see next part], divide by

1) I accept Wang Hsien-ch'ien's obvious emendation "積" to "積食".

12. The remainder is counted exclusively from the eleventh month to yield the month of the eclipse which precedes the eleventh month of the year previous [to that with which the calculation began].

$\frac{7912}{235}$ = 33 Rule Cycles, and 157 months elapsed in current Rule Cycle up to new moon preceding eclipse. 157 months minus 4 intercalations in Rule Cycle to date (see below) leaves 153 months reckoned on the basis of a 12-month year. $\frac{153}{12}$ = 12 years elapsed in current Rule Cycle and 9 months in current year (reckoned from A. N. Y.). Counting nine months beginning with the twelfth civil month of A. D. 98 (the month after A. N. Y.), the last eclipse of A. D. 99 will fall in the eighth month of the civil calendar (4 September - 2 October).

To find the number of intercalations in the current Rule Cycle: Multiply months elapsed in current Rule Cycle by intercalations per Rule Cycle (7) and divide by months per Rule Cycle, which gives the number of intercalations in the current Rule Cycle. If the remainder falls between 224 and 231, the eclipse will fall in an intercalary month.

$157 \times \frac{7}{235} \frac{\text{intercalations}}{\text{month}} = 4\frac{159}{235}$ intercalations to date. As Li Jui explains, the remainder increases by 7 per month. When it reaches 235 ($\frac{235}{235}$ intercalations) an intercalary month must be added. Since we are working with the month before the one in which the eclipse takes place, the remainder rule is equivalent to beginning the month in which the eclipse occurs at a remainder between 231 and 238, or centering it upon 235; the spread defines the leeway of the course of the month itself.

In some cases the intercalation will be predicted early or late. If so, it is to be determined by the procedure for [comparing *ch'i*-centers with] lunation dates.

The same stipulation is made in the ordinary intercalation procedure given earlier in the treatise (p. 3444; Eberhard and Mueller, p. 214, take "進退" ["advanced or retarded, early or late"] to mean "calculating backwards and forwards"). For the *ch'i*-center technique, the standard to which the other methods can only approximate, see *Han shu, 21B:* 1738, and Ch'en Chen-hsien in *Kuo wen chou-pao*, 1934, *11*. 26: 8-9. Eberhard, Mueller and Henseling, p. 940, understand the passage correctly in their commentary, but their translation is based upon a full stop before rather than after "若一日."

To find later eclipses, add $5\frac{20}{23}$ months [1] [to Accumulated Months and Month Remainder]. If the fraction adds up to unity, count it as a full month. Whenever there is no remainder, an eclipse is counted.

$7912\frac{4}{23}$ months to eclipse preceding A. N. Y.
$+ 5\frac{20}{23}$ months to next eclipse
$= 7918\frac{1}{23}$ months to eclipse following A. N. Y.
$\frac{7918}{235}$ = 33 Rule Cycles and 163 months
163 months - 4 intercalations to date = 159 months reckoned on the basis of a 12-month year.

---

[1]) Wang emends "百" to "月."

$\frac{159}{12}$ = 13 years elapsed in current Rule Cycle and 3 months in current year (reckoned from A. N. Y.).
Counting three months beginning with the twelfth civil month of A. D. 99, the eclipse is predicted for the second month (28 February - 27 March) of A. D. 100.

Method for calculating the date of the conjunction which begins the month in which a lunar eclipse takes place: Take Accumulated Months to last eclipse and multiply by 29 to give Accumulated Days. Also multiply by 499 and divide by months per Obscuration Cycle (940) [1]), adding the result to Accumulated Days. Divide the aggregate [integer] by 60 and count the remainder exclusively from the beginning of the current Obscuration [Coincidence] Cycle. The result is the first day of the month in which an eclipse takes place preceding the Astronomical First Month.

The number of months up to the new moon preceding the elipse is converted to days when it is multiplied by $29\frac{499}{940}\frac{\text{days}}{\text{month}}$. Sexagenary day cycles are counted off, the remainder being days elapsed in the current cycle. The purpose of the sexagenary count is to provide an actual date for the eclipse; in Chinese astronomy dates are necessarily sexagesimal. But the text is clearly corrupt, for the count must begin from the inception of one of the cycles by which eclipses are computed, or the procedure will not work. In emending "Obscuration Cycle" to "Obscuration [Coincidence] Cycle" I follow Shigeru Nakayama's elegant solution to this far from transparent problem (private communication).

To find the day of the eclipse, add the Great Remainder 14 to the integer and the Minor Remainder $719\frac{1}{2}$ to the numerator of the fraction. Integers which result from combining the fractions are added to the integral number [of Accumulated Days], which are counted off as before to give the date of the eclipse.

This operation amounts to adding half the days in a mean month, and thus moving from new to full moon.

To find the next conjunction preceding, and the next day of, a lunar eclipse, [take Accumulated Days and] add 27 to the integer and 615 to the numerator of the fraction. If the Month Remainder was less than 20, add another 29 to the integer and 499 to the remainder. The fractional remainder is [converted into decimal parts of a day and] compared with the number of graduations of the clepsydra indicators for that part of the year. If it does not amount to the number of graduations on one night indicator, a day is added [to the number of Accumulated Days].

Here one adds to the number of days a figure equivalent to either five ($27\frac{615}{940} = [5 \times 29\frac{499}{940}] - [2 \times 60]$) or six months as determined by the rule. If the original Month Remainder (the fraction of a month from new moon to the last eclipse) was less than $\frac{20}{23}$, conversely the fraction from that eclipse to the *next* new moon will be greater than $\frac{3}{23}$ (since a lunation contains 23 fractional parts). When $5\frac{20}{23}$ is added to the latter fraction the sum will exceed 6.

---

1) Wang emends "四百九十" to "四百九十九."

The conversion to clepsydra night parts, for which a formula is given in the Treatise, is in most cases unnecessary; the need to add an extra day can usually be estimated by inspection. Because of an apparent corruption in the later formula, the last sentence of my translation above is tentative.

In order to illustrate these procedures for computing dates, the case of the first eclipse of A. D. 100 will be worked out.

7912 months to conjunction preceding last eclipse of A. D. 99 $\times$ $29\frac{499}{940}$ $\frac{\text{days}}{\text{month}}$ = 233, $648\frac{88}{940}$ days to lunation
+ $14\frac{719\frac{1}{2}}{940}$ days from conjunction to eclipse
+ $27\frac{615}{940}$ days to next eclipse
+ $29\frac{499}{940}$ additional days as required by rule
= 233, $720\frac{41\frac{1}{2}}{940}$ days to first eclipse of A. D. 100
$\frac{233720}{60}$ = 3895 sexagenary cycles and 20 days.

We have already seen (p. 28) that the current Obscuration Coincidence Cycle is the sixth. Counting back the 640 years which have elapsed by A. D. 100, we find that the cycle begins 2, 759, 940 years after the Superior Epoch. Converting to days and casting out sixties, we are left with forty-five days elapsed in the current sexagesimal cycle. The current cycle thus begins with day no. 46, and the eclipse is predicted for day no. 6 (46 + 20 = 66, casting out sixties). Since the second month of 100 begins with a sexagenary day 51, the eclipse is predicted for the fifteenth (66-51), the full moon.

Plainly, this involved procedure serves no practical pupose, for once one knows the month of an eclipse no calculation is needed to set its date at the full moon of that month. The function of this formula is, if anything, metaphysical, in the sense that it fills out a complete system based on counting cycles.

Quarter Day Technique, Second Method (pp. 3452-3453)

Another method [for months of lunar eclipses]: Take years elapsed since the Superior Epoch and divide by the Year Number. The remainder [is multiplied by $\frac{235}{19}$; the integral part of the result] is Accumulated Months. Multiply by 112 and divide by the Month Number, discarding the integral result. The remainder is divided by the Eclipse Rule to give the eclipse following Astronomical New Year.

Cf. Eberhard and Mueller, p. 218. "Accumulated Months" has a different significance here than in the first method; it refers to integral months elapsed in the current Year Number Cycle up to A. N. Y. If Accumulated Months were multiplied by $\frac{23}{135}$ (eclipses per month), the integral dividend would be eclipses to A. N. Y., and the remainder would be fractional parts of an eclipse between the last eclipse and A. N. Y. Using the multiplier $\frac{112}{135}$ instead amounts to subtracting those fractional parts from the interval between two successive eclipses ($\frac{112}{135} = 1 - \frac{23}{135}$), leaving fractional parts of an eclipse between A. N. Y. and the *next* eclipse. Since 23 of these parts accumulate each month, division by 23 gives the number of months they have been accumulating. The remainder is the fraction of a month intervening between the new moon and the eclipse. This second method allows the first eclipse of A. D. 100 to be calculated directly:

$\frac{2760580}{513} = 5381$ Eclipse Months and 127 years.
$127 \times \frac{235}{19} \frac{\text{months}}{\text{year}} = 1570\frac{15}{19}$ months to A. N. Y.
1570 months to conjunction preceding A. N. Y. $\times \frac{112}{135} =$
$1302\frac{70}{135}$ eclipses, of which the fraction represents eclipse parts from A.N. Y. to the next eclipse.
$\frac{70}{23} = 3$ months from A. N. Y. to conjunction preceding eclipse and $\frac{1}{23}$ month from new moon to eclipse. This result, which puts the eclipse in the second civil month of A. D. 100, is identical with that obtained by the first method.

But the greater complexity of the Quarter Day eclipse techniques is due not to a fresh attack on the problem, but partly to the change of epoch and partly to a demand for predictions to the nearest day and hour that merely complicates the system without adding to its predictive power. The example illustrates this point adequately, for the lunar eclipses of A.D. 100 fell not on 14 March (the full moon of the second month) but on 13 February and 7 August, and the second was visible in China. The prediction was in principle a failure, although, preceding the eclipse of 7 August by five lunations,

TABLE VI

A Series of Lunar Eclipse predictions by the Quarter Day Technique

Year	*Calendar month predicted*	*Date of full moon*	*Nearest actual eclipse*	*Nearest visible eclipse*	*Inter-val, months*	*Evaluation*
100	2	14 Mar	13 Feb 100	7 Aug 100	5	Interpolated (unknown failure)
101	100/12	2 Feb		2 Feb 101	–	Confirmed
102	101/12	22 Jan	22 Jan 102	17 Jul 102	6	Interpolated (unknown success)
103	102/12	11 Jan		17 Jul 102	6	Interpolated
104	4	27 May		27 May 104	–	Confirmed
105	4	17 May		17 May 105	–	Confirmed
106	3	6 May		9 Nov 105	6	Confirmed
107	3	25 Apr		26 Mar 107	1	Known failure
108	2	15 Mar	15 Mar 108	5 Mar 109	12	Interpolated (unknown success)
109	1	5 Mar		5 Mar 109	–	Confirmed

it was a success in the eyes of the Han astronomers. A series of computations by the same technique (Table VI) suggests that the Quarter Day methods are inherently no more capable than their predecessor of yielding highly reliable predictions, even to the nearest month. We shall see anon that, as time passed, their failure became increasingly disastrous.

Can the Chinese scientists have been reconciled to the inadequacy of their eclipse technique? If this question is to be answered, it will be necessary to look closely at the astronomical significance of the Han cycle.

The Meaning of the Chinese Eclipse Cycle

Lunar eclipses happen only at full moon, in the middle of a Chinese month, when the moon intercepts the shadow of the earth cast by the sun [1]. If the planes of rotation of the sun and moon coincided, there would be an eclipse twice each month, a solar eclipse at new moon and a lunar eclipse at full moon. But actually the two planes are inclined at an angle of about 5°.8; they intersect at two points called the lunar nodes. The basic problem of lunar eclipse prediction is to determine what happens when the sun and moon, each travelling at its own speed, move into opposition at a given distance from a node. It can be seen from Figure 3 (which is for heuristic reasons not at all to scale) that, because the earth's shadow is larger than the moon, for a certain distance eclipses remain total. Beyond that distance the face of the moon is no longer completely obscured, so that eclipses are partial. The limit for totality is between 4°.4 and 5°.4 on either side of a node; that for partiality varies between 9°.8 and 11°.6, depending upon the distance of the moon and sun from the earth. The limits for solar eclipses are considerably greater, but the shadow cone is very narrow. It covers only a small circle on the earth's surface, and sweeps out a narrow band as it moves. A small difference in the conditions of a solar eclipse makes a greater diffence in visibility at a given place; a lunar eclipse can be seen from anywhere in the hemisphere facing

[1]) The explication which follows is not the whole story, but merely the minimum needed to make my subsequent argument intelligible to those untrained in astronomy. Further technical details are available in any textbook of elementary spherical astronomy. For the sake of simplicity and consistency I use geostatic language thoughout. Whether the sun rotates about the earth or the earth about the sun is immaterial to the mathematical prediction of lunar eclipses.

the moon. That is why, although there are more solar than lunar eclipses—in any given year the number varies from a maximum of five and two or four and three to a minimum of two and none—the number of eclipses of the sun visible from a given place (even a given continent) is considerably smaller, and the difficulty of making predictions about when a solar eclipse can be seen from any one observatory is immensely greater. It is well known that, from the T'ang on, successive groups of foreign astronomers won their entrée into the Astronomical Bureau, in the main, by predicting eclipses of the sun more accurately than the incumbents.

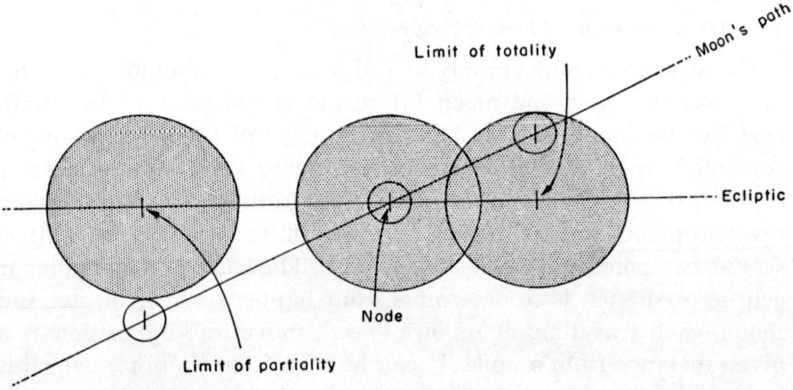

Figure 3. Conditions of a lunar eclipse. The shaded circles are a magnified projection of the earth's shadow cone upon a perpendicular plane passing through the center of the moon.

Two basic astronomical tools, neither of which was available in the Han, are required for reasonably satisfactory computations of lunar eclipses. First, one must be able to deal with variations in apparent velocity of the sun and moon, so that the moment of opposition can be predicted with greater precision than possible with mean rotation periods of the sort found in the Han systems. The major component of these variations is due to the fact that the orbits of the sun and moon are not quite circular about the earth; the resultant variation in apparent size of the moon and of the earth's shadow must also be taken into account when determining the precise conditions of interception. Second, spherical geometry or trigonometry is indispensable in order to find what angle from the node measured along the moon's orbit is equivalent to a given angle measured along the ecliptic, and to know the dependence of the moon's latitude upon nodal distance at a given moment.

Lacking all of this, it is nevertheless possible to discover cycles which have considerable predictive value for long periods.

Until modern times, the only basic eclipse cycle known to European civilization was that now called the Saros, a period of 223 lunations or about 18 years 11 days. It may be described, in the language I have applied to the Chinese system, as a concordance cycle for the eclipse year and various lunar periods, including the lunation and the nodical and anomalistic months (defined below). Since both Babylonian and Greek astronomy could deal with solar

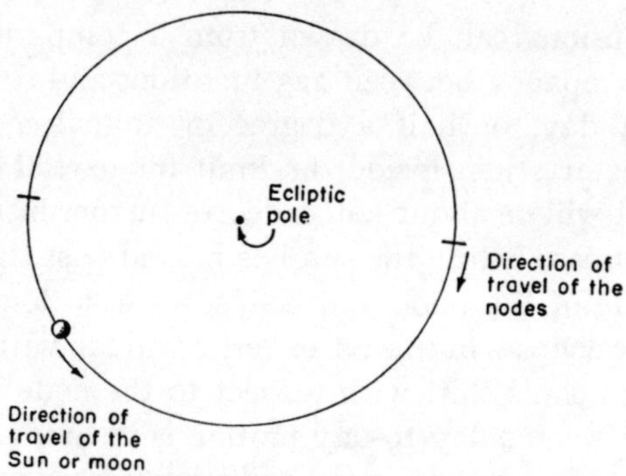

Figure 4. Movement of the lunar nodes.

and lunar inequalities and the transformation of spherical coordinates in the earliest period for which we have full records, the actual importance of the 223-lunation cycle in archaic Western astromony remains unknown [1]).

[1]) The properties of the 223-month cycle are dicussed in the first century A. D. by Pliny in his Natural History (II. x. 56), but the name is modern and based upon a misunderstanding. For a history of the legend that the Saros was the basis of Babylonian eclipse prediction, see Neugebauer, "Untersuchungen zur antiken Astronomie V. Der Halleysche 'Saros' und andere Ergänzungen zu UAA III," *Quellen und Studien zur Geschichte der Mathematik Astronomie und Physik* (Abteilung B), 1938, 4: 407-411, summarized in *The Exact Sciences in Antiquity*, pp. 141-142.

I use the term "cycle" below to denote the consecutive eclipses within a period of so many months, and "series" to denote a succession of eclipses each separated from the next by an interval of so many months. A Saros cycle is 223 months or all eclipses within that period. A Saros series is all successive eclipses 223 months apart.

The lunar nodes—which T'ang astronomers learned from India to reify as the "planets" Rahu 羅睺 and Ketu 計都—are by no means static; they move westward about the ecliptic with a period of 18.6 years [1]). The interval between successive passages of the sun through the same node (the eclipse year of 346.62 days) is thus less than a tropical year, and the corresponding period for the moon (the nodical or draconitic month of 27.2122 days) is less than a lunation (see Figure 4).

223 lunations	=	6585.32 days
19 eclipse years	=	6585.78 days
242 nodical months	=	6585.35 days

Several conclusions can be drawn from a comparison of these figures. The discrepancy between 223 lunations and 19 eclipse years is about half a day, or half a degree of annual motion of the sun. If the sun starts just inside the limit for partial eclipses, 223 lunations later it will be about half a degree further inside. Forty or fifty Saroses will pass before the sun has moved past the outer limit on the other side of the node and eclipses are no longer possible. About half of the eclipses in this Saros series will fall within the limits of totality. The moon's shift with respect to the node is much less, but because the moon's day-to-day motion is so much swifter than that of the sun, this circumstance has little effect except upon the time of day the eclipse takes place. In the odd 0.32 day of 223 lunations, the sun travels about 115°, so one eclipse will be central at a longitude which averages 115° from the next, and should still be visible, although no longer central, at the point on earth from which the first was observed. The precise shift depends upon other factors, of which anomaly (position with respect to apogee) is the most important. All of these points are illustrated concretely in Table VII [2]).

The time for the moon to return to apogee, the point in its orbit where it is furthest from the earth and its apparent motion is

[1]) Willy Hartner, "The Pseudo-Planetary Nodes of the Moon's Orbit in Hindu and Islamic Iconographies. A Contribution to the History of Ancient and Medieval Astrology," *Ars Islamica*, 1938, 5: 113-154, is to this date one of the very few monographs which studies the history of astronomy on a world-wide basis with impartially critical authority.

[2]) The average shift in longitude shown in Table VI is somewhat less than 115°, but other Saros series would diverge on the high side. In the series beginning with the eclipse of 19 September 358 B. C., for instance, the shift varies between 134° and 104°.

II

TABLE VII

Centrality and Visibility of All Eclipses in a Saros Series

(26 May 445 B.C.-21 July A.D. 259)

m	*Julian day*	*longitude,* λ	*shift,* Δλ	*latitude,* φ	*Visible ?*[1])
1	1559 033	-40		-20	No
2	1565 618	-138	98	-22	No
3	1572 203	124	98	-23	4.2
4	1578 788	25	99	-24	No
5	1585 374	-72	97	-23	No
6	1591 959	-173	101	-22	9.5
7	1598 544	85	103	-20	10.9
8	1605 130	-19	104	-18	No
9	1611 715	-126	107	-15	No
10	1618 300	126	108	-11	Total
11	1624 885	15	111	-7	No
12	1631 471	-99	114	-3	No
13	1638 056	145	116	2	Total
14	1644 641	29	116	6	Total
15	1651 227	-90	119	10	No
16	1657 812	151	119	15	Total
17	1664 397	32	119	18	Total
18	1670 983	-89	121	21	No
19	1677 568	152	119	23	Total
20	1684 153	33	119	24	Total
21	1690 739	-86	119	24	No
22	1697 324	157	117	22	Total
23	1703 909	40	117	21	Total
24	1710 495	-76	116	19	Total
25	1717 080	170	114	15	Total
26	1723 665	57	113	12	Total
27	1730 251	-53	110	8	No
28	1736 836	-162	109	4	Total
29	1743 421	92	106	-1	Total
30	1750 007	-14	106	-4	No
31	1756 592	-117	103	-8	No
32	1763 177	141	102	-12	Total
33	1769 762	41	100	-15	Total
34	1776 348	-58	99	-19	No
35	1782 933	-157	99	-21	No
36	1789 518	104	99	-23	7.8
37	1796 103	6	98	-24	No
38	1802 689	-94	100	-23	No
39	1809 274	165	101	-22	2.4
40	1815 859	61	104	-21	0.8
TOTAL ECLIPSES VISIBLE AT YANG-CH'ENG					19-20

[1]) Magnitude of visible eclipses is indicated

consequently minimal, is not equal to the lunation. Because of the gravitational effect of the sun and to some extent of the planets, the anomalistic month is only 27.5546 days long, and is connected with the Saros by the relation

239 anomalistic months = 6585.55 days.

Since the discrepancy between 223 lunations and 239 anomalistic months is small, the role of the changing apparent size and speed of the moon in altering the conditions of eclipses a Saros apart is both minor and gradual. This factor is evident not only in gradual rise and fall of latitude (φ), but especially in the regular increase in magnitude between early eclipses, and a correspondingly regular decrease between late eclipses, in a Saros series (Figure 5) [1]. An

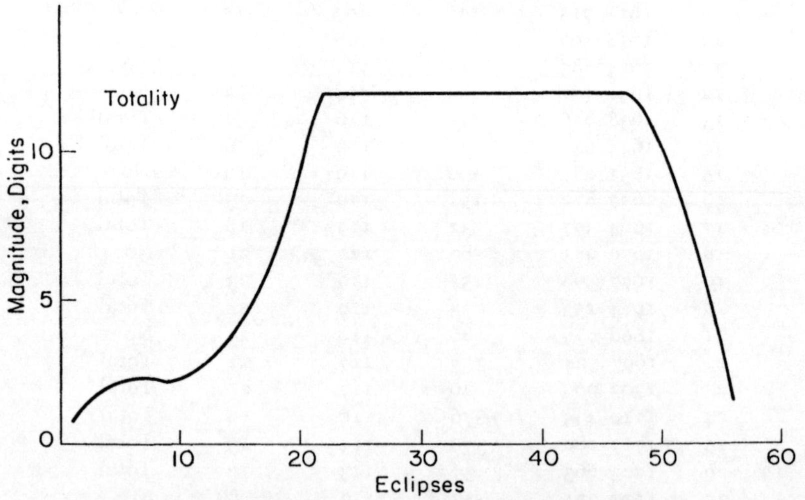

Figure 5. Magnitudes of all eclipses in a Lunar Saros series. Series begins with Julian Day 1590925, 19 September 358 B.C.

observer who notices an increase in magnitude from one partial eclipse to another 223 months later can safely conclude that a series of total eclipses will follow eventually; if the magnitude decreases, he can be sure that no more total eclipses are to come,

[1]) The series graphed was chosen to represent the *least* linear variety. The data provided in Table VII would make a much less interesting graph, since the slope of the lines connecting magnitudes of partial eclipses is nearly constant.

and that it is only a matter of time until no more eclipses can be predicted in the series.

In recent times Simon Newcomb discovered an equally fundamental cycle of 358 lunations, or twenty days less than twenty-nine years. The Inex, as this cycle has been named, has very different properties from those of the Saros [1]).

358 lunations	=	10571.95 days
30.5 eclipse years	=	10571.91 days
388.5 nodical months	=	10571.94 days
384 anomalistic months	=	10580.97 days

The discrepancy between lunations and eclipse years is only 0.04 days. If the sun's displacement from one Inex to the next is no more than 0°.04, it will take an average of 23,000 years, or almost 800 Inex periods, to work its way across the region near the nodes where eclipses are possible. Because of the odd half of the eclipse year and nodical month, the sun and moon move from one node to another each Inex, changing the direction of shadow travel across the face of the moon. The very poor correspondence of the anomalistic month with the other periods implies that the moon's longitude, latitude, and anomaly will all vary greatly from one Inex to the next. The gradual alteration in shift of longitude and the progressive change of latitude and magnitude which characterized the Saros will all be absent. Despite the tremendous longevity of the series, the magnitudes of a few successive eclipses offer no dependable clue as to what will happen later on. An

---

[1]) In my discussion of the characteristics of eclipse cycles I follow the nomenclature of George van den Bergh, *Periodicity and Variation of Solar (and Lunar) Eclipses* (Haarlem, 1955), a detailed and imaginative attempt to bring order to the multiplicy of possible prediction cycles. Several cycles which the author derives abstractly (the Tritos, described below, and others of 939, 5640, and 11045 lunations) were actually employed or proposed in China. See Kao P'ing-tzu 高平子, "古今交食週期比較論" ("A comparative study on the eclipse periods past and present"), Academia Sinica, *Li-shih yü-yen yen-chiu-so chi-k'an* 歷史語言研究所集刊, 1951, 13. 1: 1-23, which also discusses other periods merely implied by the constants of various systems. The numerous papers of Alexander Pogo on the periodicity of eclipses published in *Popular Astronomy*, Vol. 43 (1935); A. Pannekoek, "Periodicities in Lunar Eclipses," Koninklijke Nederlandse Akademie van Wetenschappen, *Proceedings* (Series B: Physical Sciences), 1951, 54: 30-41; the paper of Ch'ien Pao-tsung cited on page 6; and especially Chu Wen-hsin 朱文鑫, *Li fa t'ung chih* 曆法通志 (A general history of calendrical astronomy; Shanghai, 1934), pp. 251-269, also bear directly upon my argument.

Inex series enters both partiality and totality only by fits and starts, yielding many false predictions. An eclipse of large magnitude can be both preceded and followed at a distance of a half-dozen cycles by no eclipse at all. Even though the Inex cycle can predict the "mature" portion of each series with perfect reliability for the several millennia it takes the sun to move through the central part of the region of totality, the erratic behavior of a series in youth and old age almost entirely rules out the possibility that the cycle's overall value (its greater or lesser applicability to *every* eclipse) could have been discovered in antiquity by an analysis of eclipse records.

On the face of it, the basis of the Chinese method does not seem to be a cycle in this sense, but simply an empirical statement that there are twenty-three lunar eclipses every 135 months; the Phase Coincidence Cycle is used to predict consecutive eclipses, not eclipses 135 lunations apart. Fortunately Ssu-ma Piao preserves a statement which settles the question (and at the same time explains the cycle's name): "The lunar eclipse constants are derived from records of total eclipses 月食之既者. On the average, totality recurs every twenty-third eclipse; the [equivalent] number of months elapsed is 135 [1])."

The Tritos (to use George van den Bergh's term for the Chinese interval) is, in fact, the difference between the Inex (358) and Saros (223) cycles, and one-third of the well-known Mayan cycle of 405 lunations (about 32 years 9 months). The characteristics of the Tritos combine the limitations of the Saros and Inex.

135 lunations	=	3986.63 days
11.5 eclipse years	=	3986.13 days
146.5 nodical months	=	3986.59 days
145 anomalistic months	=	3995.42 days

The sun's shift with respect to the node is about $\frac{1}{2}^\circ$ per Tritos, so the longevity of a series will be no better than that of a Saros series. Because of the gap between 135 lunations and 145 anomalistic

---

1) *Hsu Han shu* (*chih* 3), pp. 3436-3437. I follow Ch'ien Ta-hsin's 錢大昕 (1728-1804) rejection of "食" in the last phrase.

Dubs, in *History of the Former Han Dynasty* (see page 17), I, 284, has uncovered in the Han Annals what is apparently a lunar eclipse recorded erroneously as solar, confirming that such observations were made and recorded at the court, even if not considered important enough to be cited in in the Histories. Lunar eclipse records date back to the oracle bones.

months, changes in longitude, latitude, and magnitude of successive eclipses in a series, although not as abrupt as with the Inex, will lack the predictable smoothness of the Saros (see Table VIII and Figure 6). If the likelihood of discovering a cycle is at all proportional to the amount of information it is capable of furnishing the observer, the Tritos is about as crude a cycle as ancient astronomers could be expected to know. Only in the central region of totality does it provide consistently unambiguous information. That the Chinese found it from records of total eclipses is more or less to be expected. At the same time it represents a parsimonious solution to their problem; any clearly superior cycle would have been much longer.

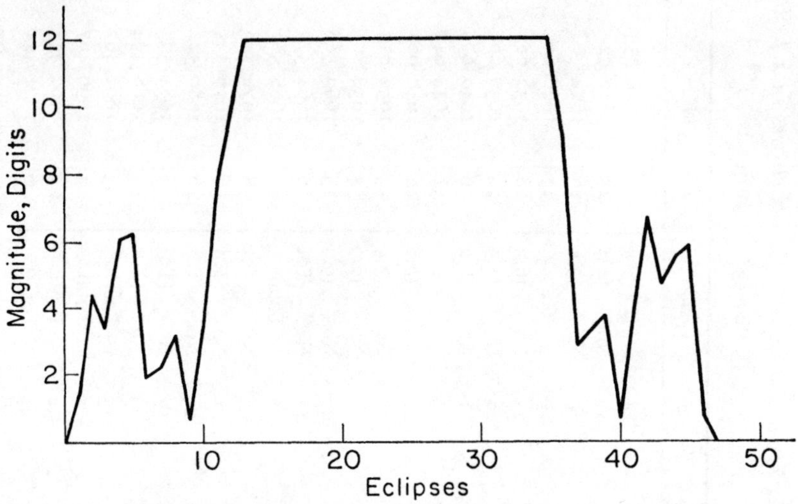

Figure 6. Magnitudes of all eclipses in a Lunar Tritos series.
Series is that of Table VIII.

The significance of the Han technique of counting off $\frac{135}{23}$ months per eclipse is still not at all clear. The Tritos implies nothing whatever about eclipses less than 135 lunations apart. It can be used to predict twenty-three consecutive eclipses only when supported by observation over a long period, in order to identify all series running within the cycle (Figure 7). This method, despite its accuracy, offers no reasonable prospect of independence from continued observation—but that independence is precisely the point of calendrical astronomy, the science of not having to look at the heavens.

Table VIII

Centrality and visibility of All Eclipses in a Tritos Series

(5 May 538 - 5 Jul 47 B.C.)

m	*Julian day*	λ	Δλ	φ	*Visible ?*	*m*	*Julian day*	λ	Δ	φ	*Visible ?*
1	1525 042	35		-14	1.4	24	1616 735	3	118	-18	No (Tot
2	1529 030	-80	115	-3	No (4.4	25	1620 722	-62	65	-8	No (Tot
3	1533 016	-142	62	9	No (3.4	26	1624 708	113	175	3	Total
4	1537 002	79	139	19	6.1	27	1628 695	14	99	15	Total
5	1540 990	-4	83	24	No (6.3	28	1632 682	-62	76	22	No (Tot
6	1544 976	-84	80	20	No (1.9	29	1636 668	-141	79	23	Total
7	1548 962	-143	59	11	No (2.2	30	1640 655	146	73	17	Total
8	1552 949	143	74	0	3.1	31	1644 641	29	117	6	Total
9	1556 936	-11	154	-12	No (0.7	32	1648 628	-8	37	-6	No (Tot
10	1560 922	-58	47	-21	No (3.4	33	1652 615	-100	92	-16	No (Tot
11	1564 909	-170	112	-24	No (7.9	34	1656 601	27	127	-23	No (Tot
12	1568 896	-86	84	-20	No (9.9	35	1660 588	-14	41	-23	No (Tot
13	1572 882	-112	26	-11	No (Tot	36	1664 575	-112	98	-17	No (9.1
14	1576 869	-179	67	1	Total	37	1668 562	-68	44	-6	No (2.8
15	1580 856	-135	44	12	No (Tot	38	1672 548	-53	15	5	No (3.2
16	1584 842	-94	41	21	No (Tot	39	1676 535	-63	10	16	No (3.7
17	1588 829	-73	21	23	No (Tot	40	1680 522	-33	30	23	No (0.7
18	1592 816	-46	27	19	No (Tot	41	1684 507	25	58	23	3.6
19	1596 801	13	59	9	No (Tot	42	1688 494	87	62	15	6.7
20	1600 788	82	69	-3	Total	43	1692 481	82	5	4	4.7
21	1604 775	44	38	-15	Total	44	1696 467	114	32	-9	5.6
22	1608 761	40	4	-22	Total	45	1700 454	-164	82	-19	No (5.9
23	1612 748	121	81	-23	Total	46	1704 441	100	96	-24	0.7

TOTAL ECLIPSES VISIBLE AT YANG-CH'ENG 17-18

Be that as it may, the figure of twenty-three eclipses per Tritos is too high. Table IX, which lists all eclipses in five non-consecutive Han dynasty cycles, shows the average to be slightly over seventeen; even this is higher than the long-term figure (16.8) implied in Oppolzer's remark that there is an average of 154.3 lunar eclipses per Julian century [1]). In other words, twenty-three eclipses is almost one and a third times as many as actually take place, and close to two and a half times as many as could have been noted by the Chinese observers unless they had a branch observatory in Boston or thereabouts.

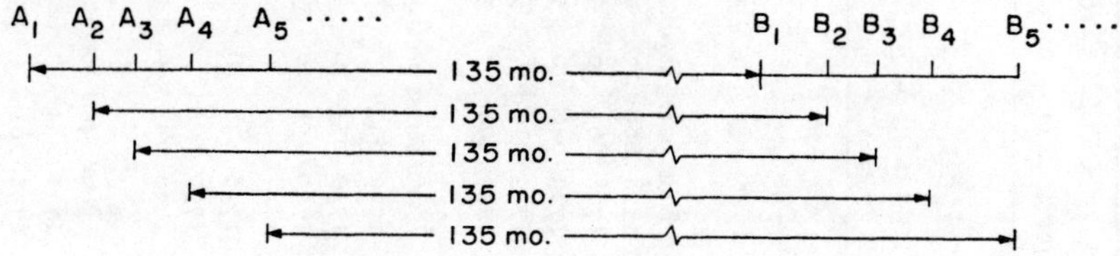

Figure 7. Use of a cycle to predict all eclipses.

Nonetheless, the number twenty-three is reflected in Table IX in a particularly striking way. It is the number of eclipses *plus blank spaces* in every cycle. Blank spaces occur when the vertical interval between consecutive eclipses is more than six months. It is a consequence of the structure of the chart that the number of blank spaces is one or two (= n-1), depending upon whether the interval is eleven or seventeen (=6n-1) months [2]).

Table IX is very condensed, since I have found it expedient to exhibit only every fifth cycle of a total of twenty-one cycles; each horizontal row gives five eclipses in the same series (unless there are empty spaces) over a period of more than two hundred years. Over this period we see the beginnings and ends of several series. The beginning of Series 15, for instance, appears on the table, since further reference to Oppolzer discloses that there is no earlier eclipse exactly 135m lunations earlier, where m is an integer less than the minimum of eclipses in a series. The beginnings of all series which commence in Table IX are indicated. These are of course approximate beginnings, as we see from Table X, since in Table IX only every fifth cycle is examined. It is evident that the tail of one series does not overlap horizontally with the head of the new one assigned

[1]) P. xxxii.

[2]) Intervals of 23 months (n = 4) are not uncommon in modern times.

TABLE IX

Five Tritos Cycles, 147 B.C. - A.D. 83

Series	CYCLE #1 *Julian day*	*Vertical Interval, lunations*	*Horizontal Interval, lunations*	CYCLE #6 (cycle #1 + 19933 days) *Julian day*	*Vertical interval*	*Horizontal interval*	CYCLE #11 (cycle #6 + 19933 days) *Julian day*	*Vertical interval*	*Horizontal interval*	CYCLE #16 (cycle #11 + 19933 days) *Julian day*	*Vertical interval*	*Horizontal interval*	CYCLE #21 (cycle #16 + 19933 days) *Julian day*	*Vertical interval*	*Horizontal interval*
1	1668 029	6	5×135	1687 963	6	5×135	1707 896	6	5×135	1727 829	6	5×135	1747 763	6	5×135
2	1668 207	6	5×135	1688 140	6	5×135	1708 073	6	5×135	1728 006	6	5×135	1747 939	6	5×135
3	1668 384	6	5×135	1688 318	6	5×135	1708 250	6	5×135	1728 183	6	5×135			
4	1668 562	6	5×135	1688 494	6	5×135									
5	1668 738	6	(10×135)							*New series begins* →			1748 442	17	(20×135)-1
6				*New series begins* →			1708 752	17	(15×135)-1	1728 686	17	5×135	1748 619	6	5×351
7	*New series begins* →			1688 996	17	(15×135)-1	1708 930	6	5×135	1728 863	6	5×135	1748 796	6	5×135
8	1669 240	17	5×135	1689 174	6	5×135	1709 107	6	5×135	1729 040	6	5×135	1748 973	6	5×135
9	1669 418	6	5×135	1689 351	6	5×135	1709 284	6	5×135	1729 217	6	5×135	1749 150	6	5×135
10	1669 595	6	5×135	1689 528	6	5×135	1709 461	6	5×135	1729 394	6	5×135			
11	1669 772	6	5×135	1689 705	6	5×135	1709 638	6	5×135						
12	1669 949	6	5×135	1689 882	6	5×135				*New series begins* →			1749 653	17	(15×135)-1
13	1670 126	6	5×135				*New series begins* →			1729 897	17	(15×135)-1	1749 829	6	5×135
14				*New series begins* →			1710 141	17	(20×135)-1	1730 073	6	5×135	1750 007	6	5×135
15	1670 451 (*New series begins*)	11	(15×135)-1	1690 385	17	5×135	1710 317	6	5×135	1730 251	6	5×135	1750 184	6	5×135
16	1670 629	6	5×135	1690 561	6	5×135	1710 495	6	5×135	1730 428	6	5×135	1750 361	6	5×135
17	1670 805	6	5×135	1690 739	6	5×135	1710 672	6	5×135	1730 605	6	5×135	1750 538	6	5×135
18	1670 983	6	5×135	1690 916	6	5×135	1710 849	6	5×135	1730 782	6	5×135			
19	1671 160	6	5×135	1691 093	6	5×135	1711 027	6	5×135						
20	1671 337	6	5×135							*New series begins* →			1751 041	17	(20×135)-1
21							*New series begins* →			1731 284	17	(20×135)-1	1751 217	6	5×135
22	*New series begins* →			1691 595	17	(15×135)-1	1711 528	17	5×135	1731 461	6	5×135	1751 395	6	5×135
23	1671 840	17	(10×135)	1691 772	6	5×135	1711 705	6	5×135	1731 639	6	5×135	1751 571	6	5+153
Total Eclipses			18			17			17			17			17

the same number. Series 5 and 20, for example, can be said meaningfully to end before Cycle 6 (actually in Cycles 4 and 5 respectively), since the eclipses given in Cycle 21 are not 135m months later. The latter in turn begin new series, to which I have arbitrarily (but naturally) assigned the same numbers as the old. The horizontal interval between the end of one series and the beginning of another in the same row is always of the form (135m-1) lunations, as we see from the notations in Table IX. Accordingly, the vertical interval between consecutive eclipses in a cycle is of the form (6n-1) lunations when it is not 6 lunations.

TABLE X

Beginning and End of Series

End of a series (A is Cycle 1, Series 5, 3 Oct 145 B.C.)				*Beginning of a series* (B is Cycle 21, Series 5, 22 Dec, A.D. 74)			
Cycle	*Julian day*	*Eclipse ?*	*Visible ?*	*Cycle*	*Julian day*	*Eclipse ?*	*Visible ?*
A-8	1636846	Yes	No	B-8	1716547	No	
A-7	1640832	Yes	No	B-7	1720534	No	
A-6	1644819	Yes	No	B-6	1724521	No	
A-5	1648805	No		B-5	1728508	No	
A-4	1652792	Yes	No	B-4	1732495	Yes	Yes
A-3	1656779	Yes	No	B-3	1736482	Yes	Yes
A-2	1660766	Yes	No	B-2	1740468	Yes	Yes
A-1	1664751	Yes	Yes	B-1	1744455	Yes	Yes
A	1668738	Yes	Yes	B	1748442	Yes	No
A+1	1672725	Yes	Yes	B+1	1752428	Yes	No
A+2	1676711	Yes	Yes	B+2	1756415	Yes	No
A+3	1680698	Yes	Yes	B+3	1760402	Yes	No
A+4	1684685	No		B+4	1764387	Yes	No
A+5	1688672	No		B+5	1768375	Yes	No
A+6	1692659	No		B+6	1772362	Yes	No
A+7	1696646	No		B+7	1776348	Yes	No

We can conclude, then, that at any given time there are 23 different series running within an eclipse cycle of 135 lunations. This is not always a meaningful statement, because of the one-month jump in the horizontal interval between the tail of an old series and the head of a new one. The ambiguity need not interfere with practice, because of the continuity between tail and head expressed by the relation B = A + (135m-1), and the substantial average length of each series—on the order of five hundred years, as we saw earlier.

If the number 23 is the number of series per cycle, what physical event corresponds to both an eclipse and a blank space in a series? Here, in the words of the great scholiast K'ung Ying-ta 孔穎達 (574-648), is an explanation familiar to educated Chinese for well over a millennium:

> The Treatise on Harmonics and Calendrical Astronomy of the History of the Former Han preserves the Triple Concordance method of Liu Hsin, according to which every $5\frac{20}{23}$ months there is one transit [of the sun across the lunar node] 交. Also according to it, if the transit precedes full moon there will be a solar eclipse at new moon and a lunar eclipse at full moon; if the transit follows full moon, there will have been a lunar eclipse at full moon, and there will be a solar eclipse at the next new moon; if the transit coincides with the lunation, there will be a total eclipse of the sun, but no lunar eclipse at the preceding or following full moon; if the transit coincides with full moon, there will be a total eclipse of the moon, but no solar eclipse at the preceding or following new moon.[1])

In other words, $5\frac{20}{23}$ months (173.331 days) approximates the time required for the sun to move from one node to the other, half an eclipse year (173.310 days). The problem which I have set out at such length becomes trivial, in fact, if, armed with hindsight, we reinterpret the graph "*shih* 食" so that when used alone it means "transit of the node" and not "eclipse," and so that it means "eclipse" only in "*jih shih* 日食" (solar eclipse), "*yueh shih* 月食" (lunar eclipse), and similar compounds.[2])

That will not do, even from the philological point of view. In the heading "Method for calculating the date of the conjunction which begins the month in which a lunar eclipse takes place" (see above, p. 30), for example, "*yueh shih*," which I have translated "lunar eclipse," also really means "transit of the node" if it has any

---

[1]) Annotation to *Ch'un ch'iu*, Hsiang 24 (*Tso chuan chu su* 左傳注疏, in *Shih-san ching* 十三經 *chu su*, Wu pen shu-chü 務本書局 reprint of 1892, *35*: 21b, cited in part in *Hsu Han shu* (*chih* 3), p. 3429; see also K'ung's comment upon *Tso chuan*, Chao 21 (*idem.*, *50*: 5b-6a), The same interpretation has been advanced in modern times in Kao P'ing-tzu (see page 39), p. 22, and in Eberhard and Henseling (see page 25), "Beiträge. . .I," *Sitzungsberichte*, 1933, 23: 217, 221.

[2]) I have preserved the distinction between "*shih*" and its compounds in my translations from *Hsu Han shu* (pp. 28-32 and 40 above), in which the character when it stands alone is rendered "eclipse" without specification of solar or lunar.

physical meaning at all; the procedure, as we have seen, has at best an indirect connection with eclipse prediction. Again, in the procedure which begins "Method of calculating lunar eclipses" and ends "... to yield the month of the eclipse which precedes the eleventh month of the year previous," there is no question that "*yueh shih*" and "*shih*" are synonymous (pp. 28-29).

Nor is it possible to retreat to the position that "*shih*" is merely ambiguous. The Han astronomers could hardly have explicitly formulated the concept "transit of the node" without distinguishing the physical circumstances of a transit from those of an eclipse. An eclipse of the moon and a transit of the node coincide only once every 135 months (to use the Chinese figure); the interval between real eclipses is always an integral number of months. The sun travels a *mean* distance of 30.67° with respect to the slowly moving nodes per month. In $5\frac{20}{23}$ months its mean travel is 180°, or the distance from one node to the next; but in six months it has gone an average of 184°.06. If six months ago the opposition of sun and moon and the sun's transit of the node coincided, today there will be a total eclipse about four degrees past the other node. Six months from today the opposition will take place eight degrees past the first node, so the eclipse will be partial and will follow the transit by more than a week. Twelve months from today opposition will happen about twelve degrees from the second node, so there will be no eclipse until the sun and moon are opposed within the eclipse limits once again. This long interval must necessarily be of the form (6n-1) months. As the distance from the second limit to the place of opposition keeps increasing at the rate of four degrees per six months, sooner or later 30.67°, or one mean lunation, less will put the opposition just inside the first limit and a new sequence of eclipses at six-month intervals will begin.[1])

There is no basis for supposing that the astronomers of the first century had formed the concept "transit of the node" (or for that matter the concept "lunar node"), since nodal transits are never spoken of as events distinct in time and space from eclipses. K'ung Ying-ta's rules of thumb, which yield much more information than the Han methods were capable of providing, do not appear in the Han treatises. There is no point in romancing about some lost source, for the Han methods are complete in themselves.

[1]) For the sake of simplicity I ignore the fact, crucial for accurate predictions, that the apparent speed and the mean speed of the sun coincide only occasionally (twice per revolution) and momentarily.

Only a feat of great exegetical violence could clear a space into which K'ung's rules might fit. The nearest thing to them one finds among the meager remains of early astronomical literature is a set of rules given in the Luminous Inception 景初 system (in use under various names 237-444) for determining whether a lunar eclipse will take place north or south of the ecliptic [1]). By the third century the concepts "lunar node" and "eclipse limit" were well understood.

But even if the idea that Han astronomers used K'ung's rules were unobjectionable, we would be no closer to solving the crucial problem of how lunar eclipses were kept sufficiently under control to avoid recurring astrological catastrophes. The official Han methods do not in fact predict transits of the nodes accurately, so that no set of simple auxiliary rules could be used with them to forecast eclipses with complete reliability. We have seen, for instance, that the Quarter Day technique predicted what in K'ung's view would be a passage of the sun across the node on 26 March 100. The nodal transit actually took place, according to modern computations, on the twenty-first of February. That is why the lunar and solar eclipse which K'ung's rules predict for 14 and 28 March (the full moon of the second month and the new moon of the third month) occurred instead on 13 and 28 February respectively.

We are now prepared to appreciate the practical significance of the Tritos cycle. Since consecutive eclipses are predicted by counting off $5\frac{20}{23}$ lunations and rounding off to an integral number of months, the net interval will be either five or six months in every case. Because of the design of the rounding-off operations, the fraction determines that, of every twenty-three intervals, twenty will be rounded off to six months and three (more or less evenly spaced) will be rounded off to five months. An eclipse was predicted to fill each of the blank spaces in the columns of Table IX, so that, for example, a seventeen-month vertical interval would contain

---

[1]) "In cases when a lunar eclipse takes place after the node is crossed, if the new moon takes place to the south of the ecliptic [lit., "outside"], the full moon will also occur to the south; if the new moon takes place to the north [lit., "inside"], the full moon will occur to the north. In cases when the lunar eclipse precedes the nodal transit, if the new moon of the lunation containing the eclipse takes place to the north, the full moon will occur to the south; if the new moon takes place to the south, the full moon will occur to the north." *Chin shu* 晉書 (*Chin shu chiao chu* 斠注 of Wu Shih-chien 吳士鑑 and Liu Ch'eng-kan 劉承幹, 1928), *18*: 12a-12b; Sung shu 宋書 (Palace ed.), *12*: 16a-16b.

two eclipses. On this understanding the average of three eclipses per cycle for which the interval is not six months is an excellent figure. The even, repetitive spacing generated by the numerical procedures of the Han treatises, however, does not correspond to astronomical reality.

The astronomers believed that there were precisely twenty-three eclipses in every cycle of 135 lunations because, using the Tritos cycle, they predicted just that many [1]. Their belief would be justified so long as they could successfully predict every eclipse which ultimately appeared. It was well known by that time that eclipses are visible in some places and not in others; there are a number of Han records of solar eclipses invisible in the capital regions [2]. Since astronomers could not distinguish a lunar eclipse which they could not observe from one which did not take place at all, it would be natural to infer that all twenty-three took place somewhere.

Why, then, did the Han procedure work as well as it did? In essence, its reliability depends critically upon its epoch. K'ung was perfectly correct in his belief that the Han astronomers were computing solar transits of the node. He was mistaken only if he meant to imply either that they were aware of what they were doing or that they were doing it accurately. Since the maximum eclipse limit is about twelve degrees, and the sun travels about a degree a day, the nodal transit and its associated lunar eclipse can never differ by more than about twelve days (or about six days if the eclipse is to be total). They necessarily fall in the same lunar

---

1) The pattern of recording solar but not lunar eclipses as omens was already set in the *Ch'un ch'iu*, the earliest extant Chinese annals, which covers the period 722-481 B. C. That the possibly still earlier *Shih ching* 193 calls a solar eclipse "ugly 醜" (= ominous?) but a lunar eclipse "ordinary 常" (= regular) has been taken more than once to mean that eclipses of the moon were under control before Confucius' time. For a translation, see Bernhard Karlgren, *The Book of Odes* (Stockholm, 1950), pp. 137-140. Cf. Han shu, *26*: 2220-2221. Maspero, "L'astronomie chinoise avant les Han" (see page 3), p. 294, discusses interpretations of lunar eclipses falling on various days of the month (!) in *Kan Shih hsing ching*, which preserves astrological traditions of the fourth century B. C.

While in theory one might say (as some of the Ch'ing commentators have done) that using the Tritos one can predict twenty-three eclipses each of the moon and the sun per cycle, I see no reason to believe that the Han astronomers knew that solar eclipses could also be systematically forecast. The number of confirmations possible, lacking a truly worldwide network of observatories, was simply too small for the cycle to have any pragmatic value.

2) See page 26 above.

month, since the lunar eclipse always falls within a day of the fifteenth. If one were to calculate nodal transits, in other words, rounding off predictions to the nearest full moon would forecast eclipses quite reliably, if (as we have seen) somewhat generously.

Since the mean interval between passages of the sun across a node is almost precisely $5\frac{20}{23}$ months, nodal transits can indeed be reckoned simply by counting increments of $5\frac{20}{23}$ months from a nodal transit. What the Chinese lacked, in short, was a nodal transit to begin counting from. Neither 9 January 104 B.C., the full moon from which eclipses were in effect counted in the Triple Concordance system, nor 8 January 28 B.C., which was the proximate "working epoch" of the Quarter Day technique (see p. 59), was a transit; the nearest passages of the sun through a node were 20 January 104 and 25 December 29 respectively.

Table XI

A Series of Eclipse Predictions by Counting Intervals of $5\frac{20}{23}$ Months from the Nodal Transit of 20 January 104 B.C.

(Predictions are for every other interval, beginning 319 intervals from the initial nodal transit)

Julian day of transit, predicted	*Nearest full moon*	*Full to transit, days*	*Nearest actual eclipse*	*Nearest visible eclipse*	*Interval, months*	*Evaluation*
1738 749	15 Jun 48	- 6		15 Jun 48	—	Confirmed
1739 095	6 May 49	+14	6 May 49	29 Oct 49	6	Interpolated
442	26 Apr 50	+ 7		26 Apr 50	—	Confirmed
789	15 Apr 51	0		15 Apr 51	—	Confirmed
1740 135	3 Apr 52	- 8		8 Oct 51	6	Interpolated
482	21 Feb 53	+14		21 Feb 53	—	Confirmed
829	11 Feb 54	+ 6	11 Feb 54	1 Feb 55	12	Interpolated (unknown success)
1741 175	1 Feb 55	- 2		1 Feb 55	—	Confirmed
522	21 Jan 56	- 9	27 Jul 55	1 Feb 55	12	Interpolated
869	11 Dec 56	+13		11 Dec 56?	—	Confirmed?

Table XI shows what accuracy the Han astronomers could have achieved had they begun counting from an actual nodal transit. The predictions evaluated are directly comparable with the second series in Table V. The first prediction of Table XI was obtained by counting 319 intervals of $5\frac{20}{23}$ months ($319 \times 173{\cdot}33 = 55292$ days) from 20 January 104 B.C. (Julian day 1683 457). In this century and a half, the discrepancy between the Chinese interval and the actual mean lapse of time between nodal transits (173.310 days) amounted to six days. Each nodal transit was thus forecast an average of six days late, but in rounding off to the nearest full moon the error happened to disappear in every case. That every one of these ten predictions succeeded according to the Chinese criteria is, of course, a matter of luck. Since predictions were always rounded off to the nearest full moon, the Chinese were implicitly assuming a maximal interval of fifteen days' solar travel between nodal transit and eclipse—which amounts to a maximal eclipse limit of roughly fifteen degrees. This is about three degrees too wide; the error could accumulate to a total of three days without ever moving a prediction over into the wrong month. Accordingly, we can say that counting from an actual passage of the sun across a node would predict lunar eclipses with unblemished accuracy for slightly less than a century (until the discrepancy amounted to more than three days), and that if the same count were used much longer, there would be occasional failures, their frequency increasing proportionately with time. In Table XI, for instance, the prediction for 21 January 56 falls on the borderline, for the correct interval between nodal transit and full moon is fifteen days, not nine days as noted. If the transit had fallen one day earlier it would have taken place on the last day of the preceding lunar month (5 January), but due to the cumulative error the Chinese astronomers would still have predicted an eclipse of the moon for 21 January.

The sun's passage through a lunar node is not an observable event. One computes it by first locating the node as the point on the ecliptic where the moon's latitude becomes zero, and then, taking the slow periodic shift of the node into account, finding the moment when the momentary longitudes of the sun and the node coincide. In the absence of the concept "lunar node" this calculation cannot in any case have been carried out.

The Han astronomers could, nonetheless, have begun their count propitiously without knowing what they were at. They had only to choose for the working epoch a total lunar eclipse of long duration, for such an eclipse would take place very close to a node. If the duration of totality were greater than fifty minutes, for instance, the nodal transit would occur within a day or two. But they did not, as we know, begin the count from a total eclipse in either system, but rather from an "invisible" eclipse which was verified by interpolation.

Inability to choose a favorable initial point meant that the Chinese techniques could not perform very well, but, because the spacing generated by the rounding-off operations corresponded to the median about which the actual situation fluctuated, even in the short run failures would tend not to be cumulative. Suppose that one of the Chinese techniques predicted that the sixth eclipse of a cycle would be immediately preceded by an interval of only five months, and that when the time came the short interval preceded the eighth eclipse of the cycle instead. In no case could this count as more than two failures, and chances are that it would count only as one, or would not be considered a failure at all. True five-month intervals between eclipses are much rarer than eleven- or seventeen-month lapses (Table IX), so that the sixth and seventh eclipses in the schema would in all likelihood not take place. Even if they were to occur, the probability that the three consecutive events would all be visible in China is small. If my reconstruction of the rules of the game is correct, the procedure is capable of masking many of its failures. The rule that an eclipse predicted but not seen is no failure, in particular, combines with what I have called the interpolation rule—which validates all predictions which yield a date five, six, eleven, or twelve months from an observed eclipse—to hide errors. In Table VI, for example, one eclipse fell only one month from an eclipse visible outside China, but could have been verified by interpolation, thereby raising the score from eight out of ten to nine out of ten.

How Crises Might Have Been Averted

Nine successes out of ten predictions is not good enough. We are not dealing with quaint Oriental gentlemen playing primitive mathematical games for their own amusement, but with scientists at work on problems of practical consequence. We are able to appreciate that they were led to an eclipse prediction method of low accuracy by their metaphysical commitment to simple linear techniques. Still, no matter how imperatively the Chinese wanted their astronomical systems to be self-maintaining cosmological mechanisms, it is difficult to conceive that a Han Emperor could have ruled with equanimity while the heavens were consistently displaying unpredicted lunar eclipses. According to the dominant theory of monarchy, a breakdown of forecasting would have been a grave crisis, certainly grave enough to have political ramifications. The correspondence of the computed ephemerides to the celestial phenomena was, after all, an objective sign that the dynasty's mandate to rule, conferred by heaven, had not been withdrawn. But while the annals record many cases in which a few hailstorms or some equally trifling omen prompted the Emperor to compose an edict on the theme "We are unworthy and sore afraid," there is no contemporary sign of corresponding unease prompted by unpredicted lunar eclipses.

The failure of this crucial scientific inadequacy to have its expected effect in the political sphere might be explained by three hypotheses:

1. That the theory of portents was a mere instrument for the manipulation of power and could be disregarded at will. One might prefer a less cynical formulation to the effect that the ominousness of omens had not been denatured away but had merely been muffled by a thick layer of convention; on this reading the frequent Imperial expressions of consternation prompted by celestial prodigies are still to be interpreted as insincere formalities. The stronger forms of this hypothesis are perhaps more conformable to the thought patterns of twentieth-century intellectuals than to those of first-century statesmen, but it would be unwise to deny them even as a possibility. Wolfram Eberhard has documented a chaotic lack of consensus in the interpretation of portents, and has pointed to widespread exploitation of them by officials for political ends and by historians to perpetuate a didactic view of the past [1]. One

---

[1]) "The Political Function of Astronomy and Astronomers in Han China," in *Chinese Thought and Institutions* (ed. John K. Fairbank; Chicago, 1957), pp. 33-70. Although this article is indispensable to any serious student because of the erudition and profound understanding of politics which it reflects, I cannot help feeling that it is vitiated by a tendency to treat as mutually exclusive factors which the sources seem to consider complementary; the overriding concern with a choice between these factors rules out what is most needed—an exploration of the balance between them and how it was maintained. The major conclusion—"that the function of astronomy, astrology, and meteorology, as defined in [relevant chapters of the Histories dealing with the Former Han] was purely political" rather than philosophic or scientific (p. 70) —is an example of the tendency to which I refer. Much of the argument is based upon characterizations of both Chinese and Western astronomy greatly at variance with those noted by historians of science. In particular, I am unable to comprehend the reasoning behind Eberhard's express denial that "as time went by, better and better data were supplied by observation and the calendar thus improved as a more useful tool step by step" (p. 65) and his unqualified generalization that "improvement of calendars was regarded as a revolutionary act and was punished" (p. 66).

might argue that lack of unanimity about meaning is not evidence of meaninglessness, but there is ample reason to explore further the possibility that, in some cases at least, manipulation bred contempt.

2. That failures were not reported to the Throne or incorporated in the historical record (which established or reinforced their ritual pertinence). The incompleteness of the Histories' records of astronomical and meteorological prodigies, as compared with the fullness of registers of observations kept in the archives of the Astronomical Bureau, is well known, and a number of detailed studies have begun the work of uncovering the patterns behind failures to report [1]).

3. That, as implied by the constants on which the official methods are based, the Han astronomers were perfectly capable of predicting every lunar eclipse with no computation whatever. The method of prognostication to which I refer was not a matter of counting off cycles from a recent epoch, as represented in the Triple Concordance system, nor from the beginning of time, as in the Quarter Day system. Instead, it consisted of counting 135 lunations

---

Finally, in order to evaluate the argument that scientific progress was prevented because astronomers "did not spend time in developing abstract laws or in studying the process of thinking . . . [and] also were not interested in applied technical sciences, e.g. in developing theoretical tools which could be used to control the flight of a cannon shell or to direct ships safely across the sea" (p. 66), the reader must be aware that the very same deficiencies were universal among astronomers, even the most scientifically progressive, of the classical period in the West.

The findings of sinologists who have examined in detail the probity of astrological documents have not tended to support Eberhard's contention that in early times many celestial omens were fabricated. See Ho Peng Yoke [Ping-yü], *The Astronomical Chapters of the Chin Shu. With Amendments, Full Translation and Annotations* (Paris & The Hague, 1966), p. 22. Ho's book, the first of its kind, conveys clearly (and with complete reliability) the great complexity of early astrology—a complexity which would be otiose if my hypothesis 1 were a total explanation.

[1]) In addition to the article of Eberhard just cited, the reader is directed to Bielenstein, "An Interpretation of the Portents in the Ts'ien Han Shu," *Bulletin of the Museum of Far Eastern Antiquities*, 1950, 22: 127-143, and to Dubs's appendixes to the various chapters of *The History of the Former Han Dynasty*.

Bielenstein has stressed (private communication) the importance of a statement in the biography of Wang Mang that at one time Wang sent to prison unauthorized persons found to be reporting auspicious omens for their own advantage. *Han shu*, 99*b*: 5737; *The History of the Former Han Dynasty*, vol. III (Baltimore, 1955), pp. 307-308.

(or, for predictions of longer range, 135m lunations) from each observed and recorded eclipse (Figure 7). Not every eclipse predicted would be confirmed by observation, but it would be possible to forecast all eclipses which were actually seen in China. After a sufficiently long initial period of observation and recording so that at least one eclipse in each series had been seen in China, twenty-three eclipses could be predicted within each cycle of 135 lunations. It is clear from the vertical columns of Table IX that sixteen to eighteen of these would take place somewhere on earth *at the time expected*, and four to seven would not, since several old series would be over and the new ones not yet established. Of course, considerably fewer than sixteen to eighteen eclipses would be confirmed.

The pattern of succession of eclipses visible in China could be expected to vary considerably within any series. While the ratio of visible to invisible eclipses would tend to unity over a long period, the life span of a series is not nearly long enough (see Table VIII). Two consecutive *visible* eclipses within a single series would ordinarily be 135m lunations apart. Roughly once every other cycle (once every twenty years or so, on the average), one eclipse would occur a month earlier than predicted. Table IX shows one to three new series beginning in each cycle, but that is due to the fact that the cycles examined are not consecutive. The *actual* beginning of any one series could fall up to four cycles earlier (cf. Table X). A separate examination of Cycle #2, for instance, reveals that it is identical with cycle #1; each of the 18 eclipses in cycle #2 falls 135 lunations later than its counterpart, so all could be successfully predicted. Taking the problem of visibility in China into account, the *average* of one anomaly every other cycle would still hold, although several eclipses at the head of a new series might not be seen.

The anomaly could be very simply brought under control by use of an auxiliary rule. In our search for the most probable formulation of this rule, let us take up where we left off. In the first eclipse of a series the sun has just come within one of the ecliptic limits, and in succeeding eclipses advances toward the node and then across it by a fraction of a degree per eclipse until it has passed outside the other limit. Both before and after a central run of total eclipses there is a run of partials, in which the moon at opposition is too far from the node, and thus too far above or below the ecliptic, to be fully obscured by the earth's shadow.

If partial eclipses of very low magnitude always mark the end of an old series and the beginning of a new one, one is naturally led to infer that the magnitude is zero, in other words that there is no eclipse, in the interval between tail and head. The Saros series exhibits this desirable characteristic, as can be seen from a graph showing magnitudes of all eclipses in a series, Figure 5. Despite the very slight auxiliary hump which sometimes appears in such curves, the beginning and end of the series would not be difficult to recognize.

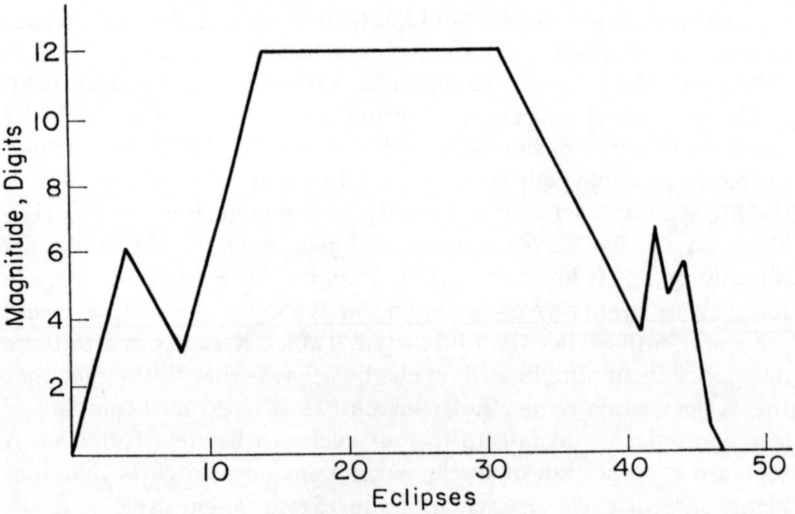

Figure 8. Magnitudes of eclipses visible at Yang-ch'eng in a Lunar Tritos series. Data are the same as those graphed in Figure 6.

A graph for a series of eclipses 135 months apart is strikingly different, as we have seen (Figure 6). The values of magnitude do not describe a smooth curve; if the jagged edges are rounded off, there are three maxima instead of one. The situation does not change radically if we graph visible eclipses alone (Figure 8). Since the Han astronomer's cycle lacked the Saros' tidiness with respect to magnitude, the likelihood that he would recognize the beginning and end of an individual series is smaller. He might equally well have inferred that between more or less continuous runs of total eclipses there were runs in which partial eclipses predominated, with the backward shift of one lunation occurring in the region of small magnitude, between two consecutive eclipses which never happened to be visible in China. A transition within the series from totality to partiality would, in this view, be a sign to look out for an eventual anomaly. Since there could be more than one such transition within what we consider an individual series, however, a more positive indicator would have been needed.

Regardless of the jagged configuration of magnitudes, there would invariably be a long gap between the last visible eclipse in an old series and the first visible eclipse in a new series. This phenomenon provides an adequate basis for a reliable auxiliary rule: If (say) 15 eclipses predicted 135 lunations apart fail successively to appear, the next eclipse is to be predicted by the interval (135m-1) lunations instead. Succeeding eclipses in the new series are to be predicted 135m lunations after the first visible eclipse forecast by the auxiliary rule.

This empirical method of prediction is simple and self-consistent. No more extensive records of lunar eclipse observations would be needed to formulate it than were required as a basis for the method given in the Histories, so it is also feasible. Whether it was known and used in the Later Han is the issue on which this hypothesis stands or falls.

The first hypothesis suggests that failures were reported to the Throne but ignored. While uncongenial to our understanding of the Han mind—unless we are willing to consign much of early intellectual life to the status of mere superstructure camouflaging political realities—it is difficult to disprove categorically. The second, which asserts that failures were not reported, is supported by the incompleteness of astrological records in general. The third, according to which there were no failures to report, is attractive *a priori*, but implies that the eclipse method of the official systems was an outright misrepresentation, maintained because it satisfied a metaphysical demand for models based upon simple repetitive intervals.

Some light can be thrown on the simple empirical questions of whether there were failures, and if so, whether they were brought to the Emperor's attention. One solid item of evidence may very well date from, or at least be derived from, the period when the Quarter Day system was still in force: a category entitled "Lunar eclipse in the wrong month 月蝕非其月" in the Treatise on the Five Elements of the Continuation of the Han History. Under that rubric are included a grand total of two lunar eclipses which defied prediction in the Later Han period [1]).

Until we understand much better how the technical treatises in the Histories were compiled, data of this sort will remain very difficult to evaluate. A report of two failures is at any rate less ambiguous than no report at all (as in the History of the Former Han). It suggests that such reports existed but were the exception rather than the rule, and thus casts doubt on the general validity of the first and third hypotheses. There is no indication in the Annals of Emperor Huan that the two unpredicted eclipses were paid special attention, but this is not necessarily significant in a period of recurrent portents, continual political catastrophes, and periodic desperate measures [2]).

But was this flagrant under-reporting necessarily due to political motives? Or was it simply meant to conceal the fact that a problem of prediction was out of control? After all, an astronomical system was meant to be like the gear train of a well-functioning machine, requiring no human intervention. As in the American dream of planned obsolescence, when a component went awry the whole machine was, at least ideally, to be replaced immediately with an improved late version, rather than being repaired.

To carry the analogy one step further, a manufacturer is usually much less reluctant to admit the defects of a product once it has been designated last year's model. It is only to be expected that, if we turn to documents which are concerned explicitly with changes of model, we should find spread before our eyes much that once had been solicitously hidden. The Treatise on Harmonics and Calendrical Astronomy of Ssu-ma Piao's Continuation preserves a remark-

---

[1]) (*Chih* 18), p. 3792. The eclipses cited are those listed in Oppolzer for 2 January 158 (the Chinese text reads "twelfth month, "but "eleventh month" is clearly meant) and 13 February 165, nos. 2110 and 2121. The latter would be only marginally visible in Yang-ch'eng (0650 hours local time, 14 February), and was likely reported from further west. The maximum phase could not, however, have been seen on 13 February anywhere east of Ferrara. That the Chinese date unambiguously corresponds to 13 February is, I think, better explained by bad timekeeping or careless recording than by reports from Western Europe to the Astronomical Bureau.

Computation by the first Quarter Day method for the last eclipse of 157 gives the date 6 September; calculating the first eclipse of 165 by the second method gives the date 15 March. Both of the observations noted thus represented known failures of prediction. I can offer no suggestion as to why these two should have been singled out from the many other known failures.

[2]) *Hou Han shu chi chieh*, 7: 286-287 and 294.

able report on various attempts to modify the eclipse prediction technique [1]). It appears to reproduce verbatim portions of documents in which the modifications were proposed. As this excerpt shows, the report not only makes no effort to cover up failures, but reveals that the Quarter Day technique was used without modification for only five years:

Many lunar eclipse predictions by the Grand Inception system were failures. The Quarter Day technique was based upon that of the Grand Inception system, but took sexagenary year 30 in the Fluvial Tranquillity 河平 period (28 B.C.) as epoch [2]). After the Quarter Day system had been in use for five years, in the first year of the Enduring Epoch 永元 period (A. D. 89), an eclipse was observed in the sky in the intercalary month following the seventh month, although it was predicted for the eighth month [3]). On the twelfth day of the first month of the second year (1 March 90), Tsung Han, Eighth-grade Meritorious Noble of Meng 蒙公乘宗紺, submitted a report saying that on the sixteenth of the current month the moon would be eclipsed, notwithstanding the official system's prediction that the event would take place in the second month. When the time came, it was as Han said [4]). The Astronomer-Royal subsequently reported that it would be advantageous to adopt Han ['s method] for official use, and Han was given an appointment awaiting edict. On sexagesimal day 41 an edict decreed that Han's technique would be employed on a provisional basis. It was so used for fifty-six years [5]).

Then, in the first year of the Root Inception 本初 period (146), an eclipse took place in the sky in the twelfth month, although it was predicted for the first month of the next year [6]). This was the beginning of a systematic discrepancy; in the twenty-nine years until the third year of the Radiant Tranquillity 熹平 period (174), there were 16 cases of eclipses occurring sooner than expected [7]). Liu Hung, Chief Administrator of Ch'ang-shan

---

[1]) *Hsu Han shu* (*chih* 2), pp. 3416-3422.

[2]) This is merely a working epoch. It represents the nearest previous year whose ordinal number (counted from the beginning of time) can be divided by the Year Number (513) without remainder (see above, p. 28). Adoption of a working epoch obviates repeating each time the determination of years elapsed in the current cycle.

The precise date from which eclipses were counted is 8 January 28, the first full moon after the Astronomical New Year.

[3]) At 1: 15 A. M., Yang-ch'eng time, 8 September 89, magnitude 3.4.

[4]) At 1: 00 A. M., 5 March 90, total. Li Jui's emendation ("十二年" to "二年") is confirmed by Oppolzer's *Canon*.

[5]) It could not officially replace the old technique without a complete "calendar reform."

[6]) On 3 February 147, magnitude 3.5.

[7]) I doubt that there is any significance in the statement that the eclipses were early. Whether an eclipse is recorded as seen four months later than one prediction or two months earlier than the next is a matter of Astronomical Bureau practice.

常山長史劉洪, submitted to the Throne the Seven Luminaries technique which he had worked out. On day 41, an edict commanded Liu Ku, Gentleman of the Palace in the Astronomical Bureau 太史部郎中劉固, Feng Hsun, Secretary in the Astronomical Bureau [?] 舍人馮恂, *et al.*, to test the efficacy of the new method against observation. An Eight Epoch technique was also worked out [at this time, and] Liu Ku and his associates worked out a lunar eclipse technique of their own. When these were all compared, Liu Ku's technique and the Seven Luminaries technique agreed that among the failures of the official system to predict lunar eclipses would be a case in sexagesimal year 56 (179), when an eclipse would take place in the fourth month. According to Hsun's method it would fall in the third month; according to the official system, in the fifth month [1]). The Astronomer-Royal proposed that when the time came these predictions be tested against observation, and that the [technique] which was confirmed be adopted. On day 54 an edict replied giving permission. In the fourth year (175), Tsung Han's grandson Ch'eng 誠 submitted a report saying that he had been taught Han's method, and that further modifications [in the official system] were indicated. There would be an eclipse in the twelfth month of the current year, although the official system forecast it for the first month of the next year. When the time came, it was as Ch'eng said [2]). He was appointed Bureau Secretary. On day 33, an edict gave permission to use Ch'eng's method. In the second year of the Glorious Harmony 光和 period (179), the fifty-sixth year of the current cycle, [the full moons of] both the third and fifth moons were cloudy.

The notation of a systematic discrepancy deserves special examination. Oppolzer lists forty-three eclipses for the period up to 174. Of these, only nineteen would be definitely visible at Yang-ch'eng. Two more would be near the limit of visibility. In twenty-nine years, that is to say, only three to five eclipse predictions were confirmed! This is a very different picture from that provided by the contemporary records in the Treatise on Five-elements Phenomena. Even with supplementary reports from outlying observers—less likely to be systematic in this period of disunity—the number would not have been significantly greater.

The report goes on to describe a series of attempts to settle finally on the most satisfactory technique, culminating in a far from clear-cut decision to adopt Tsung Ch'eng's modified Tritos approach pending further confirmation, rather than Chang[=Feng ?] Hsun's 張恂 new method based on the period relation 961 eclipses = 5640 lunations. As this choice is explained, we are provided with a frank statement, a hundred years after the Quarter Day technique,

---

[1]) No eclipse occurred during the seventeen months preceding 2 November 179.

[2]) At 6: 30 A. M., 14 January 176, magnitude 2.7.

that the theoretic basis of mathematical astronomy was inadequate:

Considering now the motions of the sun and moon, the sun moves along the ecliptic, while the moon follows the Nine Roads [1]). According to [measurements with] a declination ring, the sun at winter solstice is 115 degrees distant from the celestial pole [2]). As to right ascension, the equator passes through the twenty-first degree of the lunar mansion Southern Dipper 斗, and the ecliptic through the nineteenth degree of the same mansion. Comparing the two circles, [we see that] because the motions of the sun and moon are not parallel, they become advanced and retarded [with respect to each other when measured equatorially]. When the moon is passing through the mansions Eastern Well 井 and Herdboy 牛 [its daily motion measured along the equator] exceeds fourteen degrees; when in Horn 角 or Mound 婁, twelve degrees and a fraction. Neither case corresponds [to the mean], and so the ratio [between the mean daily motions of the sun and moon] does not hold [3]).

In view of this [limited feasibility of a mathematical analysis], there is no point in rejecting any method which does not conflict with observation, nor in adopting any method whose utility has not been practically demonstrated. The Way of Heaven is so subtle, precise measurement so difficult, computational methods so varying in approach, and chronological schemas so lacking in unanimity, that we can never be sure a technique is correct until it has been confirmed in practice—nor that it is inadequate until

---

[1]) These stand for the moon's path and eight successive positions of apogee. See Needham, *Science and Civilisation in China*, III, 392-393.

[2]) The Chinese degree (*tu* 度) is defined as one day's mean solar motion, and thus equals $\frac{360}{365^1/_4}$ °. On the Han value of obliquity, see Willy Hartner, "The Obliquity of the Ecliptic According to the *Hou-Han-shu* and Ptolemy," in *Silver Jubilee Volume of the Zinbun-Kagaku-Kenkyusyo*, Kyoto University (Kyoto, 1954), pp. 177-183.

[3]) In the first two lunar mansions the moon is far from the equator and moving in a nearly parallel direction; in the second two, near it and moving obliquely. Clearly the problem of transforming equatorial to ecliptic coordinates is not yet under control. The two-degree discrepancy between the position of the winter solstice point on the equator and on the ecliptic represents an attempt to reconcile tropical and solar years which was to lead in a century and a half to the discovery of the first Chinese constant of precession. An earlier note of the need for reconcilement is translated in Needham, III, 355-356.

The location of the solstice is given elsewhere more precisely as $19\frac{1}{4}$ degrees in Southern Dipper. The same passage, which discusses the transformation problem in some detail, has the moon's daily motion varying between thirteen and fifteen degrees (*Hsu Han shu* [*chih* 2], pp. 3396-3398). The modern value for mean speed would be 13.4 Chinese degrees per day.

My translation of "*lou*" as "Mound" is tentative. The most important early senses of the word are verbal. "Mound", the only meaning I have found in ancient sources which is clearly nominative (as are all the names of lunar mansions), is commonly differentiated as "塿."

discrepancies have shown up. Once a method is known to be inadequate, we change it; once it is known to be correct, we adopt it: this is called "sincerely holding to the mean [1])."

There is thus no dearth of evidence both that the official eclipse computation method was recurrently in a state of crisis and that the most predictable consequences of such a crisis failed to materialize. This document moreover directly indicates that the Throne was involved in the crisis, to the extent of appointing experts to official posts and sponsoring trials by observation. The center of gravity is thus shifted abruptly toward the hypothesis that the Emperor was quite aware of the crisis but was free to ignore its implications. Are we then to discard as irrelevant the report of only two failures in the Treatise on the Five Elements? If we could be sure that the relation of the two documents was a mere matter of a credibility gap, it would be simple enough to conclude, at least for the second half of the second century A.D., that the first hypothesis is reasonable and the second is not. (We would not, of course, be so crude as to extend the first hypothesis to all astrology or to the whole Later Han without considerable further investigation). My preference that the matter be left in suspense derives from a conviction that we know far too little about what might be called bureaucratic epistemology. I doubt that we sufficiently comprehend the early Chinese official's patterns of thought to answer such questions as: Does knowledge for purposes of institutional adjustment necessarily amount to knowledge for purposes of royal ritual? Did a memorial reporting the necessity for expert consultations or new techniques have the same force and the same cosmologic consequences as a memorial whose purport was the appearance of an omen? The first alternative conventionally belongs to the calendrical function, and the second to the astrological function; these two were never entirely distinct, but their separation was, as I have indicated, much more than an expedient of historiographic format. I regret being unable to offer clear-cut, unambiguous answers, but then the major aim of this paper is not to provide definitive solutions, but to exhibit the complexity of the problems as a guide to more fruitful and more profound investigations.

Nor will it do to let the third hypothesis wither away without a backward glance. There is no positive evidence whatever that the

---

[1]) Analects, XX, 1, where the Mandate of Heaven is represented by the celestial regularities 曆數 which will later become the cyclical constants of the calendar.

empirical method of eclipse prediction was ever used. At the same time, its simplicity, the ease with which astronomers could have discovered it and used it to deal with the eclipse problem, and its possible value as an alternative explanation for the absence of reported failures until the second half of the second century, justify its incorporation in this analysis.

One might argue, in fact, that had the empirical method been used it could not have been used openly, for counting off within individual series rather than within cycles involved a commitment to continued observation, which Chinese calendrical astronomy was dedicated to transcend. To posit its use requires a conspiracy theory of sorts; if it was a trade secret, not every astronomical official learned it. If the empirical method was known and put into practice for however long or short a period, during the time it was used the official methods must have been sedulously ignored, except for periodic tinkering which it was hoped would improve them. The only function they could have succeeded in performing during such periods would be to serve as a link in the morphologically consistent and all-embracing schema which the Chinese at this time required.

One might speculate more plausibly—so long as we are capable only of speculation at this point—that the Han astronomers never did find the empirical method, precisely because they were incapable of looking for it so long as their formal postulates ruled it out. One might just as well expect medieval Europeans, who were convinced of the immutability of the heavens, to have seen sunspots or novae when they looked at the sky.

Finally it is necessary to record that this empirical method may not have been altogether unique in Chinese astronomy. It is equally conceivable, and equally remains to be proven, that the unsatisfactory methods for computation of planetary phenomena in the early systems could have been ignored. A detailed critical study remains to be done, but we have a provisional evaluation of the Triple Concordance and Great Patrimony 大業 (608) planetary techniques by Yabuuchi Kiyoshi, the greatest living expert on Chinese astronomy: "As far as our present thinking is concerned, the Chinese method appears to have been not a calculation based on theoretical considerations, but rather a mere rearrangement, on an apparently suitable basis, of the actual observations." [1]

---

[1] Yabuuti, "Astronomical Tables in China, from the Han to the T'ang Dynasties" (see page 7), p. 492. See also his "Chūgoku tenmongaku ni okeru

One notes as a possible direction for future research that accurate planetary predictions without computation were in principle also possible even in the first century. One of the approaches of Seleucid astronomy, seen in what A. J. Sachs has called "Goal-year Texts," involved counting off by concordance cycles for years, synodic revolutions, and sidereal revolutions, corresponding to the Chinese Synodic Cycles (p. 15 above), from one observed phenomenon to the next. For Jupiter, as an example, the Mesopotamian astronomers used the relation 65 mean synodic periods = 6 sidereal revolutions = exactly 71 lunisolar years; and for Mars, 22 synodic periods = 25 revolutions = 47 years + 2 days. Since the intervals between planetary phenomena varied with respect to position along the ecliptic—which recurs at the end of a sidereal period—theoretic deficiencies were empirically short-circuited [1]). If this should prove to have been the case in China too, the purpose of Yabuuchi's "mere rearrangement" will have been not fudging the records, but making accurate predictions by a simple counting-off method closely analogous to that I have outlined for eclipses.

The Demise of the Cosmos

We have seen that in the early calendrical treatises the large cycles which tied a system together were determined by finding lowest common multiples for mean periodic intervals of constant value.

Some aspects of the ephemerides could be treated quite satisfactorily in this way. The basic calendrical functions—determination of lunations and years, and intercalation to reconcile them—were very early brought to a pitch of perfection which more than satisfied any practical need (unless we are to consider "practical" in this sense the insatiable demands of the historical chronologist). The twenty-four seasonal divisions (*ch'i* 氣) of the tropical year were incorporated in the ephemerides in such a way that the Imperial responsibility for regulation of agriculture was discharged with far more than adequate precision.

---

gosei ondō ron 中国天文学における五星運動論" (On the theory of planetary motions in Chinese astronomy), *Tōhō gakuhō* (Kyoto), 1956, 26: 90-103.

[1]) A. J. Sachs, "A Classification of the Babylonian Astronomical Tablets of the Seleucid Period," *Journal of Cuneiform Studies*, 1950, 2: 282-285; Antonie Pannekoek, *A History of Astronomy* (London, 1961), pp. 54-57. I am indebted to Asger Aaboe and William B. Stahlman for discussions on this problem.

But the times of consecutive eclipses and major planetary phenomena cannot be predicted with high accuracy by counting off constant intervals. Open conflict gradually developed in these areas between the desire for a simple cyclical model and the demand for a reliable ephemerides. Until the first consideration could be disregarded or modified in favor of the second, truly successful methods of prediction were unattainable.

It was in this sense inevitable that the classic calendrical treatise should eventually become something less than a complete cosmological entity, incorporating a first-order linear program on one hand and every sort of cosmo-numerological system on the other. The first step was abandonment of the Jupiter cycle for numbering years. As we have seen, this cycle had been applied to give 145 sexagenary year-numbers in each 144 (tropical) years—that is, it defined the "sexagenary year" as one-twelfth of the sidereal period of Jupiter. In essence, the Jupiter cycle was dead by the end of the first century A.D. when Li Fan and his collaborators, in the Quarter Day system, rejected the planetary constants of Liu Hsin.

The Supernal Manifestation 乾象 system, compiled about 180 by Liu Hung 劉洪 and used in the Shu kingdom from 223 to 280 [1]), was no longer tied to the numerological apocrypha which had given the constants of the earlier treatises so much more than merely astronomical significance. This same Liu Hung was the last great Chinese astronomer to take eclipse cycles seriously. He worked out an improved cyclical relation, 1882 eclipses = 893 years = 11045 months, which corresponds within one and a half minutes to the modern value for half an eclipse year (using Liu's value for year length, $\frac{893}{1882} \times 365\frac{145}{589}\ \frac{\text{days}}{\text{year}} = 173.3075$ days, modern value 173.3100). The procedure for applying this cycle is in no way more sophisticated than those of the Han treatises. It is, in fact, a conflation of the two Quarter Day methods, discarding the irrelevant and useless methods for forecasting date and hour of the eclipse [2]). Despite the theoretically better cycle, and use of a much more recent Superior Epoch (7173 B.C.) to minimize the cumulative effect of residual error in the cycle constants, Liu's technique is incapable of drastically increasing the proportion of confirmed pre-

[1]) Yabuuti, "Astronomical Tables in China, from the Han to the T'ang Dynasties," departs from his source (Chu, *Li fa t'ung chih* [see page 39]) in giving the date of adoption as 222, but according to *San kuo chih* 三國志, *Wu* 吳 *chih* (Palace ed.), *2*: 14b, Chu is correct.

[2]) *Chin shu*, *17*: 10a, 15b-16a. See Appendix B below.

dictions. Liu was able to incorporate elsewhere in the system his knowledge that the lunar motion is not constant, and that both the point on the moon's orbit where the speed is greatest, and the intersection of the moon's path with the ecliptic, move. The pertinence to eclipse prediction of these periodic changes could not be realized, however, until the model of eclipses as simple cyclical phenomena was given up. Liu's increased comprehension of lunar phenomena could only add to the pressure for abandonment of the cyclical model, which was not long delayed. In the Luminous Inception system (used 237-444) of Yang Wei 楊偉, the concept of (solar) distance from the node 去角度 was clearly defined, with a maximum value of fifteen (Chinese) degrees for partial eclipses. This value is too large to have been derived empirically; there can be little doubt that it springs from the old practice of rounding off eclipse predictions to the full moon of the month in which they fell (see p. 51 above). Nodal distance was also taken in Yang's system as a measure of eclipse magnitude 虧蝕分 [1]).

By the eighth century, capping a development which had begun about 600, it was possible to abandon the mean motion of the sun as well as that of the moon, thereby first attaining in principle a sophistication comparable to that of Babylonian astronomy in the Seleucid period (last three centuries B.C.). I do not mean to assert that the approach of the two civilizations was ever identical. The modern reconstruction of Mesopotamian celestial kinematics is based entirely upon Seleucid documents, which represent the culmination of the tradition but yield only indirect clues about its development. The comparative crudity of the Chinese techniques, as well as the basic differences of approach in certain important areas, suggest that if there was a transmission it occurred before the Western art had settled upon many of the characteristics with which we are familiar.[2])

The abandonment of the Jupiter cycle and of simple eclipse

---

1) *Chin shu, 18*: 11b-13a; *Sung shu, 12*: 15b-16a.

2) The reader is referred to Needham, *Science and Civilisation in China* (see page 5), vol. III, esp. pp. 232-259, where comparisons are carefully drawn. Perhaps the most significant divergence was the Chinese tendency to use meridian transits of circumpolar stars to indicate the positions of invisible lunar mansions; the Babylonians generally relied on horizon phenomena when they wanted to refer phenomena to the zodiac. At the same time, there are indications from non-mathematical tablets that until past the middle of the first millennium B. C. the Babylonians, like the Chinese, located celestial events equatorially rather than by reference to the ecliptic.

cycles, and the incorporation of lunar and solar inequalities, are more than technical improvements. They mark fundamental changes in the Chinese approach to the calendrical art, and represent the astronomical tradition's final line of evolution, which was away from counting off by mean intervals. The advance of observational and computational techniques and the gradual improvement in values of astronomical constants kept the movement going until the time of Kuo Shou-ching 郭守敬 (1231-1316), when the proto-trigonometrical aspect of Chinese mathematics and the precision of armillary-type instruments were sufficiently developed to allow apparent positions, derived indirectly from observation, almost completely to replace mean positions computed from cycles. The achievement of Kuo Shou-ching remains the climax of the Chinese astronomical tradition. Here we see an instance of the law that the highest point of the *yang* is the inception of its decline, for after Kuo's time the tradition lost its vitality, and soon his work was no longer comprehended [1]).

The demand for precision had to win out, once it had been maneuvered into confict with the goal of metaphysical consistency and unity. With the aid of hindsight, we might propose that the Chinese had formulated their classic conception of the universe as a congeries of cyclical time relationships on the basis of too primitive a model. The assumption of simple cyclical behavior could not have survived for long. In the Han it was maintained because it made mathematical astronomy possible, but at the cost of compromising the integrity of the system. When this cost became intolerable, the assumption was discarded. It was never replaced by new assumptions more conformable to the complexity of the celestial motions, for by the time of its rejection the technical tasks of astronomy could be carried out without such assumptions. Later Chinese calendrical science was marked by an indifference toward cosmology —but this was the indifference of the disenchanted, not that of the inexperienced.

In the final creative phase of Chinese astronomy, from the time of Shen Kua 沈括 (1031-1095) [2]) to that of Kuo Shou-ching, as its

---

[1]) For a sketch of Kuo's life and work (in particular his accomplishments in hydraulic engineering, which deserve to be more widely known) see Li Ti 李迪, *Kuo Shou-ching* (Shanghai, 1966).

[2]) For a full-scale critical biography of Shen, not entirely satisfactory from the astronomical point of view, see Chang Chia-chü 張家駒, *Shen Kua* (Shanghai, 1962).

mathematics moved further in the direction of geometry, astronomers began to develop a willingness to think in terms of physical models which might have led to a new transformation if calendrical science had not lost its vitality (see Appendix C). The brilliance, much later, of Wang Hsi-shan's 王錫闡 (1628-1682) conservative attempt to provide a cosmological basis for traditional astronomy by a critical overhaul of Tycho Brahe's world system—the best available in China at the time—is ample proof that a new growth was not inherently impossible [1]). But the tradition Wang wished to resuscitate was long dead, and his labors were abortive.

One sees rough parallels in the history of classical Western astronomy for certain aspects of the developments I have reconstructed [2]). There was, indeed, a long period of fixation on the idea that the motions of the celestial spheres, in order to be eternal (or, to use an unclassical word, inertial), must be circular and constant. From the time of Ptolemy (ca. A.D. 150) on, there arose also a gradually increasing tension between philosophical rigor and kinematic accuracy which could be resolved only by a revolution. The Western conception, however, was not only much more concrete and physical, envisioned in terms of spatial relations within orbits rather than as predominantly concerned with time cycles, but its form was also comparatively advanced. It demanded merely that apparent speeds be resolvable into *combinations* of constant velocities, and was thus prepared from the start to envision extremely complex aggregate motions. If Eudoxus' (ca. 370 B.C.) experiments with counter-rotating concentric spheres were only partially successful, no matter; Apollonius' (ca. 200 B.C.) eccentrics and epicycles and Ptolemy's equants were the foundation for a set of geometrical models which stood the test of new observations without a major breakdown for fifteen hundred years.

But the Aristotelian first principles whose authority Ptolemy accepted had been formulated originally to fit a much more innocent estimate of the subtlety which astronomy would need. Ptolemy's

---

[1]) See Wang's *Wu hsing hsing tu chieh* 五星行度解 (Explications of the Motions of the Five Planets; in *Shou shan ko ts'ung-shu* 守山閣叢書), of which I have prepared a translation. For a provisional evalution of Wang Hsi-shan's work, see Hsi Tse-tsung 席澤宗, "試論王錫闡的天文工作," *K'o-hsueh-shih chi-k'an*, 1963, 6: 53-65.

[2]) The best one-volume introduction to classical astronomy is still J. L. E. Dreyer's *A History of Astronomy from Thales to Kepler*; reprint, New York, 1953. Considerably more elementary and topical, but still generally reliable, is Thomas Kuhn, *The Copernican Revolution* (Cambridge, 1957).

adherence to the circularity postulate, in particular, had to be qualified—certain motions could not be quite centered on the earth (as center of the universe), or even be quite uniform about their own centers—if the model were to predict locations with great accuracy. At the same time, as a consequence of his geocentric frame of reference, Ptolemy's techniques for the various luminaries were connected by no apparent mathematical necessity; they were a cosmos only by custom.

The Islamic and European reformers of Ptolemaic astronomy, of whom Copernicus considered himself one, were motivated by a desire to restore metaphysical rigor while maintaining or improving astronomical precision, in short to perfect the relation of cosmology and mathematical astronomy. This motivation, so decisive in the gestation of modern science, seems to have been in China an early casualty of a premature and unworkable relation between form and content.

* * *

I acknowledge with gratitude the comments and criticisms of many distinguished colleagues—too many to list here—and the financial support of the History of Science Department, Harvard University; the Department of Humanities, M.I.T.; National Institutes of Health; and National Science Foundation.

Appendix A

Eclipse prediction technique from the Triple Concordance System [1]

Take months elapsed in the current Coincidence Month, multiply by 23 and divide by 135. To the remainder, add 23 at a time, each time counting one month, until 135 is reached. When the number of months obtained is counted exclusively from the Astronomical First Month, the result is the month in which the eclipse occurs. The time of the eclipse is given by the hour 辰 of opposition at full moon [, computed normally].

Appendix B

Eclipse prediction technique from the Supernal Manifestation System [2]

Take the number of the year for which the eclipse is wanted, counted exclusively from the Superior Epoch, and divide by the

[1]) See p. 25, note 1.
[2]) See p. 65, note 2.

Coincidence Year 會歲 (893). The remainder is multiplied by the Coincidence Factor 會率 (1882) and divided by the Coincidence Year to give Accumulated Eclipses.

The first remainder represents years elapsed in the current Coincidence Year cycle. It is multiplied by $\frac{1882}{893}\frac{\text{eclipses}}{\text{year}}$ to give the number of eclipses up to Astronomical New Year of the year wanted.

If there is a remainder, add 1 [to Accumulated Eclipses]. Then multiply [Accumulated Eclipses] by the Coincidence Month 會月 (11045) and divide by the Coincidence Factor to give Accumulated Months; the remainder is the Month Remainder.

Rounding off the remainder to the next higher unit extends the count past A. N. Y. to the time of the next eclipse so that, like the second Quarter Day method, this procedure forecasts the first eclipse of the year wanted rather than the last eclipse of the previous year. Accumulated Months thus represents lunations in the current Coincidence Year cycle up to the month in which the eclipse occurs. The Month Remainder is, strictly speaking, the fraction of a month elapsed from new moon to eclipse.

Multiply Intercalations per Rule Cycle (7) by the Year Remainder [from the first operation], dividing by Years per Rule Cycle (19) to give Accumulated Intercalations, which is to be subtracted from Accumulated Months.

This procedure corresponds to that of the first Quarter Day method.

The residue is divided by Months per Year(12), and the remainder counted [exclusively] from the Astronomical First Month.

Appendix C

Shen Kua (1031-1095) on planetary motions

The beginning of Chinese thought on physical models for astronomical phenomena is to be seen, juxtaposed with indications of attitudes which were to contribute to the eventual death of the native tradition, in an important but hitherto untranslated passage from Shen Kua's Dream Creek Essays (*Meng ch'i pi-t'an*) 夢溪筆談 [1]). The model proposed in the first part is instructive in its use of a figure out of nature to perform the function which the very abstract epicycle had been performing for twelve hundred years in

[1]) *Meng ch'i pi t'an chiao cheng* 校證 of Hu Tao-ching 胡道静 (rev. ed., Shanghai, 1959), *8*: 334-335. For another example of Shen's concrete astronomical imagination see Needham, *Science and Civilisation in China*, III, 415-416.

the West. The career of the epicycle was on the one hand a metaphysically significant attempt to apply the circularity postulate, and on the other was rigorously supported by a mathematical analysis. The problem of planetary retrogradation had not been approached previously in China, to my knowledge, in terms of real spatial configurations. Shen's schema is a characteristically original but offhand suggestion which he knew he would never have the data to confirm, for his extremely ambitious program of data-gathering (which anticipates that of Tycho Brahe five hundred years later) came to nothing in his lifetime.

The illustrations are my own tentative reconstructions.

I have determined that, according to ancient and modern astronomical methods, the anomalies 差 of the five planets are greatest in magnitude near the stationary points. If the direct motions [of the planets] are from within, their retrogradations must face outward [Figure 9 (a)]. If their direct motions are from without, their retrogradations must be on the insides [of the orbits; Figure 9 (b)]. Their

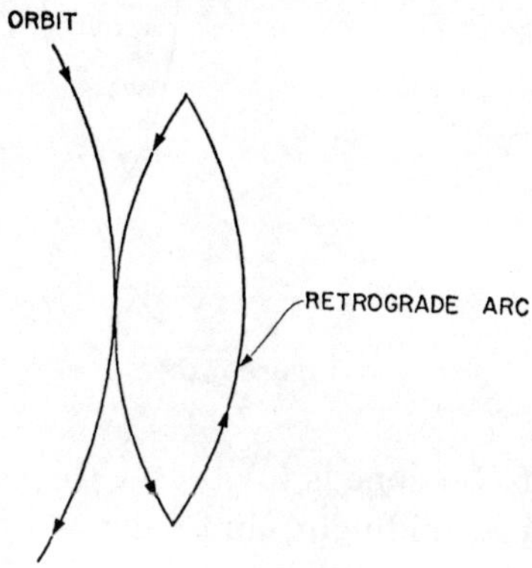

Figure 9 (a)

orbits must be shaped like a willow leaf, with the two ends [i.e. the stations] pointed and, in between, the paths followed in passing back and forth separated by a considerable distance. Thus the slight retardation of a planet when moving in [the neighborhood of] the two ends is due to the fact that its motion is oblique [to the line of sight]. That the angular motion is slightly accelerated in between is because its path is perpendicular 絕 [to the line of sight]. Astronomers have been aware only that there are divergences from the mean speed, but have not seen that there is a variation in the inclination of the path [to the observer].

In the Splendid Peace 熙寧 period (1068-1077) I held the office of Astronomer-Royal. Wei P'u 衛朴 prepared a calendrical treatise in which the solar and lunar elements were corrected, but when it came to the motions of the five planets, there were no registers of observations [from which the appropriate elements] could be verified. In previous generations most calendrical treatises had been compiled by merely modifying values taken from older treatises, without checking them against observed celestial positions.

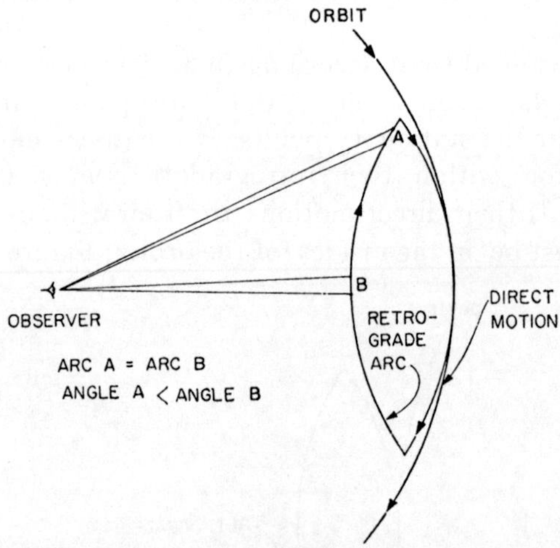

Figure 9 (b)

Now what has to be done is to observe the positions of the moon and planets at dusk, midnight, and dawn, to the nearest fraction of a degree, and to establish a register in which they are to be recorded. When five full years have passed, subtracting cloudy nights and day-time appearances, one would have the apparent motions for three years [i.e. for three-fifths of the time]. Subsequently, the constants could be mathematically derived [lit., "threaded"]. This is what was called in ancient times "*chui shu* 綴術 "[lit., "the technique of threading"] [1]).

[1]) Shen Kua implies even more strongly elsewhere that he is thinking of the Method of Finite Differences: "[The method by which] one seeks the motions of the planets and variations in the lunar and solar periods is called 'the technique of threading,' since it can be found only by mathematical

At that time the positions of the officials in the Bureau of Astronomy were hereditary sinecures, so that in fact none of them knew anything about astronomy. Piqued that P'u's expertise exceeded theirs, they joined together in a campaign of slander, repeatedly bringing serious charges against him. Although in the end they were unable to shake him, to this day the register of observations has not been completed. The techniques used in the Oblatory Epoch 奉元 calendrical treatise [official 1075-1093] to account for the motions of the five planets employ values which are mere modifications of those in earlier treatises, correcting the very worst errors, but only five or six discrepancies in ten could be dealt with. P'u's mastery of technique is unequalled and unprecedented: How sad that the backbiting of that bunch of calendar-makers could have kept him from bringing his art to fruition!

'threading,' not by examination of figures [in space]." *Ibid.*, *18*: 572; cf. Needham, *Science and Civilisation in China*, III, 123. Whether the "*chui shu*" of the fifth century was in fact the Method of Finite Differences is another matter; the former had not been passed down to Shen's time, according to Ch'ien Pao-tsung, *Chung-kuo shu-hsueh shih* 中國數學史 (History of Chinese mathematics; Peking, 1964), pp. 85-86.

Retrospect

This essay began with a remarkable seminar at Harvard on the comparative study of chronology, taught by Willy Hartner. Professor Hartner, visiting from the University of Frankfurt, was an immensely learned astronomer-historian. He had published valuable studies in the mathematical astronomy of a number of civilizations, using primary sources in all their languages. Most of these essays were eventually gathered in the two volumes of his *Oriens–Occidens* (Hildesheim: Georg Olms, 1968–1984). He encouraged me to study the time reckoning of China, which I had not previously explored. The result, an eight-hour oral report to the seminar, raised technical questions that I eventually settled, as far as I could, with uncounted hours of computation on an abacus.

There is a deeper point to this essay that is not mentioned in the Abstract. It is about a scientific revolution, a transformation in the central questions of a science and what constitutes an answer to them. That is what the new indifference to cosmology amounted to. No one had found a revolution in Chinese science because no one had gone looking for one. This point was perhaps made a little too subtly, since I drew on the problematique of Tom Kuhn's *The Structure of Scientific Revolutions* without using its jargon.

Perhaps I should mention something that I learned from this essay. It was often extremely difficult to reconstruct the technical decisions in back of the concise and stereotyped mathematical formulations in the documents. I was greatly aided by the work of several of the great scholars of the last three centuries such as Li Jui 李銳 (1765–1814; see page 21, note 2), who were as interested in early astronomy as in every other aspect of antiquity. When our understanding of a text differed, I was at first inclined to suppose that this was because they did not have the benefit of the excellent textbooks from which I was teaching myself spherical astronomy. I gradually came to realize that such differences generally meant that I did not understand the text well enough. I did not consider the paper finished until I had resolved all these differences, generally in favor of my ancient colleagues.

Since this essay was published there has not been a great deal of new work on such general questions in early astronomy. Christopher Cullen's "Motivations for Scientific Change in Ancient China: Emperor Wu and the Grand Inception Astronomical Reforms of 104 B.C.," *Journal for the History of Astronomy, 24:* 185–203, deepens our understanding when it emphasizes the role of religious cosmography and the emperor's obsessions.

On page 9 above, at the end of the third line from the end of the main text, the word "specify" was omitted.

III

SHEN KUA

Shen Kua 沈括 was the most prominent of the polymathic statesmen who flourished in the Northern Sung. He was born in 1031, with registration at Ch'ien-t'ang 錢塘 [now Hangchow, Chekiang province], China, and died at Ching-k'ou 京口, Jun prefecture 潤州 (now Chinkiang, Kiangsu province) in 1095.

Shen was the son of Shen Chou 沈周 (ca. 978–1052) and his wife, whose maiden name was Hsu 許. Shen Chou came of a gentry family with neither large landholdings nor an unbroken tradition of civil service. He spent his life in minor provincial posts, with several years in the capital judiciary. Shen Kua apparently, like a number of other important southern intellectuals of his time, received his early education from his mother. A native of Soochow, known for its flourishing manufactures, commerce, and agriculture, she was forty-four or forty-five years old when he was born. Among other things, she taught him and his brother P'i 披 the military doctrines of her elder brother Hsu Tung 許洞 (ca. 976–ca. 1016).[1]

Shen's background made it possible for him to enter the imperial bureaucracy, the only conventional road to advancement for educated people of his time. Unlike colleagues who came from the ancient great clans, he could count on few advantages save those earned by his striving and the full use of his talents. Shortly after he was assigned to the court, he became a confidant of the emperor and played a brilliant part in resolving the crises of the time. Within slightly over a decade, his career in the capital was ended by impeachment. After exemplary service in a provincial appointment and five years of meritorious military accomplishment, he was doubly disgraced and politically burnt out. The extremes of Shen's career and the shaping of his experience in science and technology become comprehensible only if the pivotal circumstances of his time are first considered.

1. See Hu Tao-ching, "Shen Kua chün-shih ssu-hsiang t'an yuan." Full citations for important studies of Shen are given in the Bibliography at the end of this chapter. Works pertinent only to a given point are fully documented in the notes.

III

2

Historical Setting

Shen's time was the climax of a major transition in the Chinese polity, society, and economy.

Three centuries earlier the center of gravity in all these respects still lay in the north, the old center of civilization of the Han people. Wealth and power rested in the hands of the old aristocratic land-owning families. Governmental institutions could not resolve the tension between civil servants' private interests and the emperor's inevitable desire to concentrate authority. The civil service examination system was beginning to let the central government shape a uniform education for its future officials. Since birth or local recommendation determined who was tested, the mass of commoners were not involved. The social ideals prevalent among the elite were static; the ideal past was cited to discourage innovation; and the moral example of those who ruled, rather than responsive institutions or prescriptive law, was held to be the key to the healthy state. The classicist's ideal of a two-class society—self-sufficient agriculturalists ruled and civilized by humane generalists, with land as the only true wealth—did not encourage commerce, industry, or the exploitation of natural resources. The wants of the great families, whose civil servant members were becoming city dwellers by the middle of the eighth century, nonetheless gave momentum to all of these activities. The rural majority of the population still took no part in the rudimentary money economy.

The T'ang order began a long, slow collapse about 750, until in the first half of the tenth century the empire of "All under Heaven" was reduced to a jumble of ephemeral and competing kingdoms. When the Northern Sung rebuilt the universal state (960–1126), its foundations were in many important respects different from those of the early T'ang. A new dynasty was not only, as classical monarchic theory had it, a fresh dispensation of the cosmos; it was also the occasion for institutionalizing a new distribution of power in society. Changes in taxation made the old families accountable for their estates as they had not been earlier, and encouraged smaller landholdings. The cumulative result was a wider diffusion of wealth.

The center of vitality moved southeast to the lower Yangtze valley, which had long before emerged as the major rice-yielding region. By this time its fertility, combined with fewer social restrictions, had bred a new subculture that was more productive in industry than elsewhere. It encouraged the growth of commerce and stable markets, the beginnings of a uniform money economy, and the great broadening of education that printing had just made possible. The new southern elite was, on the whole, small gentry, and lacked the military traditions of the ancient northern clans and of power holders in the period of disunion. Their families were often too involved in trade for them to despise it. Although conservative, as all Chinese elites have been, they were prepared to think of change as a useful tool. The novelties of attitude and value were often slighter or subtler than such a brief account can convey, but within the established limits of Chinese social ideals their consequences were great.

In Shen Kua's time the old families of the north and west still provided many of the very highest officials and thus wielded great influence, positive or obstructive, in discussions about the future of China. But they had become merely influential members of a new political constellation that brought a variety of convictions and interests to that perennial debate. An especially obvious new element was that many southern small gentry families like Shen's established traditions of civil service, either as a main means of support or to protect and further their other concerns. Once a family's social standing was achieved, one or more members could enter the bureaucracy freely because of experience as subordinates in local administration or because they were amply prepared by education for the examinations. Their sons could enter still more freely because offspring of officials were given special access to both direct appointment and examination.

Not sharing the old vision of a social order fixed by precedent, men of the new elite were willing to sponsor institutional renovation in order to cope directly with contemporary problems. Dependent on their own talents and often needing their salaries, they were dedicated to building a rational, systematic, and in most respects more centrally oriented administration. They were willing to make law an instrument of policy, and insisted that local officials be rated not only on the moral example they set but also quantitatively—on how effectively they made land arable and collected taxes. In the name of efficiency, they devoted themselves to removing customary curbs on imperial authority and (with only partial success in the Sung) to dismantling the structures of privilege that underlay regional autonomy. Only later would it become clear that they were completing the metamorphosis of the emperor from paramount aristocrat to autocrat. At the same time they were successfully demanding more policymaking authority as the emperor's surrogates, although at the cost to themselves of greater conformity than officials of the old type had willingly accepted.

This irreversible transition did not lead to a modern state, but only to a new and ultimately stagnant pattern. The most accelerated phase of change was the activity of what is called the New Policies group (*hsin fa* 新法,

actually a shifting coalition) between 1069 and 1085. In 1068 the young emperor Shen-tsung, who had just taken the throne, brought its leader, Wang An-shih 王安石 (1021–1086), to the capital. Within two years Wang had become paramount Grand Councillor. He resigned for nine months in 1074, when pressure from his antagonists persuaded the emperor to be less permissive, and returned to private life in 1076. The New Policies continued to be applied and extended, but with less and less attention to their founding principles, until Shen-tsung's death in 1085. Under the regency of the Empress Dowager, enemies of the reform attempted for eight years to extirpate Wang's influence and take revenge upon his adherents. When Emperor Che-tsung came of age in 1093, the New Policies were revived, but were so bent toward factional ends and administered so disastrously that the word "reform" is hardly applicable.

Wang An-shih's opponents were many: the old aristocrats, career bureaucrats of the sort who would oppose any change as disruptive, officials whose individual or group interests ran in other directions—and men of high ideals who found his proposals ill-advised and his personal style too intolerant.

No institution had evolved through Chinese history to work out and resolve conflicts of political viewpoint. This lack was filled by cliques, intrigues, and appeals to imperial intervention. Division and corruption among active supporters of the New Policies had been a problem from the start. Wang's program was so ambitious that he had to take competent support where he found it. The new access to power that he offered attracted enterprising men. Many had little sympathy for his convictions, and dedicated themselves to manipulation and graft. Once Wang was gone, the leadership of his group tended to become a battleground for aspirations of this kind.[2] The internal and external enemies of the New Policies left the program a shambles by the time the Chin Tartars drove the Sung south in 1127.

A primary aim of the reforms was financial security of the state, which prompted initiatives in water control and land reclamation, encouragement of extractive industries and agriculture, intervention in commerce, and ration-

---

2. In the successive reform movements of the Northern Sung there were considerable differences in the alignment of men with different beliefs and backgrounds. See the discussion in James T. C. Liu, "An Early Sung Reformer: Fan Chung-yen," in John K. Fairbank (ed.), *Chinese Thought and Institutions* (University of Chicago Press, 1957), 105–131, esp. 107–109. Robert M. Hartwell has examined regional variations of change in "Demographic, Political, and Social Transformations of China, 750–1550," *Harvard Journal of Asiatic Studies*, 1982, *42*. 2: 365–442. I am grateful to him for counsel.

alization of taxes. Another goal, particularly at the emperor's insistence, was military strength. There had been a long confrontation along the northern border between the Chinese and the powerful Khitan empire, pastoral masters of mounted combat (renamed Liao in 1066). Seventy years of fitful peace were punctuated by humiliating Chinese failures to recapture territory south of the Great Wall. The peace was maintained by large annual bribes. For three decades the Tangut people of the northwest had posed an almost equally unpalatable demand for appeasement. The wealth that the New Policies could generate from man's exploitation of Nature,[3] the emperor hoped, could buy victory, or détente through strength, on both fronts. Either goal demanded expertise in cartography, strategic theory and tactical doctrine (both of which contained cosmological elements), design and manufacture of war matériel, fortification, troop organization and training, and development of a stable economy in border regions.

Shen Kua contributed to nearly every field of New Policies activity, both civil and military. His social background and political commitments cannot be considered responsible for his scientific talent or curiosity; the antecedents and loyalties of other major contemporary scientific figures were very different from his. But a review of his career and of his work will show how regularly his involvement with particular technical themes and problems grew out of his activities in government.

Life[4]

From about 1040 Shen traveled with his father to successive official posts from Szechwan in the west to the international seaport of Amoy. He was exposed not only to the geographical diversity of China but also to the broad range of technical and managerial problems in public works, finance, agriculture, and maintenance of waterways, that were among the universal responsibilities of local administrators. Because his physical constitution was weak, he became interested in medicine at an early age.

---

3. There was no Chinese word that exactly corresponded to "Nature," a protean and culture-bound term, until the end of the nineteenth century. I use the English word simpy as a convenient label for the sum of all objects of empirical study outside of social relations. In this sense it is equivalent to *t'ien-ti* 天地, often abbreviated as *t'ien.*

4. I do not give references for the well-documented facts of Shen's life found in Chang Chia-chü's and other standard biographies.

Late in 1051, when Shen was twenty, his father died. As soon as the customary inactivity of the mourning period ended in 1054, Shen received the first of a series of minor local posts. His father's service exempted him from the prefectural examination. His planning ability became almost immediately apparent when he designed and superintended a drainage and embankment system that reclaimed some hundred thousand acres of swampland for agriculture. This was the first of a series of projects that established his reputation for skill in water control. This ability seems to have run in his family. Shen recorded a visit to his brother P'i, subprefect in Ning-kuo 寧國 (now Fu-hu, Anhwei province), in 1061. After a cartographic survey and a historical study of previous earthworks in the region, Shen P'i applied the labor of fourteen thousand people to another massive land reclamation scheme that won the recognition of the emperor.[5]

In 1063 Shen Kua passed the national examinations in the highest category. Posted to Yangchow, he impressed Chang Ch'u 張芻 (1015–1080), the Fiscal Intendant (a post then equivalent to governor), who recommended him for a court appointment leading to a career in the professional financial administration.[6]

Shen apparently used the time not occupied by his early metropolitan appointments, which were conventional and undemanding, to study astronomy. In reply to the informal questions of a superior he set down clear explanations, still extant, of the sphericity of the sun and moon as proved by lunar phases, of eclipse limits, and of the retrogradation of the lunar nodes. They demonstrate an exceptional ability to visualize motions in space, which were at best implicit in the numerical procedures of traditional astronomy and seldom were discussed in technical writing. In 1072 Shen was given an additional appointment as Supervisor of the Directorate of Astronomy. With the collaboration of his remarkable commoner protégé Wei P'u 衛朴 and the aid of other scholarly amateurs, using books gathered from all over the country, he undertook a major calendar reform. He planned an ambitious series of daily observations to extend over five years, using renovated and redesigned instruments.

---

5. The standard sources credit this accomplishment to Shen Kua, but Teng Kuang-ming has corrected this error.

6. The succession of fiscal posts that could lead to a seat on the Council of State in the eleventh century has been documented by Hartwell in "Financial Expertise, Examinations, and the Formulation of Economic Policy in Northern Sung China," *Journal of Asian Studies*, 1971, *30:* 281–314.

The incompetent career officials who staffed the bureau stymied him. He forced the dismissal of six whom he caught falsifying records of phenomena. Those who remained doomed his program of observations and kept his new system of ephemeris computation from being among the two or three most securely founded before modern times. Shen's personal involvement in later stages of the reform undoubtedly was limited by his gradual movement into the vortex of factional politics.

Shen was early known to Wang An-shih, who composed his father's epitaph while a young provincial official. Shen eventually was publicly identified by enemies of the New Policies as among the eighteen members of Wang's intimate clique. In late 1072, in support of Wang's program, Shen used an original technique to survey the silting of the Pien Canal near the capital, dredged it, and demonstrated the value of the silt as fertilizer. Until mid-1075 he spent much time traveling as a troubleshooter of sorts, inspecting and reporting on water control projects, military preparations, and local administrations—and, it has been conjectured, providing encouragement to Wang's provincial supporters. Shen was put in charge of arsenal activities and, in 1075, was sponsored by Wang (then head of government) to revise defensive military tactics, a task the throne had proposed for Wang himself.

In 1074 the Khitan were pressing negotiations to move their borders further south. Incompetent and timorous Chinese negotiators were conceding unfounded Liao claims about the language and substance of previous agreements. Shen built a solid Chinese case by going to the archives, as no one before him had bothered to do. His embassy in mid-1075 to the camp of the Khitan monarch on Mt. Yung-an 永安 (near modern P'ing-ch'üan, Hopei) was triumphant. He described himself, encircled by a thousand hostile onlookers, calling on his staff, who had memorized the old documents of the Khitan themselves, to cite without pause or flurry the exact reference to refute one historical bluff after another.[7]

Shen returned to China (with biological specimens and maps of the territories he had passed through) to become a Han-lin Academician, to take charge of a large-scale water control survey in the Yangtze region, and then to become head of the Finance Commission. While in this very powerful position he untangled a variety of contradictory policies, producing in the process some of the most penetrating writings before modern times on how

7. Chang Ya-ch'in 张雅琴, "Shen Kua yü Sung Liao hua chieh chiao-she 沈括与宋辽划界交涉" (Shen Kua and the border negotiations between Sung and Liao), *Shih i* 史绎, 1975, 12: 10–25.

to affect and regulate supply and demand, how to forecast prices in order to intervene effectively in the market, and how hoarding, counterfeiting, and melting affect the supply of currency as the price of the metal in it fluctuates about its controlled monetary value. In the autumn of 1077, just as he had launched his revision of critical fiscal measures, the corrupt and vindictive censor Ts'ai Ch'ueh 蔡確 (1036–1093) impeached him. The charge was that Shen had opposed a New Policies taxation measure in an underhanded, inconsistent, and improper way. Historians credited it for centuries, but modern Chinese research has refuted it in every detail.[8] His protector Wang An-shih had just left government. By threatening an established budget item in order to ease the burdens of the poor, it seems, Shen became an easy victim of factional maneuvering.

The emperor was not only the ritual synapse between the political and natural orders. He was a human being whose likes and dislikes courtiers indulged within broad limits that could be further widened by force of his charisma and will. The closer to him an official penetrated, the more achievement and even survival became subject to imperial whim and the intrigue of colleagues. Wang An-shih maneuvered Shen Kua into the proximity of the throne because of Shen's brilliance, judgment, and effectiveness at complicated tasks. He was not adept at protecting himself. He attracted the most damaging animosity not from opponents of the New Policies but from designing members of the coalition to which he belonged. Once the emperor qualified his support of the New Policies in 1074, the risk of debacle remained imminent. Many officials who had risen with Wang fought furiously for the power that would keep them afloat even though the program sank. They did not wish to be deterred by a colleague who judged issues on their own merits. They probably also felt, as others did, that a man in his early forties and low in rank did not merit the emperor's increasing confidence.

Ts'ai Ch'ueh was rising into the vacuum that Wang's retirement had left. The emperor depended increasingly on Ts'ai's monetary counsel and could not easily disregard what he insisted upon. For three years it was impossible to overcome his objections and those of another censor, and to rehabilitate Shen. In 1080 he was sent to Yen-chou 延州 (now Yenan, Shensi province), on the inevitable route for military operations by or against the Tanguts, as Commissioner for Prefectural Civil and Military Affairs. The Tanguts were then divided and weakened, minor Chinese conquests around 1070 had set the stage for a war, and the treasury had ample funds. Shen played an impor-

8. Chang Chia-chü, 121-127.

tant part in organizing and fortifying for the victorious offensive of the autumn of 1081. In extending Sung control he showed a practical as well as a theoretical mastery of the art of warfare. He was cited for merit and given several honorary appointments. It was probably at the same time that he was ennobled as State Foundation Viscount. In his sixteen months at Yen-chou, Shen received 273 personal letters from the emperor. His standing at the court was in principle reestablished. Whether he had become shrewd enough to survive there was never tested.

Shen and a colleague followed up the victory by proposing fortifications to close another important region to the Tanguts. The emperor referred the matter to an ambitious and arrogant official who, ignoring the proposal, changed the plan to provide defenses for what Shen argued was an indefensible and strategically useless location. Shen was commanded to leave the vicinity of the new citadel so as not to share in the credit for the anticipated victory. When the Tangut attack came, the emissary's force was decimated while Shen, with imperial permission, was successfully defending a key town on the enemy invasion route to Yen-chou. He denied the Tanguts an opening for advance.

Be that as it may, Ts'ai Ch'ueh was now a Grand Councillor. As titular military commander Shen was held responsible for the defeat and considerable loss of life. At the age of fifty-one he found his career was over. The anti-New Policies regime later, to no advantage, abandoned the towns he protected, just as another negotiator lost the lands he had saved from the Khitan through diplomacy.

Shen spent six years in fixed probationary residence, forbidden to engage in official matters. He used at least two of these years to complete a great imperially commissioned atlas of all territory then under Chinese control. He had been working on this atlas intermittently since, as Finance Commissioner a decade earlier, he had had access to court documents. His reward included the privilege of living where he chose.

Ten years earlier Shen had bought, sight unseen, a garden estate on the outskirts of Ching-k'ou. In 1086, visiting it for the first time, he recognized it as a landscape of poignant beauty that he had seen repeatedly in dreams, and named it Dream Brook *(Meng ch'i* 夢溪, also read *Meng hsi).* He moved there in 1088. With a pardon and sinecures to support him, he spent his last years in leisure, isolation, and illness.[9]

---

9. For a translation that conveys the flavor of Shen's autobiography see Holzman, "Shen Kua," 275–276.

III

10

Writings

Shen's writings, of which only a few are extant even in part, included commentaries on Confucian classics, two atlases, reports on his diplomatic missions, a collection of literary works, and monographs on rituals, music, mathematical harmonics, administration, mathematical astronomy, astronomical instruments, defensive tactics and fortification, painting, tea, medicine, and poetry. Of three books compiled during his last years at Dream Brook, one, Good Medicinal Formulas *(Liang fang* 良方), was devoted to medical therapy, doctrine, and philology. The other two belong to particularly Chinese genres.

Records of Longings Forgotten at Dream Brook *(Meng ch'i wang huai lu* 夢溪忘懷錄), a collection of notes on the life of the gentleman farmer in the mountains, contains useful information on implements and agricultural technique and, unlike more conventional agricultural treatises up to that time, on the culture of medicinal plants.

Brush Talks From Dream Brook *(Meng ch'i pi t'an* 夢溪筆談) survives and has been well edited in modern times. It is by any reckoning one of the most remarkable documents of early science and technology, not to mention philosophy, art and literary criticism, diplomacy, occultism, linguistics, archaeology, and not a few other topics. It is a collection of about six hundred recollections and observations, ranging from one or two sentences to about a page of modern print. "Because I had only my writing brush and ink slab to chat with, I call it Brush Talks." The jottings are loosely grouped under topics (seventeen in all current versions), of which seven are most pertinent to Nature and man's use of it: "Regularities Underlying the Phenomena" (mostly astronomy, astrology, cosmology, divination), "Technical Skills" (mathematics and its applications, technology, medicine), "Philology" (including etymology and meanings of technical terms), "Strange Occurrences" (incorporating various natural observations), "Artifacts and Implements" (techniques reflected in ancient objects), "Miscellaneous Notes" (greatly overlapping other sections), and "Deliberations on Materia Medica" (most of it devoted to untangling historic and regional confusions in identities of medical substances).

Under all these rubrics, notices of the highest originality stand cheek by jowl with trivial didacticisms, court anecdotes, and ephemeral curiosities. Other sections are given to topics conventional in collections of jottings—memorable people, wisdom in emergencies, and so on. Shen's theoretical discussions of scientific topics employed the abstract concepts of his time—yin-yang, the Five Phases *(wu-hsing* 五行), *ch'i* 氣, and so on.

A large fraction of the book's contents is devoted to fate, divination, and portents. Historians seeking in him the prototype of the modern scientist have ignored Shen's belief in such things, normal among people of his time.[10] The author of Brush Talks has been compared with Leibniz. In an era of happy relations between China and the Soviet Union, Hu Tao-ching 胡道靜, the foremost authority on Shen, called him the Lomonosov of his day. But Shen was writing for gentlemen of universal curiosity and humanistic temperament. Custom, wisdom, language, and spiritual anomaly were as important themes as nature and artifice.[11]

Because Shen's interests were multifarious, the record unsystematic, and its form too confining for anything but fragmentary insight, only accumulation can fairly convey why he is important. What follows is a mere sample of his attempts to deepen the contemporary understanding of Nature, his observations that directed the attention of his educated contemporaries to important phenomena or processes, and his own technical innovations. I have grouped

---

10. Brenier et al. perceptively contrast the modern view of Shen as a scientist and the traditional view of him as a man of letters ("Shen Gua," 350). It is necessary to qualify this contrast in two respects. First, for a thousand years, while littérateurs have seen Shen as a distinguished author, Chinese astronomers have considered him one of their own predecessors. "Man of letters" is itself a Western stereotype, and ignores Shen's service in the highest technical post in the empire. Second, the view of Shen as a scientist in the modern mold has faded out of historical writing since the 1950's. It hardly survives outside the People's Republic of China, as the innovative work of Teraji Jun in Japan and Fu Daiwie (Ta-wei) in Taiwan attests. In China it subsists in large measure because history is an ideologically sensitive profession. A rejection of positivism is correctly understood to be a challenge to the Marxist-Leninist view of science.

Historians of science will also find odd the judgments in this essay that yin-yang, *ch'i,* and so on are not "real" theoretical concepts, but mere labels for correspondence schemata. The authors define "theory" anachronistically as "an exclusive and coercive explanatory system," which makes it possible to deny that there were scientific theories anywhere before the nineteenth century. Equally oddly, the authors find yin-yang et al. unimportant because such terms appear in only an estimated five per cent of the jottings in Brush Talks (p. 345, n. 31). This conclusion (aside from inflating the significance of simple counting) ignores the commonplace observation that in *belles lettres* theory is more often implicit in notions of order and harmony than explicit.

11. Fu Daiwie has studied systematically the jottings on marvels and oddities. He argues that their character is directly related to the "newly emerging official-intellectual culture of the Northern Sung," which involves a rejection of "the old aristocratic high culture." "A Contextual and Taxonomic Study of the 'Divine Marvels' and 'Strange Occurrences' in the *Mengxi bitan,*" *Chinese Science,* 11: 3–35, esp. 22.

them to bring out contiguity of subject matter without interposing the radically different, and quite irrelevant, disciplinary divisions of modern science. On the basis of these samples I discuss—tentatively, given the state of research—the epistemological underpinnings of Shen's work, and the unity of his scientific thought with elements that today would be considered unscientific, primitive, or superstitious. That done, I evaluate Shen's life as a case study in the reconcilability of Confucianism and science, which the conventional wisdom among Sinologues for over a generation has tended to place in opposition.

Quantity and Measure

Mathematics was not the queen of sciences in traditional China. Chinese natural philosophers, unlike most in the postclassical West, did not dismiss the possibility that terrestrial phenomena could conform to mathematical regularities. But given the strength of Chinese quantitative sciences in numerical rather than geometric approaches, the very late and partial development of mathematical generalization, and the complete absence of notions of rigor, it is only consistent that much of the effort to discover such regularities produced numerology (that is, the use of numbers to label qualities, mainly in order to establish associations, meanings, or hierarchies). This practice was firmly founded on the "Great Commentary" to the Book of Changes.

Mathematics was, at least until the seventeenth century, embodied in specific problems about the physical world. Abstract thought about numbers was always concerned with their qualities rather than their properties, and thus remained numerology. This art, although it blended into arithmetic, was only partly distinct from other symbolic[12] means for exploring the inherent patterns of Nature and man's relation to it. Computation, on the other hand, was applied in a coherent tradition of textbooks to a great variety of mensurational, accounting, and other everyday tasks of the administrator. They were categorized by the type of practical problem to be solved: surveying, computing land tax, estimating requirements, and so on. Occasionally curiosity and skill pushed beyond these pragmatic limits, but, in Shen's time, never very far. Some of the problems that Shen presented in Brush Talks had no application, but his enthusiasm for them was undiminished.

In addition to this accumulation of individual problems, there were two exact sciences in which mathematics served to advance knowledge of the patterns underlying the phenomena. One was mathematical harmonics *(lü lü*

---

12. In the anthropologist's sense: ritual, religious, divinatory.

律呂), which explored the relations between musical intervals and the dimensions of instruments that produced them, in ways analogous to the Pythagorean art. Its appeal was much the same in both China and Greece: it demonstrated how deeply the power of number was rooted in Nature. For this reason in China mathematical harmonics was often put into the same category as mathematical astronomy, which also had foundations in cosmology. Astronomy, by far the more technically elaborate of the two exact sciences, was normally employed on behalf of the monarch. Successfully predicting a phenomenon fitted it into the dynamic cosmic and social order that the emperor maintained on behalf of his people. Unpredictable phenomena and failed predictions were either good or bad omens. Each challenged the established order, and each had a meaning. Bad omens warned that the emperor's mediating virtue, which maintained concord between the cosmic and political orders, was deficient. Good portents generally signalled, and approved, a step in a new direction. Successfully predicting celestial events neutralized their ominousness, preserving the charisma of the ruling dynasty.

The annual calendar (or almanac) issued by the throne was thus important in the solemn ceremonies that asserted the emperor's authority, as the ancient motto put it, to "grant the seasons." The calendar encompassed all predictable events, including planetary phenomena and eclipses. The utilitarian calendrical aspects—lunar months and solar years—had long since been refined past any practical demand for accuracy, but astronomical reinforcement of the Mandate of Heaven called forth endless attempts to improve constants. As it became conventional to institute a complete new system for computing these ephemerides when a new emperor was enthroned, technical novelty was at a premium. When new ideas were unavailable, the Directorate of Astronomy tended to trivially repackage old techniques.

Repeated failures of prediction were another motive to reform the astronomical system. In such cases too, the system was in principle replaced as a unit rather than repaired. Most systems survived or fell on their ability to predict eclipses, particularly solar eclipses. These were the least suited of all celestial phenomena to the algebraic, non-geometric style of mathematics. Prior to Shen's time little effort had gone into predicting the apparent motions of the planets, which lacked the significance of solar and lunar phenomena. This was, in fact, an omission that Shen seems to have been the first to confront.

General mathematics. As wood-block printing became widespread, the government, eager to control the content of education, recognized its value. The court used it to propagate carefully chosen and edited collections of ancient textbooks. It was doing so for medicine at the time Shen entered the capital bureaucracy. In 1084 it printed a collection of ten mathematical manuals, made four centuries earlier and reconstituted as well as extant texts allowed. The authority of these projects served both to fix textual traditions, preserving selected treatises from further attrition, and to consign to oblivion the books that government committees decided not to include. Shen thus lived at a decisive period in the development of mathematical traditions. His judgments on lost techniques and disused technical terms (e.g., 300, 306)[13] have played an important part in later attempts to interpret them. Brush Talks is also an essential source for the study of pre-Sung metrology, currency, and other subjects related to computation.

Shen used mathematics in formulating policy arguments more consistently than most of his colleagues. Examples are his critique of military tactics in terms of space required for battle formations (579) and his computation that a campaign of thirty-one days is the longest that can feasibly be provisioned by human carriers (205). But of the computational methods discussed in his "Technical Skills" chapter, those not related to astronomy are almost all abstractly oriented.

This original bent emerges most clearly in two problems. One is a new method for computing the frustum of a solid rectangular pyramid. Shen worked out the volume of the same figure if composed of stacked articles (he mentioned flat round chess pieces, bricks, and wine vats) that leave interstices (301). Since Shen intended this "volume with interstices" (*hsi chi* 隙積) method to be applicable regardless of the shape of the objects stacked, what he gave is a correct formula for the number of objects, which are thus to be considered of unit volume. His presentation has several interesting features. Needham has suggested (III, 142–143) that the concern with interstices (and, one would add, unit volumes) may have been a step in the direction of geometric exhaustion methods, although it was tentative and bore fruit only in seventeenth-century Japan. Second, instead of the worked-out problem with actual dimensions that is conventional in early textbooks, Shen simply gave a generalized formula: "double the lower length, add to the upper length, multiply by the lower width," and so on. Third, this was the earliest known case in China of a problem involving higher series. Built on earlier numerical

13. Numbers in parentheses are item numbers in the Hu Tao-ching 1960 ed. of *Meng ch'i pi t'an* (or "Brush Talks"). Roman volume numbers followed by page numbers refer to translations in Joseph Needham et al., *Science and Civilisation in China.* Where my own understanding differs considerably from Needham's, an asterisk follows the page reference. All translations are my own.

approaches to arithmetical progressions, it provided a basis for more elaborate treatment by Yang Hui 楊輝 (1261) and Chu Shih-chieh 朱世傑 (1303).

The second problem of interest was said "in a story" to have been solved by one of China's greatest astronomers, the Tantric Buddhist patriarch I-hsing 一行 (682–727). It sought the number of possible situations on a *go* board, with nineteen by nineteen intersections on which any number of black or white stones may be placed. Whether I-hsing actually solved this problem we do not know; Shen's single paragraph was the first and last known discussion of permutations in early Chinese mathematics. It stated the order of magnitude of the answer—"approximately speaking one must write the character *wan* 萬 (10,000) fifty-two times in succession." This amounts to an order of magnitude of roughly 10^{208}. Shen added to this solution exact answers for smaller arrays, three methods of solution, and a note on the limited traditional notation for very large numbers (304).[14]

Mathematical harmonics. The Pythagoreans were fascinated by the relations of concordant intervals to the plucked strings that produced them, since the lengths between stops were proportionate to simple ratios of integers. By about 200 B.C., the Chinese built up a similar science on a gamut of standard pipes. Beginning with a pipe eight inches long and 0.9 inch in diameter, they generated the lengths of subsequent pipes by multiplying the previous length alternately by 2/3 and 4/3, making twelve pipes within an approximate octave. The dozen were then related to such categories as the twelve divisions of the tropical year, in order to provide a cosmic basis for the system of modes that the pipes determined. A pentatonic scale, which could be used in any of the twelve modes, provided similar associations with the Five Phases This basis was extended to metrology by defining the lengths and capacities of the pipes in terms of millet grains of standard dimensions. Shen lucidly and concisely explained these fundamentals of mathematical harmonics, and corrected grotesque complications that had crept into a canonic source through miscopying of numbers (143, 549).

---

14. Needham has translated part of this jotting (III, 139*). The extant text, even in Hu's edition, is very corrupt. Ch'ien Pao-ts'ung has edited and considerably emended it in *Sung Yuan shu-hsueh-shih lun-wen-chi*, 266–269. For a general study of the treatment of large numbers in Chinese mathematics see Joël Brenier, "Notation et optimisation du calcul des grandes nombres en Chine. Le cas de l'échiquier de go dans le Mengqi bitan de Shen Gua (1086)," in *Nombres, astres, plantes et viscères. Sept essais sur l'histoire des sciences et des techniques en Asie orientale,* ed. Isabelle Ang & Pierre-Étienne Will (Memoires de l'Institut des Hautes Études Chinoises, 35; Paris: Collège de France, 1994), 89-111.

He also studied stringed instruments. By straddling strings with paper figures, he showed that strings tuned to the same notes on different instruments resonate, as do those tuned an octave apart on the same instrument (537; cf. IV.1, 130). His two chapters on music and harmonics are also a trove of information on composition and performance.

Astronomy. Shen's major contributions in astronomy were his attempts to visualize celestial motions spatially, his arc-sagitta methods that for the first time moved algebraic techniques toward trigonometry, and his insistence on daily observational records as a basis for his calendar reform. The first had no direct application in computation of the ephemerides, although it may well have inspired (and at the same time have been inspired by) the second, which evolved from traditional mensuration. It has been suggested that the clarity of Shen's cosmological explanations led to his appointment to the Directorate of Astronomy, which provided opportunity for his contributions in the second and third areas. But circumstances that arose from the bureaucratic character of mathematical astronomy made these contributions futile in his lifetime.

Shen's discussions of solar, lunar, and eclipse phenomena (130–131; excerpts, III, 415–416) have been mentioned (p. 6). By far the most remarkable of his cosmological hypotheses attempted to account for variations in the apparent planetary motions, including retrogradation. This concern is not to be taken for granted, since traditional astronomers, in addition to their disinterest in planetary problems, preferred purely numerical approaches to prediction, unlike the spatial geometric models of Greek antiquity. Constructing explanatory models was not part of their work. Noting that the greatest planetary anomaly occurred near the stationary points, Shen proposed a model in which the planet traced out a figure like a willow leaf attached at one side to the periphery of a circle. The change in direction of the planet's motion with respect to the stars was explained by its travel along the pointed ends of the leaf (148).[15]

The willow leaf, in other words, served one of the same functions that the epicycle served in Europe. It is characteristic that, having taken a tack that in the West was prompted entirely by geometric reasoning, Shen's first resort should have been a familiar natural object with rich poetic connotations. Use of a pointed figure would hardly have survived a mathematical analysis of observational data, but this simile remained an offhand suggestion.

15. Translated above in Chap. II, Appendix C; see its Figure 9.

Another early outcome of Shen's service at the court was a series of proposals to redesign major astronomical instruments: the gnomon, which was still employed to measure the noon shadow and fix the solstices; the armillary sphere, with which angular measurements were made; and the clepsydra, used to determine the time of observations as well as to regulate court activities. Shen's improved versions of the armillary and the clepsydra apparently were not built until late 1073, after he had taken charge of the Directorate of Astronomy. The armillary at least was discarded for a new one in 1082, a casualty of his personal disgrace.

Shen's clepsydra proposals represent a new design of the overflow-tank type (Needham's Type B; III, 315–319, 325), but the most significant outcome of his work on this instrument was a jotting on problems of calibration. Day and night were by custom separately divided into hours, the length of which varied with the season. The time was read off graduated float rods, day and night sets of which were changed twenty-four times a year. Shen pointed out that this crude and inadequate scheme amounted to linear interpolation, "treating the ecliptic as a polygon rather than a circle," and argued for the use of higher-order interpolation (128).

The best armillary sphere available in the central administration when Shen first worked there was based on a three-hundred-year-old design "and lacked ease of operation" (150). Shen's most interesting improvement was in the diameter of the naked-eye sighting tube. At least from the first millennium B.C. a succession of stars had been designated the polestar until each strayed unacceptably far from the celestial pole, the still point in the sky about which the stars rotate in the course of a day and night. In the late fifth century A.D. Tsu Keng 祖暅 discovered that the current polestar, 4339 Camelopardi, rotated about a point slightly more than a degree away. This determination of the true pole was incorporated in subsequent instruments by making the radius of their sighting tubes 1.5 Chinese degrees (each of which is 360/365.25°). The nightly excursion of the polestar just inside the field of view thus provided a check on whether the instrument was pointing true north. Six hundred years later Shen found that the polestar could no longer be kept in view throughout the night. He gradually widened the tube, using plots of the polestar's position made three times each night for three months to adjust aim, until his new calibration revealed that the distance of the star from "the unmoving place at the celestial pole" was now slightly over three degrees (127; III, 262). Shen's successors followed him in treating the distance as variable. The relation of this secular change to what is now called the precession of the equinoxes remained unexplored. Aware of the periodic retrogradation of the lunar nodes, Shen also discarded the armillary ring representing the moon's path, which did not reflect this motion. It was never used again.

Calendar reform. On the accession of Shen-tsung in 1068, a new computational system was expected. The inability of the incumbent specialists to produce one left Shen with a clear mandate when he took over the Directorate of Astronomy in 1072. The situation became even more awkward when he could dislodge few of the timeservers already in the bureau, and was forced to bring in Wei P'u and others from outside the civil service in order to begin work on the calendar reform. It is not yet possible to tell what part of the work was done by Shen and what part by his assistants. Wei surely took responsibility for compiling the system as Shen became increasingly occupied elsewhere in government. Wei bore the brunt of fervent opposition within the bureau. He was even formally accused of malfeasance.

Shen knew that previous Sung astronomical systems had suffered greatly from reliance on old observations. He clearly conceived what new data were needed for the first major advance in centuries. Unabating opposition within the bureau and his own demanding involvements outside it limited the number of important innovations in his Oblatory Epoch (Feng-yuan 奉元) system. It was the official basis of calendar computation from 1075, the year of its completion, to 1094, a period very close to the average for systems of the Northern Sung. That the system was not used longer has little to do with its merits, since except in cases of spectacular failure, Sung astronomical systems changed as rulers changed. Shen's was replaced when the coming of age of Che-tsung began a new era. The immediate vicissitudes and long-term influence of three special features will give an idea of how historical actualities limited Shen's astronomical ambitions.

The boldest aspect of Shen's program was the attempt to predict the apparent positions of the planets—not merely their mean speeds and prominent phenomena—for the first time. The computational tools available did not permit this to be done with a few observations of stationary points, occultations, and maximum elongations. Shen and Wei therefore planned a series of observations of a kind not proposed in Europe until the time of Tycho Brahe, five centuries later: exact coordinates read three times a night for five years. Similar records were to be kept for the moon's positions, since previous Sung systems had still used the lunar theory of I-hsing, which after 350 years had accumulated considerable error. This program of observing was the most unfortunate casualty of the obstruction within the Bureau. Shen and Wei had no recourse but to produce a conventional planetary theory based mainly on old observations. They were able to correct the lunar error, but even this proposal provoked such an outcry that it could be vindicated only by a public demonstration using a gnomon (116).

A second issue was the central one of eclipse prediction. Previous attempts to add or subtract correction factors had shown the futility of tinkering. It was Wei P'u who "realized that, because the old eclipse technique used the mean sun, [the apparent sun] was ahead of it in the accelerated phase of its motion and behind it in the retarded phase." He therefore incorporated apparent solar motion into the eclipse theory (139). This had been done centuries earlier but abandoned.

A major obstacle in eclipse prediction, as well as in such workaday problems as projecting observations in equatorial coordinates onto the ecliptic, was the absence of spherical geometry. Shen's evolution of arc-chord-sagitta relations out of some inferior approximations for segment areas given in the arithmetical classics was a first step toward trigonometry. He made it possible in effect to apply sine relations and a fair approximation of cosine relations (301; III, 39, with diagram). The great remaining lack, as in planetary theory, was a mass of fresh observations on which to base new parameters. That this weakness could threaten the continuance of the system became clear in 1076, the year after it was adopted, when the failure of a predicted lunar eclipse to occur left Shen and his associates open to attack. Shen parried with a successful request that students in the Palace Library's Bureau of Astronomy be ordered to carry out his observational program "for three or five years" and to communicate the results to the original compilers. Whether this expedient could have bypassed the stalemate at the Directorate of Astronomy's observatory remains unknown, for in the next year Shen's impeachment aborted it.

In sum, the immediate outcome of the Oblatory Epoch calendar reform was undistinguished, and within half a century the official documents embodying it had been lost. It is impossible to be sure, for instance, to what extent Shen, after inventing arc-sagitta relations, had incorporated them. But enough information survived in proposals, reports, Shen's writings, and compendia of various sorts for his astronomical system to play a considerable part in the highest achievement of traditional Chinese mathematical astronomy, the Season-Granting *(Shou shih* 授時) system of Kuo Shou-ching 郭守敬 (1279). Kuo carried out a sustained program of observation using instruments that incorporated Shen's improvements. He took up Shen's arc-sagitta formula, greatly improving the cosine approximation, and applied it to the equator-ecliptic transform. Aware of Shen's emphasis on the continuous variation of quantities in Nature and his criticism of linear interpolation in clepsydra design, Kuo used higher-order interpolation to an unprecedented extent in his calendar reform.[16]

Shen recorded another scheme for reform of the civil calendar that was most remarkable for his time and place. It almost certainly occurred to him in the last decade of his life. The traditional lunisolar calendar was a series of compromises in reconciling two incommensurable quantities. The modern value for the tropical year is 365.2422 days, and that for the synodic month 29.53059 days, so that there are 12.3683 lunar months per solar year. The practical problem was to design a civil calendar with an integral number of days each month, and an integral number of months each year, in such a way that the long-term averages approach the astronomical constants. Hardly two of the roughly one hundred computational systems recorded in early China solved this problem in exactly the same way, just as there was endless tactical variety in other traditional societies, but the strategy was generally the same. Months of twenty-nine and thirty days alternated, with occasional pairs of long months to raise the average slightly. Intercalary thirteenth months were inserted roughly seven times every nineteen years. That comes to 0.3684 additional months per year, an excellent approximation.

By a millennium before Shen's time the calendar was more than adequate in these respects for every civil need, although attempts to further refine the approximation led to endless retouching. The rhythms of administration, and to some extent of commerce, were of course paramount in the design of the lunisolar calendar, despite pieties about imperial concern for agriculture. Chinese farmers did not need a printed almanac by which to regulate their activity. What they consulted was its notations of lucky and unlucky days. Division of the year by lunar months is, in fact, useless for agriculture, since the seasons that pace the farmer's work vary with the sun alone. The Chinese calendar also incorporated twelve equal divisions of the tropical year *(ch'i*, like the Babylonian *tithis)*, further subdivided into twenty-four periods with such names as Spring Begins, Grain Rains, and Insects Awaken. These provided a reliable notation for seasonal change in the part of northern China in which the series originated, but were useless elsewhere. They were included for their antiquity, not because they filled an agricultural need.

Shen's suggestion was a purely solar calendar, based on the twelve divisions of the tropical year (average 30.43697 days in his system) instead of on the lunation. The civil calendar would thus alternate months of thirty and

---

16. An excellent biography is Li Ti 李迪, *Kuo Shou-ching* (Shanghai: Shang-hai Jen-min Ch'u-pan-she, 1966).

thirty-one days, with pairs of short months as necessary to approach the average. This would provide truly seasonal months and at the same time do away with "that goitrous excrescence" the intercalary month. "As for the waxing and waning of the moon, although some phenomena such as pregnancy and the tides are tied to them, they have nothing to do with seasons or changes of climate; let them simply be noted in the almanac" (545). Shen was aware that because the lunisolar calendar went back to hoary antiquity "it is by no means appropriate to criticize it." He predicted that his discussion "will call forth offense and derision, but in another time there will be those who use my arguments." This proposal was in fact considered by later scholars the greatest blemish on Shen's astronomical talent. His posterity appeared in the mid-nineteenth century, with the even more radical solar calendar enacted for a few years by the T'ai-p'ing rebels.[17] His work was cited to justify more respectable proposals between that time and the adoption of the Gregorian calendar in 1912.

Configuration and Change

Despite Shen's exceptional interest across the board in mathematical quantity and his conventional fondness for numerology, the most obvious of his contributions to understanding the earth and its phenomena are qualitative. I group them below, for convenience, by modern category.

Magnetism. For more than a millennium before Shen's time, south-pointing objects carved from magnetite were used from time to time in ceremonial and magic. In 1044 a book on military arts recommended, as aids to pathfinding, objects cut from sheet iron and magnetized by thermoremanence. Shen took up the matter of needles rubbed against lodestone by contemporary magi, discussed floating and other mountings, recommended suspension, noted that some needles point north and some south, and asserted that "they are always displaced slightly east rather than pointing due south"—all in about a hundred characters (437; IV.1, 249–250). Noticing the result of what is today called magnetic declination depended not only on trials with a sus-

17. Kuo T'ing-i 郭廷以, *T'ai-p'ing t'ien kuo li-fa k'ao-ting* 太平天國曆法考訂 (Critical study of the calendrical methods of the T'ai-p'ing Heavenly Kingdom; Peking: Commercial Press, 1937, reprinted Taipei: idem, 1963); Lo Erh-kang 羅爾綱, *T'ien li k'ao chi t'ien li yü yin yang li jih tui-chao-piao* 天曆考及天曆與陰陽曆日對照表 (On the T'ai-p'ing calendar, with a concordance table for the lunar and Gregorian calendars; Beijing: San Lien Shu-tien, 1955).

pended needle but also on the improved meridian determined by Shen's measurement of the distance between the polestar and true north.[18]

Shen may have been anticipated by geomancers, who practiced a sophisticated proto-science of land configuration and siting, but the dates of texts on which such claims have been based are questionable. The use of compass needles in navigation is recorded shortly after Shen's death, and later descriptions provide enough detail to show that the twenty-four-point rose that Shen substituted for the old eight compass points (perhaps also under the stimulus of the better meridian, if not of geomantic practice) was widely adopted. He apparently was unaware of the polarity of magnetite itself, since in another article he explained the difference between north-pointing and south-pointing needles as "perhaps because the character of the stone also varies" (588; IV.1, 250).

Cartography. It has been conjectured that Shen was the first to use a compass in mapmaking, although traditional methods would have sufficed. Neither his early maps of Khitan territory nor the atlas of China completed in 1087 have survived to answer this question. But in an enclosure to the latter he did separately record bearings between points using his twenty-four-point compass rose. He also listed rectilinear distances rather than, as customary, distances along established routes. He called the use of distances "as the bird flies" ancient, but we have no earlier record. "Thus although in later generations the maps may be lost, given my book the territorial divisions may be laid out according to the twenty-four directions, and the maps speedily reconstructed without the least discrepancy" (575; III, 576). His great atlas included twenty-three maps drawn to a uniform scale of 1:900,000. The general map was ten by twelve Chinese feet. There is no evidence that the handbook outlasted the maps.

Three-dimensional topographic maps go back at least to Hsieh Chuang 謝莊 (421–466), who had a demountable wooden model carved, apparently on the basis of an ancient map. In 1075, while inspecting the Khitan border, Shen embodied information gathered from the commander and the results of his own travels in a series of relief maps. He had them modeled, for the sake of portability, in plastic media (wheat paste and sawdust until the weather turned freezing, then beeswax) on wooden bases. These were carried to the

---

18. Needham and Peter J. Smith, "Magnetic Declination in Mediaeval China," *Nature*, 17 June 1967, 1213–1214. See the historical table in *Science and Civilisation in China*, IV.1, 310. Declination in Shen's part of China at the time has been estimated as between five and ten degrees.

capital and duplicated in wood. The government thenceforth required similar models from other frontier regions (472; III, 580).

Shen regularly used both historical research and special on-the-ground surveys to solve such cartographic problems as tracing changes in watercourses (431). Typical of his ingenious topographic survey methods were those used in 1072 to measure the slope of the Pien Canal near the capital. He built a series of dikes in temporary, narrow parallel channels to measure incremental changes in water level (457; III, 577*).

Formation of the earth. In 1074, in the T'ai-hang 太行 mountain range (Hopei), Shen noticed strata of "bivalve shells and ovoid rocks running horizontally through a cliff like a belt. This was once a seashore, although the sea is now hundreds of miles east. What we call our continent is an inundation of silt. . . . This mud year by year flows eastward, forming continental land." A similar stratum had been observed long before by Yen Chen-ch'ing 顏眞卿 (708–784), who vaguely suggested its origin in the sea; but Shen—whose duties had made him intimately familiar with the process of silting—opened a new line of investigation by proposing a cause (430; III, 604).

Probably on his southward drought survey earlier in the same year, Shen saw the Yen-tang 雁蕩 range (Chekiang), a series of fantastic rock formations "invisible from beyond the ridgeline [opposite], but towering to the sky when seen from the valleys. If we trace the underlying pattern, it must be that great waters in the valleys have attacked and washed away all the sand and earth, leaving only the great rocks erect and looming." His explanation proceeded to generalize the shaping role of erosion, and then to apply it to the hills that divide streams in the loess country of northwest China—"miniatures of the Yen-tang mountains, but in earth rather than stone" (433; III, 603–604).

Shen reported diverse contemporary finds of petrified plants and animals (373–374; III, 614–618). He remarked particularly on a stony formation he identified as originally a grove of interconnected bamboo roots and shoots, found dozens of feet below ground level at Yenan 延安 (Shensi). He knew from his military service there that the climate was too dry to grow bamboo: "Can it be that in earliest times [literally, 'before antiquity'] the land was lower and the climate moister, suitable for bamboo?" (373).[19] About a century

19. Fu Daiwie observes that Shen did not consider these objects to be anything that could be called fossils in the modern technical sense. He understood them by applying the contemporary idea that one living creature metamorphoses into another (29–31).

later the great philosopher and polymath Chu Hsi (1130–1200), who knew Shen's jottings well and often extended ideas from them in his teaching, suggested that the stone of certain mountains was itself petrified silt deposits. But Shen's notion of prehistoric climatic change, like that of the reshaping of land by erosion, was not pushed further soon after his lifetime.

Atmospheric phenomena. Although Shen did not report important original discoveries of his own, he preserved a number of interesting observations not recorded elsewhere. Perhaps the most important is a vivid description of a tornado (385; translated in Holzman, "Shen Kua," 286). Modern meteorologists questioned its veracity until, in the first decade of the twentieth century, the Sikawei Observatory in Shantung reported phenomena of the same kind, previously thought restricted to the western hemisphere.

Shen transmitted an explanation of the rainbow by Sun Ssu-kung 孫思恭, an elder contemporary in the court who was also considered one of the best mathematical astronomers of his era. "The rainbow is the image [literally, 'shadow'] of the sun in rain, and occurs when the sun shines upon it." This sentence does not, as often claimed, adduce refraction (pinhole or mirror images were regularly called "shadows"; see 44). Sun's observations prompted Shen to determine by experiment that the rainbow is visible only opposite the sun (357). Later Chu Hsi, aware of Shen's account, added that by the time the rainbow appears "the rain *ch'i* has already thinned out; this in turn is because sunlight has shone on and attenuated the rain *ch'i.*" *Ch'i* must mean vapor here; the notion of reflections off individual drops is hardly implicit. Shen also recorded the fall of a fist-sized meteorite in more detail and with less mystification than in previous reports. The particulars of its fall came from a careful account by another of Wang An-shih's associates. The object was recovered and exhibited, but Shen did not claim that he himself had observed that "its color is like that of iron, which it also resembles in weight" (340; III, 433–434).

Products of the earth. Shen's responsibilities with respect to fiscal policy gave him a detailed knowledge of important commodities, their varieties, and the circumstances of their production, as may be seen from his descriptions of tea (208) and salt (50, 221, 224, 422). Inflammable seepages from rock had been known a millennium before Shen's time, and for centuries had been used locally as lamp fuel and lubricant. While Commissioner near Yen-chou in 1080-1081, he noted the blackness of soot from petroleum. Good ink was then made by burning pine resin, but Shen knew that North China was being rapidly deforested. He remarked that, in contrast with the growing scarcity of trees, "petroleum is produced inexhaustibly within the earth." He began an industry to manufacture the solid cakes of carbon ink used for writing and painting throughout China. The name Shen coined for petroleum is the one used today, and the source in Shensi province that he developed is still exploited. In the same article he quoted a poem of his that is among the earliest records of the economic importance of coal, then beginning to replace charcoal as a fuel (421; III, 609, partial).

Optical phenomena. Shen did not directly link his interest in image formation with his worldly concerns. The play of his curiosity over old artifacts, not the improvement of naked-eye astronomical instruments, motivated him.

In the canons of the Mohist school (ca. 300 B.C.) is a set of propositions explaining the formation of shadows and of optical images (considered a kind of shadow) in plane, convex, and concave mirrors. Most scholars believe that one proposition concerns pinhole images, although textual corruption and ambiguity make this uncertain. These propositions are in many respects physically correct, although very schematic. They do not presuppose rays of light.

Shen asked himself why a concave mirror forms an inverted image. He suggested that an "obstruction" *(ai* 礙*)*, analogous to an oarlock, constricts the "shadow" to a shape like that of a narrow-waisted drum (more or less two cones apex to apex, the second constituting the inverted image). Like the Mohists, Shen clearly believed that inversion takes place *before* the image is reflected. He expressly likened the inverted image to that of a moving object formed on the wall of a room through a tiny opening in a paper window. Aware for the first time that there is a range of distances from a concave mirror within which no image is formed (that is, between the center of curvature and the focal point), he explained that this blank region, corresponding to the pinhole, is the locus of "obstruction" (44).[20] Shen's pinhole observation was adventitious, but his approach to the burning-mirror settled the details by experiment.

Two other observations of optical interest are found under the rubric "Artifacts and Implements." The first, in the Sequel to Brush Talks, noted that when the ancients cast bronze mirrors, they made the faces just convex enough that, regardless of size, every mirror would reflect a whole face. By

---

20. This jotting and the Mohist texts are translated in A. C. Graham and Sivin, "A Systematic Approach to the Mohist Optics," in S. Nakayama and Sivin (ed.), *Chinese Science: Explorations of an Ancient Tradition* (Cambridge, MA: MIT Press, 1973), 105–152.

Shen's time this refinement had been abandoned and the reasoning behind the curvature forgotten, so that collectors were having the faces of old mirrors scraped flat (327; IV.1, 93).

The second jotting is the oldest record of a Far Eastern curiosity still being investigated: "magic mirrors," or, as Shen called them, "transparent mirrors." Shen described a bronze mirror with the usual smooth, polished face and an inscription integrally cast in relief on its back. When this mirror was used to reflect the sun onto a wall, the inscription appeared within the image. Shen cited with approval an anonymous explanation: "When the mirror is cast, the thinner parts cool first. The raised design on the back, being thicker, cools later and the shrinkage of the bronze is greater. Although the inscription is on the obverse, there are imperceptible traces of it on the face, so that it becomes visible within the light." He noted that this explanation is incomplete, because he had tried mirrors in his own and other collections that were physically indistinguishable from the "transparent" ones and found that they did not cast images (330; IV.1, 94*). His doubt was justified, although the approach taken by his informant was at least as good as those of some modern metallurgists.

Manipulation of the cooling rate can only accentuate the effect. The variation in thickness appears to be primarily responsible for the image in this sort of mirror, the most common among several types extant. Filing considerable bronze off the face of the mirror after casting is the key. This releases tensions in the metal and gives rise to slight deformations that produce the image.[21]

Productive Techniques and Materials

The technologies of Shen's time, like their European counterparts, were not cumulative and linked to science, but independent artisanal traditions transmitted from master to pupil. Shen left so many unique and informative accounts of ancient and contemporary processes among his jottings that Brush Talks has become a major source for early technology. Shen's interests in contemporary techniques can in most cases be linked to broad concerns of his official career; but the exceptional richness of his record bespeaks a rare

21. For Chinese research see Julia K. Murray & Suzanne E. Cahill, "Recent Advances in Understanding the Mystery of Ancient Chinese 'Magic Mirrors': A Brief Summary of Chinese Analytical and Experimental Studies," *Chinese Science*, 1987, 8: 1–8. This report gives more attention to temperature than to surface deformation. I am also grateful for counsel to Prof. Mao Tseng-tien 毛增滇.

curiosity, and the trenchancy of his descriptions a seriousness about mechanical detail unusual among scholar-officials. His notes on techniques that had been lost before his time reflect the application of this technical curiosity and seriousness to archaeology, which was just becoming a branch of antiquarianism in the eleventh century.

Most of Shen's cultured contemporaries keenly appreciated good workmanship, but saw no reason to notice the artisans responsible for it, except occasionally when condescending to them. Shen wrote about resourceful craftsmen and ingenious laborers with much the same admiration he gave to judicious statesmen. He never lost sight of the social distance between himself and members of the lower orders, but in his writing there is no snobbishness about the concert of hand, eye, and mind.

Contemporary techniques. The most famous example is Shen's account of the artisan Pi Sheng's 畢昇 (fl. 1041–1048) invention of movable-type printing. Shen described how ceramic types were carved and fired, and imbedded and leveled in a layer of resin, wax, and paper ash in an iron form. A second form was set up as the first was printed. As in woodblock printing, water-based ink was used. Since the porous, thin paper took it up readily, no press was needed. Shen also remarked, with his usual acumen, that the process could become faster than carving wood blocks only with very large editions (historians estimate the average printing of a scholarly book in the Northern Sung period at between fifty and a hundred copies). Unevenness of the surface and absorption of ink by the fired clay must have posed serious problems. The process was abandoned after Pi died, probably due to the lack of economic incentive that Shen noted. The long series of royally subsidized Korean experiments in the fifteenth century that perfected cast-metal typesetting still began with Pi Sheng's imbedding technique as described by Shen. Whether he knew Pi is unclear, but Shen's nephews owned Pi's original font (307).[22]

Shen described several metallurgical processes. Especially interesting are the recovery of copper from a mineral creek by replacement of iron, a process then being carried out on an industrial scale to provide metal for currency (455; II, 267); two of the three steelmaking processes used in early China (56);[23] and a little-known cold-working method that the smiths of the Ch'iang people of western China used to make extremely tough steel armor (333).

22. Translated by T. H. Tsien in *Science and Civilisation in China,* V.I, 201–202.

23. Translated in Needham, *The Development of Iron and Steel Technology in China* (London: The Newcomen Society, 1958), 33–34.

Water control techniques of which he records details include pound-locks with double slipways (213; IV.3, 351–352), piles for strengthening embankments (210; IV.3, 322–323), and sectional gabions for closing gaps after embankment repairs (207; IV.3, 342–343).

Ancient techniques. The concern for understanding ancient techniques began more than a millennium earlier, as commentators strove to understand everything mentioned in the Confucian and other classics. Exegesis remained an important activity in China, and the productive methods of golden antiquity were investigated with the same assiduity as anything else mentioned in its literary remains. Less than a century before Shen's time, for various reasons—among them the recovery of ancient artifacts in large numbers for the first time, the growth of collecting, and the elaboration of a conscious aesthetics—archaeology began to emerge from the footnotes, especially in monographs on ancient implements and ritual institutions. He was familiar with this literature and responded to it critically. Much of his writing in the "Artifacts and Implements" chapter falls squarely in this antiquarian tradition, drawing on the testimony of both objects and books.

Shen's vision of the past as a repertory of lost processes introduced an influential new theme. A constant concern in his writing was not only that the workmanship of the past be esteemed for its excellence, but also that the present be enriched through understanding what the practical arts had been capable of. Although his contemporaries generally believed that semidivine monarchs of archaic times had invented the arts that made civilization possible, in a letter Shen saw technology emergent for just the opposite reason: "How could all of this have come from the Sages? Every sort of workman and administrator, the people of the towns and those of the countryside: none failed to participate."[24]

Shen's remarks on magic mirrors are typical of his effort to understand lost processes. Another example is his reconstruction (and trial) of ancient crossbow marksmanship, interpreting a gnomic aiming formula in an ancient footnote with the aid of a graduated sight and trigger assembly that he examined after it was unearthed (331; III, 574–575). The most famous instance of Shen's use of literary sources to study techniques has to do with the remarkable modular system of architecture used in public buildings. The set of standard proportions is well known from an official compilation printed about a decade after Shen's death. By describing the earlier proportion system of the Timberwork Canon *(Mu ching* 木經), attributed to a great builder of about

24. *Ch'ang-hsing chi, 19:* 53b.

1000 and already falling out of use, Shen demonstrated the antiquity of this art (299; IV.3, 82–83).

Medicine

Medicine, from at least the second century B. C., drawing heavily upon natural philosophy for its conceptual underpinnings, accumulated a classical tradition. Not only was each new treatise consciously built upon its predecessors, but a main goal of new work was restoring an understanding that medical scholars believed was deepest in the oldest writings. The revealed truth of the archaic canons was too concentrated for ordinary latter-day minds, who could hope to recapture it only as the culmination of a lifetime of study. Writers in the intervening centuries referred to the early classics as the ultimate source of significance, even while aware that empirical and practical knowledge had considerably advanced since antiquity. The continuous tradition of medical writing fitted new experiences into the old framework and, when necessary, constructed new frameworks in the spirit of the old. As government committees compiled and disseminated standard editions of the chief classics, the curing arts became more respectable as a field of study. Large numbers of men from the scholar-official class began to take up medicine, not in competition with those who made a living by it but as a means of self-cultivation allied to cosmology and occasionally useful. The initial motivations commonly were personal ill health and the desire to serve one's sick parents.[25]

Shen, we have seen, began the study of medicine early, for the former reason. One of his two therapeutic compilations survives in altered form. Its preface is a long disquisition on the difficulty of adequate diagnosis and therapy, as well as on the proper selection, preparation, and administration of drugs. His criticisms of contemporary trends toward simplification remind us that the development of urban culture and the spread of education in Sung China had led to increased medical practice among ordinary people as well as study by the literati. As popular adaptations of elite medicine began to displace ritual folk therapies (at least in the cities), there were more half-educated physicians to be criticized by learned amateurs such as Shen.

25. Robert Hymes demonstrates conclusively for one affluent southeastern locality that "a permanent shift among elite families from exclusive concentration on the achievement of high office to a broader strategy focusing on local position" came only at the end of the N. Sung. "Not Quite Gentlemen? Doctors in Sung and Yuan," *Chinese Science*, 1987, 8: 9–76.

Shen's most characteristic contribution was undoubtedly his emphasis on his own experience, unusual in a tradition whose literature in the Sung still depended heavily on copying wholesale from earlier treatises. Shen included only medical formulas the efficacy of which he had witnessed, and noted the circumstances in which most had succeeded. He precisely described many animal, vegetable, and mineral drugs. Although like other pharmacognostic scholars he had no interest in general taxonomic schemes, his concern for exact identification and for philological accuracy gave his critical remarks enduring value. Many were incorporated into later compilations on materia medica. Shen's writing also served as a stimulus to the work a few decades later of the great pharmacognostic critic K'ou Tsung-shih 寇宗奭 in his Dilatations on Materia Medica *(Pen-ts'ao yen i* 本草衍義, 1116).

Historians are still debating whether certain medical preparations from human urine collectively called "autumn mineral" *(ch'iu shih* 秋石) contain high concentrations of steroid and protein hormones. In Good Formulas Shen gives one of the earliest accounts, in the form of detailed instructions for two such preparations that he performed in 1061 (other accounts by contemporaries are harder to date).[26]

Perhaps Shen's most famous writing on general medical matters is one in which he refuted the common belief that there are three passages in the throat. This belief was so general that three passages appear in the first book of drawings of the internal organs based directly on dissection (1045).[27] His

---

26. See Lu Gwei-djen and Needham, "Medieval Preparations of Urinary Steroid Hormones," *Medical History,* 1964, *8:* 101–121. Their claim that the oldest known formula goes back to 1025 (p. 111) is erroneous, so that Shen's account is the oldest extant. See Juan Fang-fu, "Shen Kua tui k'o-chi-shih ti yu i chung-yao kung-hsien—kuan-yü wo-kuo shih-i shih-chi ts'ung jen-niao t'i-chü hsing-chi-su ti chi-tsai 沈括对科技史的又一重要贡献—关於我国十一世纪从人尿提取性激素的记载" (Another important contribution by Shen Kua to the history of science and technology. On records of the extraction of sexual hormones from human urine in our country in the eleventh century), in *I-hsueh hsin lun,* 219–230, esp. 225–227. For an analysis of Shen Kua's account, see Meng Nai-ch'ang 孟乃昌, "Shen Kua ho Li Shih-chen tui 'ch'iu shih' ti li-lun ch'an-shu 沈括和李时珍对秋石的理论阐述" (The theoretical expositions of Autumn Mineral by Shen Kua and Li Shih-chen), *Chung-hua i-shih tsa-chih* 中华医史杂志, 1987, *17.* 3: 187–188. Attempts to repeat Shen's preparations have failed to yield steroids. See the summary in H. T. Huang et al., "Experiments on the Identity of *Chiu shi* (Autumn mineral) in Medieval Chinese Pharmacopoeias," *Pharmacy in History,* 1990, *32:* 63–65.

27. Persons untrained in medicine dissected executed bandits in 1045 and recorded what was found under the direction of an enthusiastic amateur. A similar

supporting argument is not from independent dissection but from the principle of sufficient reason: "When liquid and solid are imbibed together, how can it be that in one's mouth they sort themselves into two throat channels?" He thus saw the larynx as the beginning of a network for distributing throughout the body the vital *ch'i* carried in atmospheric air, and the esophagus as carrying nutriment directly to the stomach cavity, where its assimilation begins. This was a significant increase in clarity as well as accuracy (480).

A passage that has been praised for its simple but beautiful language takes issue with the ancient principle that medicinal plants should be gathered in the second and eighth lunar months (when they were thought easiest to identify). In a few hundred characters it epitomizes the variation of ripening time with the identity and variety of the plant; the part used in therapy; the physiological effect needed for the application; altitude; climate; and, for domesticated medicinal plants, variation with planting time, fertilization, and other details of horticulture. The sophistication of this jotting (485) reflects not only increasing cultivation (exceptional before the Sung) but also the integration of drugs from every corner of China into the expanding commercial network.

Conclusion

The Northern Sung was an important period in the history of every branch of science and technology because society was expansive and relatively open to talent, and the government increasingly invested in a broad spectrum of learning. Shen was not the first polymath this era produced. There was also Yen Su 燕肅 (fl. 1016), who designed an odometer and south-pointing chariot (in which a differential gear assembly kept figures pointing in a constant direction as the chariot turned), improved the design of the water clock and other astronomical instruments, and wrote on mathematical harmonics and the tides. In Shen's lifetime there was Su Sung 蘇頌 (1020–1101), who was paramount Grand Councillor during the last part of the reaction against the New Policies (1092–1093). Through the 1060's he played a major part in a large imperially sponsored compilation of materia medica, and in the editing

---

episode, meant to correct the earlier drawings, took place at the beginning of the twelfth century. There is no reliable account of either in any European language, but see Watanabe Kôzô 渡邊幸三, "Genson suru Chûgoku kinsei made no gozô rokufu zu no gaisetsu 現存する中國近世までの五藏六府圖の概説" (A survey of extant Chinese anatomical drawings from before modern times), *Nihon ishigaku zasshi* 日本醫史學雜誌, 1956, *7:* 88.

and printing of ancient medical classics. In 1088 a group that he headed completed a great water-driven astronomical clock incorporating an escapement device. Their detailed description of the mechanism *(Hsin i-hsiang fa yao* 新儀象法要) included the oldest star map extant in printed form, based on a new stellar survey. That Yen, Su, and Shen were all in the central administration is not surprising. The projects on which they were trained and those in which they worked out many of their ideas were of a scale that only the imperial treasury could (or at least would) support. What little is known about the career of another contemporary polymath, Shen Li 沈立, suggests that his narrower range of interests (in natural history, geography, water conservancy, etc.) reflects a career that evolved in provincial administration and moved to the capital late in life.[28]

Breadth of interest alone does not explain Shen's importance for the study of Chinese scientific thought. Another aspect is his profound technical curiosity. Many of the phenomena he recorded were mentioned by others; but even when their descriptions were fuller, they usually treated their subject matter as a mere curiosity or an occasion for anecdote rather than as a challenge to comprehension. Above all, one is aware in Shen, as in many other great figures, of a special directness. A member of a society in which the weight of the past always lay heavily on work of the mind, he nevertheless often cut past deeply ingrained structures and assumptions. This was as true in his program of astronomical observations and his audacious solar calendar as in his work on government policies.

People in the Sung were aware that society had become much more complicated since antiquity. Questioning of precedent (in the name of a return to classical principles) was inherent in the New Policies. Shen's commitment to Wang An-shih's political program can only have made him more aware of

---

28. The book on the astronomical clock has been studied and translated in Wang Ling et al., *Heavenly Clockwork* (Cambridge University Press, 1960). See Wang Chin-kuang 王锦光, "Sung-tai k'o-hsueh-chia Yen Su 宋代科学家燕肃" (Yan Su, a scientist of the Song dynasty), *Hang-chou ta-hsueh hsueh-pao* 杭州大学学报, 1979, 3: 34–38; Su K'o-fu 苏克福 et al. (ed.), *Su Sung yü Pen-ts'ao t'u-ching yen-chiu* 苏颂与本草图经研究 (Studies of Su Sung and the Illustrated canon of materia medica; Changchun: Ch'ang-ch'un Ch'u-pan-she, 1991). The essays in the latter collection vary greatly in quality. The only study of Shen Li is "Pei Sung k'o-hsueh-chia Shen Li 北宋科学家沈立" (The Northern Sung scientist Shen Li), in Li Ti, *Chung-kuo k'o-hsueh chi-shu shih lun-wen-chi* 中国科学技术史论文集 (Collected essays on the history of Chinese science and technology; Hohhot: Nei-meng-ku Chiao-yü Ch'u-pan-she, 1991), 73–83.

the unceasing need to improve techniques and to achieve better results that had driven the best Chinese astronomers for centuries. Even given these predispositions and opportunities, Shen remains in many senses an atypical figure, even in his time and among his associates.

A modern scientist or engineer finds much familiar, not only in the way Shen went about making sense of the physical world, but also in the temper of his discourse, despite the ancient concepts he used. One comes away from his writings confident that he would see much of modern science as a culmination (not the only possible culmination) of his own investigations—more confident than after reading Aristotle, St. Thomas Aquinas, or Copernicus. But does Shen's special configuration of abilities and motivations suggest that a genetic accident produced, out of time, a scientific rationalist-empiricist of essentially modern type? To answer this question it is necessary to look at Shen's larger conception of reality, of which his scientific notions are only a part but from which they are inseparable.

The relation of scientific thought to reality. The sense of cumulative enterprise in mathematical astronomy did not imply the positivistic conviction that eventually the whole pattern could be mastered. Instead, from the earliest discussions on, the attitude was prevalent that scientific explanation—whether in terms of number or of abstract qualitative concepts, such as yin-yang—merely expresses, for human purposes, limited aspects of a pattern of constant relations too subtle to be understood. No one expressed this attitude more clearly than Shen. In instance after instance he emphasized the inability of secular knowledge to encompass phenomena: the reason for magnetic declination (437), why lightning striking a house can melt metal objects without burning the wooden structure (347), the way in which every constant and every mean value obscures continuous variation of every parameter (123). This and similar evidence amount not merely to an appreciation of the role of abstraction in science, but also to the steady conviction that abstraction is a limited process incapable of producing universal and fundamental knowledge. Nature is too rich, too multivariant, too subtle (*wei* 微). This limitation did not detract from the interest or worth of theoretical inquiry, and did not lead intellectuals to question whether learning could contribute to the satisfaction of social needs; but the ambit of rationalism in traditional scientific thought was definitely circumscribed.[29]

29. I have studied this theme in "On the Limits of Empirical Knowledge in the Traditional Chinese Sciences" in J. T. Fraser et al., *Time, Science, and Society in China and the West*; for a revised version see below, Vol. II, Chap. IV.

In this light Shen's explanatory metaphors, such as the willow leaf for variations in planetary speed and the oarlock and waisted drum for optical image inversion, become more comprehensible. He likened the variation in polarity of different magnetized needles to the fact that two species of deer shed antlers in opposite seasons (588; IV.1, 250), and so on. Geometric figures, numbers, and quantities were useful for computation but had very limited value, not so great as cogent metaphors from the world of sensual experience, in understanding its inherent patterns. We have already seen that numerology and mathematics were alternate means to the same goal.

Other kinds of knowledge. Did Shen believe that other ways of knowing complemented and completed empirical and theoretical investigation? Aside from its scientific aspects, Shen's thought has been so little studied that only some tentative suggestions can be offered.

Contemplation and disciplined self-examination were ancient themes in Confucianism. By Shen's time many learned people considered illumination to be a source of knowledge complementary to experience of the external world. The domestication and secularization of Buddhist and Taoist meditation were gradually leading to more introspective and less ritualistic approaches to self-realization. This tendency was later elaborated with great variety of emphasis and weight in the schools of neo-Confucianism.

To understand what part contemplation and meditation played in the thought of Shen Kua requires a clearer view than we now have of their practice and meaning in his time, of Wang An-shih's considerable influence on his intellectual development, and of his own attitudes as indirectly expressed in his literary remains. His interest in Taoist arcana seems to have peaked in his thirties. His public remarks express sympathetic interest in illuminationist (Ch'an, Japanese Zen) Buddhism. He stated in an autobiographical fragment that Ch'an meditation was one of the things to which he turned his attention after retirement. We can hardly overlook such enthusiasms if we hope to grasp his epistemology.

Teraji Jun 寺地遵 has demonstrated this point when examining how strong a factor in Shen's motivation and individuality was his belief in destiny and prognostication. In his commentary on Mencius, Shen spoke of the necessity for choosing what is true and holding to it, and called the rule of the heart and mind by sensory experience "the way of the small man." The basis of moral choice was an autonomous inner authority defined in an original way but largely in Mencian terms, a centeredness "filling the space between sky and earth," unquestionably linked with the self-reliance that marked Shen's own unhappy career.

It is not immediately obvious why someone who so valued individual responsibility should have been fascinated by fate and divination, the themes of whole chapters of Brush Talks. Shen did not view these enthusiasms as in conflict with his scientific knowledge. His delight in strange occurrences and his tendency to place matters of scientific interest under that rubric begin to make sense under the hypothesis that he accepted the odd, the exceptional, and the affront to common sense as a piquant challenge to understanding. Phenomena that seem to be unique turn out to be similar to something else somewhere or at some other time. As Fu Daiwie has put it, "Shen was carving out new classes of things in nature, subjecting them to his own classificatory grid."[30] Other phenomena deserve a place in the chapters on oddities because Shen cannot fathom their place in the cosmic schema, and the challenge remains to be met by someone else—without assuming that explanation is inevitable. In his hundreds of jottings on people, the person he chose to praise is most often the one who did not do the obvious thing, even when it seemed the sound thing to do.

At one point Shen provided a thoroughly rational explanation of the relations between fate and prognostication. Certain people can of course know the future, he said, but it is a mistake to conclude that all matters are preordained. The vision of the future is always experienced in present time; the years in the interim also become simultaneous. One can do nothing to avoid an undesirable future so glimpsed. Authentic foreknowledge would have witnessed the evasive measures; a vision that failed to see them could not be authentic foreknowledge (350).

In addition to the visionary ability of certain minds, Shen pondered universally accessible methods of divination, which (he seems to have believed) do not describe the future or the spatially distant so much as provide counsel about them or aid thought about them. In one of his chapters, "Regularities Underlying the Phenomena," he explained why the same divinatory technique gives different outcomes when used by different people, and thus is not inherently verifiable. He quoted the Great Commentary to the Book of Changes to the effect that understanding is a matter of the clarity and divinity (in an abstract sense) within one's mind. This divinity is, for Shen's sources, the moral center of the individual. Because the mind is never without burdens that hinder access to its divinity, Shen reasoned, one's communion with it may take place through a passive mediating object or procedure (144, 145). The rituals of prognostication invoke the power of self-examina-

30. Fu, 27.

tion. Access to the future, whether by vision or by divination, is a perfectly natural phenomenon. It is related, on the one hand, to the moral faculties, which choose the future, and, on the other, to the sciences, which rationally comprehend the natural order as reflected in all sensory experience.

Thus introspection supplemented by divinatory procedures was a legitimate means to knowledge in Shen Kua's eyes, like painstaking observation and measurement of natural phenomena. He neither confused the two approaches nor attempted to draw a clear line between them. Nor was he inclined to compare the importance of these ways of knowing.

The complementarity in Shen's attitudes toward knowledge parallels another in the external world of his work. Computational astronomy and divination of various kinds (including judicial astrology) were equally weighty functions carried out by the central government on the emperor's behalf. Both kinds of activity supported his charisma. An important memorial of Wang An-shih (1072) implies that the marriage of scientific activity and imperial ritual is indispensable to government: because the monarch acts on behalf of the natural order *(t'ien* 天), he must understand it in order to safeguard the empire and strengthen the loyalty of his subjects. Ritually expressed awe of that order, without knowledge, is not enough. Teraji has acutely pointed out that this is precisely the political justification for Shen's research, and the reason that conventional bureaucratic technicians, concerned mainly with maintaining ancient practices, were not what Wang wanted.[31]

Confucianism and science. Attempts in both East and West to construct a historical sociology of Chinese science have in large part been built around a contrast between Confucianist and Taoist convictions. The values of the Confucian elite are often described as oriented toward stasis, hierarchy, bureaucracy, and bookishness. These characteristics are seen as perennially in tension with the appetite of socially marginal Taoists for novelty and change, their tendency to contemplate Nature and the individual in it as a system, and their fascination with techniques, which kept them in touch with craftsmen and made them willing to engage in manual work themselves. I examine this set of assumptions elsewhere.[32] For the moment what matters is how relentlessly this discussion has treated both ideologies as collections of vague, free-floating abstractions.

---

31. *Hsu tzu chih t'ung chien ch'ang pien* 續資治通鑑長編 (Unabridged draft of the Sequel to the Comprehensive Mirror for Aid in Government, presented to the throne 1168; 1881 ed.), *236:* 16b.

32. Vol. II, Chaps. VI, VII.

Sociology is about groups of people. Doctrines are germane to sociology to the extent that their effect on what groups of people do, or on how they form, can be demonstrated. Generalizations about people who accept a certain doctrine have no significance unless such people can be shown to acted as a group, or at least so to have identified themselves. In ancient China, whether the use of an idea reflects an intellectual commitment, or merely conventional language, is very much to the point.

The term "Confucian" is used indifferently even by specialists to refer to a master of ceremonial, a professional teacher of Confucian doctrines, a philosopher who contributes to their elaboration, someone who attempts to live by Confucius' teachings, any member of the civil service, any member of the gentry regardless of ambition toward officialdom, or any conventional person (since it was conventional to quote Confucian doctrines in support of conventional behavior). A "Taoist" can be anyone from a hereditary priest ordained by the Heavenly Master to a retired bureaucrat of mildly unconventional tastes living on a city estate. Either group, by criteria in common use, includes people who would make opposite choices on practically any issue. This being so, the proposition "Taoists were more friendly toward science and technology than Confucians" reduces to "Educated individuals who hold unconventional sentiments are more inclined to value activities unconventional for the educated than are educated people who hold conventional sentiments." That tautology is sociologically vacuous and historically uninteresting.

Unease of this sort is probably the most obvious outcome of reflection on Shen Kua's career. By sentimental criteria he can be assigned to Confucianism, Taoism, or Buddhism, to suit the historian's tastes.[33] He was a member of the elite, a responsible official, a writer of commentaries on several of the Confucian classics, and a user of the concepts of Confucius' successor Mencius to explore the depths of his own identity. He spoke well and knowledgeably of Buddhism. He practiced arcane disciplines, such as breath control, that he called Taoist.

---

33. Two essays in Anonymous, *Ju-Fa tou-cheng yü wo kuo ku-tai k'o-hsueh chi-shu ti fa-chan* 儒法斗争与我国古代科学技术的发展 (The struggle between Confucianism and Legalism and the development of science and technology in our country in ancient times; Peking: K'o-hsueh Ch'u-pan-she, 1974), 118–140, portray Shen as a Legalist and a relentless opponent of Confucianism. This odd idea was prominent in the "anti-Confucius anti-Lin Piao" campaign of the Cultural Revolution. See "Shen Kuo, the Legalist, and his Scientific Achievements," *Scientia sinica*, 1974, *17:* 717–728.

As for his allegiances, Shen was prominently associated with a powerful but shifting group of politicians whose backgrounds were generally similar to his own. Social stasis and institutional fixity were impediments to their aims in reshaping government. At the same time, the new balance of power toward which they strove was more authoritarian than the old. Underlying their common effort was an enormous disparity of motivation, from the well-intentioned (Shen) to the indefatigably manipulative and corrupt (Ts'ai Ch'ueh).

Were these Confucians more or less Confucian than their Confucian opponents? Wang An-shih, the leader of the New Policies clique, earned enduring stature for his commentaries on the classics and his thought on canonic themes. His followers seem to have found inspiration in the classics as often as their enemies and as those who avoided taking a political position. This is not to say that everyone understood the Confucian teachings in the same way. The latter were not, from the viewpoint of intellectual history, a set of tightly linked ideas that set fixed limits on change; rather, they were a diverse and fragmentary collection of texts reinterpreted in every age. They were understood differently by every individual and group who looked to them for guidance when coping with problems of the moment.

The major commentaries, which attempted to define the meanings of Confucian teachings philologically and didactically, carried enormous authority. Governments repeatedly attempted to make one interpretation orthodox. The administration of Wang An-shih, for instance, originated an important practice by requiring that candidates for the imperial examinations memorize his annotations. But the urge to pin down meanings was always in conflict with precisely what made these books classic. Their unlimited depth of significance depended more on what could be read into them than on precisely what their authors had meant them to say. That depth made them applicable to an infinity of human predicaments and social issues, unprecedented as well as perennial. Late neo-Confucian philosophers striking out in new directions demonstrated again and again how little the bounded intellectual horizons and social prejudices of the ancient canons' authors limited what may be drawn from them.

In other words, the Confucian canon had the influence it did because it set down patterns of thought that, over the centuries, educated people used and redefined in working through new quandaries, justifying action and inaction. Arguments against the New Policies often cited the classics as a pattern for static social harmony and willing subordination. Shen used them to argue for flexibility in social relations and for greater receptivity toward new possibilities than was usual in his time. Either as a social institution or as an ideology, Confucianism is too protean and thus too elusive a base for generalizations about the social foundations of science and techniques in China.[34]

Institutions also changed constantly, but at least they were tangible entities. No one seeking the motors of technical change is likely to succeed without scrutinizing them. Historians are often distracted from such scrutiny by musings on whether Taoism predisposed anyone and everyone to scientific research, or whether neo-Confucians were capable of "rational analysis."[35]

It is only sensible to consider institutions when tracing the social connections of science. Very little is known about how scientists were educated in the Northern Sung period. In Shen's case we can see a pattern that certainly was not unique. He was, so far as we know, self-educated in astronomy, but with many learned associates to draw upon. In medicine and breath control he probably received teaching in the traditional master-disciple relationship. Defined in the ages before printing enabled access to large collections of books, this relationship involved the student's memorizing the classics that the teacher had mastered. This verbatim transmission of a text was supplemented by the teacher's oral explanations. The relation was deepened by ceremonial formality; the master took on the obligation to monitor the disciple's moral as well as intellectual growth, and the disciple accepted the responsibility of becoming a link in an endless chain of transmission. Schools were largely communities of masters and disciples.[36] The scale of government-sponsored elementary schools in the provinces was small in Shen's youth. Such schools began to compete with private academies only in the New

---

34. Kidder Smith, Jr., et al. illustrate this point nicely in *Sung Dynasty Uses of the I Ching* (Princeton University Press, 1990).

35. On the latter point, see a recent argument that neo-Confucians, because of "their lack of interest in experimentation . . . were unable to examine their incorrect premises" about physical phenomena. This paper does not define "neo-Confucian." It mentions Shen but does not notice the trenchancy of his critical reasoning on such matters. Nor does it observe how often in early modern Europe avid experimention was combined with inability to examine incorrect premises. Charles E. Hammond, "The Interpretation of Thunder," *Journal of Asian Studies,* 1994, *53.* 2: 487–503, citing 498. Aside from these generalities, Hammond does well to correct the tendency among historians of an earlier generation to read a scientific mind-set into every correct description.

36. I have examined these relationships, defined in the third and second centuries B.C., in "Text and Experience in Classical Chinese Medicine," in Don G. Bates (ed.), *Knowledge and the Scholarly Medical Traditions* (Cambridge University Press, 1995).

Policies period. The two sorts together did not serve more than a small minority of youth.

In the central government, by the eighth century there were small schools to train technical functionaries. The masters, usually several in number, were officials representing the bureau that the disciples were being trained to staff. The schools for medicine and astronomy could not lead to the top of government, but guaranteed steady advancement between minor sinecures. Very few of the important physicians or astronomers of traditional China began in these schools.

In the absence of evidence to the contrary, there is no reason to believe that Shen Kua ever attended a school of any sort, nor does that make him untypical. His early education by his mother, his training in medicine by an obscure physician and others who remain unknown, and his catch-as-catch-can studies of most other matters, do not set him apart from his contemporaries. With no knowledge of particulars one cannot even guess how his personal style in technical work was formed. But to say that we are ignorant is not to say everything. The intimate relations of master and teacher and the isolation of the autodidact were themselves important institutions in the Northern Sung, institutions of a sort that did not discourage the emergence of unforeseen abilities in the small number of people who had the opportunity to be educated. Shen did not have to cope with a standard curriculum, for better or worse. If we are searching for the decisive curriculum of technological practice, it is necessary to look outside the realm of education.

The civil service and the sciences. One institution above all others influenced the mature ideas and attitudes of intellectuals, namely the bureaucracy. This was even true of those who did not belong to it. What can be said about its influence on science and technology in the life of Shen Kua? First, like every bureaucracy, it depended upon science and technology. It supported both sorts of activity on a scale otherwise unattainable, and unimaginable in Europe at the time. Shen's curiosity, experience, and skills were so largely shaped by the civil service that it is absurd to ask what he would have become had he lived as a country gentleman or a Taoist priest. On the other hand, as elsewhere, technicians were further from the priorities of the state than administrators. The responsibility of the former was to provide the emperor and his administrators with wealth and other tools for the realization of policy. Specialist positions in the sciences and in engineering did not often serve as the beginnings of great careers.

Shen's epoch was pivotal in this respect as in many others. By the New Policies period a career stream for economic experts had been established. It could assimilate people who combined technological acumen with fiscal skills, and carry them to the central councils of the empire. Shen's early technical feats were performed in general administrative posts, but his special talents came to be valued and he rose quickly through formal and informal structures. It is not irrelevant that his management of the Directorate of Astronomy was never more than a concurrent position. His attempt to combine an effective voice in the shaping of change with a presence at the frontiers of knowledge ended in personal disaster. He was ruined by men of his own faction, apparently for his political seriousness and naïveté. His astronomical work was rendered futile by subordinates because of his demand that they do their jobs. The bureaucracy was not neutral; it was a two-edged sword.

The civil service provided a form for great projects in science and technology. It practically monopolized certain disciplines, such as mathematical astronomy, observational astrology, and cartography. Printing gave it the wherewithal to determine much of the content of elementary technical education, as in medicine and mathematics. A man of Wei P'u's genius, who had not had the opportunity to enter the bureaucracy by a regular route, was looked down upon and deliberately frustrated. Had Shen himself chosen to be a mere technician, his incumbency in the Directorate of Astronomy might have been sufficient to protect him from personal attack. He would have had more time but less power. It would be rash indeed to speculate that his calendar reform would have succeeded. But there is a larger issue.

Gentlemen were generalists by upbringing. Shen's mind was shaped for the civil service, as were those of his ancestors and peers, by an early education centered in moral philosophy and letters. Only a superficial knowledge of technical matters—arithmetic and so forth, the canonical Six Arts—was expected of youths like him. The development of depth in thought and work was left to his own proclivities. This situation was not very different from that of the British gentleman of some decades ago.

Shen's growing responsibilities in fiscal affairs were the one aspect of his career that clearly encouraged him to draw coherence out of his varied experiences and studies. For this reason and others of which we are still ignorant, the great breadth of his knowledge was accompanied by enough depth to let him write important monographs and, even through his brief jottings, to reshape Chinese understanding of certain phenomena. But distraction is a theme that runs through his writings: promising studies laid aside; endless skirmishes to defend administratively measures that spoke for themselves technically and strategically; pathbreaking proposals negated by political setbacks. Regardless of his capacity for technical depth and his willingness to find his way to it, the sheer busyness of his career blocked his way. The works of his final leisure, however valuable, were all superficial in form. Was this the result of habit, of distance necessitated by disillusion, or of a style appropriate for chatting with one's brush and ink slab in a silent garden? Of all three, no doubt.

What, then, was responsible for Shen Kua's personality? We do not know the answers to all sorts of prior questions. The greatest difficulty comes in learning what these questions should be—in isolating the important issues, in coming to terms with the paucity and partiality of the sources, and in doing justice to a rich mind that, despite its absorption in a quest that belongs to all peoples and all eras, partook fully of its time and place. It is not a matter of mechanically juxtaposing the usual factors: intelligence, subjectivity, philosophical convictions, social background, career, experiences. We have already seen how problematic the last three are. The most conspicuous traits of Shen's consciousness were open curiosity, mental independence (without the intolerance for disagreement that was a great limitation of Wang An-shih), sympathy for the unconventional, ambition, loyalty, and consciousness of his place in the world. The first four are considered marks of promise among technical people today, although one often meets great scientists who lack one or more of them. Were these characteristics in Shen due to heredity, to early experiences and education, or to influences encountered in adult life? This is another example of the sort of question that bars understanding; surely Shen was the sum of all these and more. The secret of his uniqueness will not yield itself to historical method, however powerful, unless it is applied with imagination, artifice, and awareness of the springs of human complexity.

Attitudes toward Nature. When examined closely, attitudes toward Nature in the late eleventh century become as elusive as attitudes toward Confucian humanism. The richly articulated philosophic vision of man in harmony with his physical surroundings could not prevent the deforestation of northern China, virtually completed a generation after Shen's death. One cannot even speak of the defeat of that vision in an encounter of ideas; there was none. What happened? The most obvious part of the answer is that the people who were chopping down the trees for timber and charcoal were not the people who were seeking union with the ineffable cosmic Tao. That social difference was of too long standing, however, to explain the crescendo of exploitation in the Northern Sung. The coincidence of that fateful shift with the rise of large-scale industry and market networks is again obvious enough.[37] What needs to be explained, rather, is the survival of the naturalist ideal until modern times.

The dilemma emerges clearly in the attitudes of Shen Kua and Wang An-shih toward their natural environment. The orientations that pervade Brush Talks are in most respects the same as those of literati thinking about Nature a millennium earlier. Philosophical pigeonholes are largely beside the point. Some "Confucians" thought about Nature a great deal, and some, convinced that human society is the sole proper object of reflection and action, as little as possible. The notions of heaven and earth they drew on were, on the whole, the ones common to all Chinese who could read and write. Nature was an organismic system, its rhythms cyclic and governed by the inherent and concordant pattern uniting all phenomena.

With that in mind, it comes as a shock to see Shen's definition of salt in a memorial of 1077: "Salt is a means to wealth, profit without end emerging from the sea."[38] This was not a slip, nor is it difficult to find philosophical precedents. Shen expected wealth from Nature to pay for the expanding fiscal functions of the state. That mattered, because he briefly held supreme responsibility for those functions. He recommended—on the basis of his own early accomplishments—encouraging extractive industries and manufactures, and mobilizing labor for land reclamation, in order to increase the wealth of the state. In that respect he was faithful to the priorities of Wang An-shih. This is a far cry from the senior cadre in China in the 1960's designing a campaign to convince farmers that Nature is an enemy to be conquered, tamed, and remolded to social ends. But neither is it the pastoral ideal.

Why this discrepancy between Nature as the ideal pattern to which man adjusts, and Nature as an exploitable resource? Why does Shen seem not to be conscious of the contradiction? The research has yet to be done, but Shen Kua's career, considered in the round, suggests a working hypothesis. Such notions as yin-yang, the Five Phases, and certain related ideas associated with the Han commentaries on the *Book of Changes*, are sometimes called hindrances to an autochthonous scientific revolution in traditional China. This is, of

---

37. See the series of papers by Hartwell: "A Revolution in the Chinese Iron and Coal Industries during the Northern Sung, 960–1126 A.D., " *Journal of Asian Studies*, 1962, *21:* 153–162: "Markets, Technology, and the Structure of Enterprise in the Development of the Eleventh-Century Chinese Iron and Steel Industry," *Journal of Economic History*, 1966, *26:* 29–58; "A Cycle of Economic Change in Imperial China: Coal and Iron in Northeast China, 750–1350," *Journal of the Economic and Social History of the Orient*, 1967, *10:* 102–159.

38. *Hsu tzu chih t'ung chien ch'ang pien*, 280: 17b–21b.

course, an elementary fallacy.[39] The old Chinese world view had much in common with cosmological ideas (the four elements and so on) universal among the educated in Europe until the seventeenth century, ideas that readily faded away as Aristotelianism withered. Historically speaking, Chinese organismic naturalism was not a rigid framework of ideas that barred change; rather, it was the only conceptual language available for thinking about Nature and communicating one's thoughts, *new or old,* to others. Like any language, it imposed form and was itself malleable. Its historical possibilities were less a matter of original etymology or definition than of the ambiguity and extensibility that let people in later ages read new and often drastically changed import into old words. There is no true paradox in appeals to the harmony of man and Nature by Shen and others before and after him who favored the exploitation of Nature in the interests of the state. Although such activist thinkers stretched the old pattern of understanding, its fabric remained seamless. Their definition of what they wanted could not transcend it. Only the more desperate urgencies of another time could finally stretch it until it tore.

Bibliography

Primary sources

The best attempt at a complete list of Shen's writings is in an appendix to Hu Tao-ching's 胡道静 standard ed. of Brush Talks, *Meng ch'i pi t'an chiao cheng* 夢溪筆談校證 (Brush Talks From Dream Brook, a Variorum Edition, hereafter *Chiao cheng;* Shanghai: Chung-hua Shu-chü, 1956; rev. ed., 2 vols., Peking: idem, 1960), 1151–1156. There are forty titles, including lost works mentioned in early writings about Shen. Some titles belong to parts or earlier versions of larger writings. Ch'en Teng-yuan 陳登原 has suggested that the high rate of attrition was due to the campaign of Ts'ai Ching 蔡京 (1046–1126), virtual dictator during the revival of the New Policies in the first quarter of the twelfth century, to obliterate the literary remains of his predecessors as well as their enemies. See his *Ku-chin tien-chi chü-san k'ao* 古今典籍聚散考 (A study of the collection and dispersal of classical writings in ancient and modern times; Shanghai: Hai-wai T'u-shu Ch'u-pan-she, 1936), 54. Six of Shen's books are extant, although only two appear to be substantially unal-

39. This and other fallacies are analyzed in Chap. VII below.

tered, and considerable fragments of four others exist. Those of scientific interest are described below:

1. *Meng ch'i pi t'an* 夢溪筆談 (Brush Talks From Dream Brook), written over the greater part of Shen's retirement and possibly printed during his lifetime. It includes 587 jottings.

Brush Talks was first quoted in a book dated 1095. Originally it consisted of 30 ch. *(chüan* 卷). All extant versions, descended from a xylograph of 1166, follow an unknown prior editor's rearrangement into 26 ch. The editor of the 1166 reprint noted a number of errors that he could not correct for want of variants. One copy of the 1166 version survives in Japan. A reprint of 1305 has been reproduced as *Yuan k'an Meng ch'i pi t'an* 元刊梦溪笔谈 (Peking: Wen-wu Ch'u-pan-she, 1975).

There are two sequels, *Pu pi t'an* 補筆談 (Supplement to Brush Talks, 91 jottings), listed in most early bibliographies as 2 ch., but rearranged in the 1631 ed. into 3 ch. with some alteration of order, and *Hsu pi t'an* 續筆談 (Sequel to Brush Talks, 11 jottings in 1 ch., mostly on literature). Hu suggests that both sequels were edited posthumously from Shen's notes. There is even stronger evidence for this hypothesis than he adduces, for some articles appear to be rejected drafts of jottings in Brush Talks (cf. 588 with 437, 601 with 274).

The practically definitive ed. of Brush Talks and its sequels, and in many other respects the foundation of later studies, is the 1960 Hu Tao-ching recension. It includes a carefully collated and corrected text with variorum notes and modern (but occasionally faulty) punctuation, based on all important printed versions beginning with that of 1495 and on five previous sets of variorum notes. It also provides exegetic and explanatory notes and generous quotations from documents concerning Shen, from his other books, from the reflections of other early writers on his subject matter, and from modern Chinese (and to some extent Japanese and Western) scholarship. Appendixes include thirty-six additional jottings or fragments that have survived only in the writings or compilations of others; all known prefaces and colophons; notes on eds. by early bibliographers and collators; a chronological biography; a list of Shen's writings; and an index to names and variant names of all persons mentioned in Brush Talks (a tool still very rare in Chinese publications). Hu also published a 1-vol. version of the text with minimal apparatus as *Hsin chiao cheng Meng ch'i pi t'an* (Brush Talks, newly edited; Peking: Chung-hua Shu-chü, 1957). He summarized foreign scholarship up to the late 1970's in "Meng ch'i pi t'an tsai kuo-wai 梦溪笔谈在国外" (Brush Talks abroad), *Shu-lin* 书林, 1979, 1: 14–17. For a retrospect see his *"Meng ch'i pi t'an chiao cheng* wu-shih nien 梦溪笔谈校正五十年" (Fifty years of the variorum ed.), *Tu shu* 读书, 1979, 4: 120–125.

Hu Tao-ching's unflagging labors in China and Yabuuchi Kiyoshi's 藪内清 and Joseph Needham's broad use of Brush Talks in their publications have made Shen's name generally familiar to historians of science, but one can hardly discern a scholarly industry forming around him. Several of Yabuuchi's pertinent publications are gathered in *Sô-Gen jidai no kagaku gijutsushi* 宋元時代の科學技術史 (History of science and technology in the Sung and Yuan periods; Kyoto: Kyôtô Daigaku Jimbun Kagaku Kenkyûsho, 1967). Teraji Jun 寺地遵, "Shin Katsu no shizen kenkyû to sono haikei 沈括の自然研究とその背景" (The natural investigations of Shen Kua and their background), *Hiroshima Daigaku Bungakubu kiyô* 廣島大學文學部紀要, 1967, *27*. 1: 99–121, is a notable attempt at an overview. Sakade Yoshinobu's positivistic discussion of Shen's use of theory, "Shin Katsu no shizenkan ni tsuite" (On Shen Kua's conception of Nature), *Tôhôgaku* 東方學, 1970, *39:* 74–87, contains a number of stimulating insights. The only study of comparable breadth since 1975, discussed in n. 10 above, is Joël Brenier et al., "Shen Gua (1031–1095) et les sciences," *Revue d'histoire des sciences,* 1989, *42:* 333–351.

There have been a couple of useful partial commentaries: Hu & Chin Liang-nien 金良年, *Meng ch'i pi t'an tao tu* 梦溪笔谈导读 (Reader's guide to Brush Talks; Chengdu: Ba-Shu Shu She, 1988), with vernacular notes; Jean François Billeter, "Florilège des *Notes du Ruisseau des réves (Mengqi bitan)* de Shen Gua (1031–1095)," *Asiatische Studien,* 1993, *47*. 3: 389–451. Partial translations into modern Chinese include Li Ch'ün 李群, *Meng ch'i pi t'an hsuan tu. Tzu-jan k'o-hsueh pu-fen* 梦溪笔谈选读。自然科学部分 (Selected readings from Brush Talks. Sections on natural science; Peking: K'o-hsueh Ch'u-pan-she, 1975); Chung-kuo K'o-hsueh Chi-shu Ta-hsueh 中国科学技术大学 (Chinese Science and Technology University) et al., *Meng ch'i pi t'an i chu. Tzu-jan k'o-hsueh pu-fen* 梦溪笔谈译注。自然科学部分 (Brush Talks, translated and annotated. Sections on natural science; Hofei: An-hui K'o-hsueh Chi-shu Ch'u-pan-she, 1979). The former book contains 86 jottings, and the latter 138, chosen from the more than two hundred on science and technology.

2. *Hsi-ning Feng-yuan li* 熙寧奉元曆 (The Oblatory Epoch astronomical system of the Splendid Peace reign period, 1075), lost, but listed in a Sung bibliography as 7 ch. This was the official report embodying Shen's calendar reform. The usual arrangement provided lists of constants and step-by-step instructions for computation, with tables as needed, so that a clerk with no

knowledge of astronomy could calculate the complete ephemeris. Since a *Hsi-ning Feng-yuan li ching* 曆經 (Canon of the Oblatory Epoch Astronomical System . . .) in 3 ch. is separately recorded, the remaining 4 ch. may have been, as in other instances, an official critique *(li i* 曆議) outlining the observational basis of the system and reporting on trials of its accuracy. The Sung standard history also records a ready reckoner *(li ch'eng* 立成) in 14 ch., containing tables and other aids to calculation, and a detailed explanation of the mathematics with worked-out examples *(pei ts'ao* 備草) in 6 ch. The great student of ancient astronomy Li Jui 李銳 (1765–1814) gathered surviving fragments of the basic document under the title *Pu hsiu Sung Feng-yuan shu* 補修宋奉元術 (Restoration of the Sung Oblatory Epoch techniques), in *Li shih i shu* 李氏遺書 (Posthumous works of Mr. Li, printed 1823). Three proposals of 1074, in which Shen outlined his designs for armillary instruments, clepsydras, and gnomons, survive in the Sung history *(Sung shih* 宋史, Chung-hua Shu-chü 1977 ed., *48:* 954–965).

3. *Liang fang* 良方 (Good medicinal formulas), a work of 10 or 15 ch. compiled during Shen's retirement. In the Sung (probably ca. 1126) it was combined with various medical formulas recorded by the greatest literary figure of Shen's time, Su Shih 蘇軾 (1036–1101), a moderate but influential opponent of the New Policies. The conflation is called *Su Shen nei-han liang fang* 蘇沈内翰良方 (Good medicinal formulas by the Han-lin Academicians Su and Shen), often referred to as *Su Shen liang fang.* The most broadly based text is that in the *Chih pu-tsu chai ts'ung-shu* 知不足齋叢書 collection and modern reprints descended from it. One copy of an illustrated Ming ed. still exists. Shen's original compilation was lost sometime after 1500. There is some overlap between ch. 1 of *Su Shen liang fang* and jottings in ch. 26 of *Meng ch'i pi t'an.*

Hu has reconstituted Shen's *Liang fang,* so far as it can be done, by separating out the part of *Su Shen liang fang* written by Su. See "Su Shen nei-han liang fang Ch'u Shu p'an 苏沈内翰良方楚蜀判" (Attribution of the medicinal formulas in Good Formulas by Su and Shen), *She-hui k'o-hsueh chan-hsien* 社会科学战线, 1980, 3: 195–210, and Wu Tso-hsin 吴佐忻, *"Su Shen liang fang* i-wen hsi 苏沈良方佚文析" (Analytical study of lost texts from Good Formulas by Su and Shen), *Chung-hua i-shih tsa-chih,* 1982, *12.* 4: 257.

A lost collection of medicinal formulas in 20 ch., *Ling yuan fang* 靈苑方 (Formulas from the Holy Garden), is quoted in Sung treatises on materia medica. Hu Tao-ching has shown that it was written before *Liang fang (Chiao cheng*, 830–831). See the comparison between Good Formulas and medical jottings in Brush Talks, 880–882.

4. *Meng ch'i wang huai lu* 夢溪忘懷錄 (Record of longings forgotten at Dream Brook), 3 ch., compiled during Shen's retirement. It incorporates a lost book on mountain living written (or at least begun) in Shen's youth and entitled *Huai shan lu* 懷山錄 (Record of longings for the mountains). His retirement to Dream Brook satisfied his early longings, hence the title of the later collection. It was lost soon after his death. The only well-known excerpts, in the *Shuo fu* 說郛 collection, are on implements useful to the well-born mountain dweller, but Hu Tao-ching in a recent study has shown that the book was plausibly classified by early bibliographers as agricultural. Hu has also collected all known fragments. See Hu, "Shen Kua ti nung-hsueh chu-tso *Meng ch'i wang huai lu* 沈括的農學著作夢溪忘懷錄" (Shen Kua's agricultural work Record of Longings Forgotten), *Wen shih* 文史, 1963, 3: 221–225; Hu & Wu Tso-hsin, "*Meng ch'i wang huai lu* kou-ch'en—Shen Ts'un-chung i-chu kou-ch'en chih i 梦溪忘怀录钩沉—沈存中佚著钩沉之一" (Record of Longings Forgotten recovered. Shen Kua's lost works reconstituted, 1), *Hang-chou ta-hsueh hsueh-pao* 杭州大学学报, 1981, *11*. 1: 1–16.

5. *Ch'ang-hsing chi* 長興集 (Collected Literary Works of [the Viscount of] Ch'ang-hsing), originally 41 ch., almost certainly a posthumous compilation. This book includes prose, poetry, and administrative documents prized for their language. It also contains important astronomical documents and sources for Shen's intellectual formation, in particular his commentary on Mencius (*Meng-tzu chieh* 孟子解) in ch. 23. By ca. the fifteenth century, when the book was reprinted in *Shen shih san hsien-sheng wen chi* 沈氏三先生文集, only 19 ch. of the Sung version remained. *Ssu pu ts'ung k'an* 四部叢刊 reproduces this ed. The ed. of 1718 contains an additional 3 ch. collected from other sources, but was very carelessly edited; see Teng Kuang-ming 邓广铭, "Pu-hsu-yao wei Shen Kua chin-shang t'ien hua—Wan-ch'un wei ping fei Shen Kua hsing-chien hsiao k'ao 不需要为沈括锦上添花—万春圩并非沈括兴建小考" (No need to gild the lily for Shen Kua; the Myriad Springs Embankment system was not built by him), in *Shen Kua yen-chiu* 沈括研究 (Studies of Shen Kua, compiled by the Seminar on Sung History of Hangchow University; Hangchow: Che-chiang Jen-min Ch'u-pan-she, 1985), 16–26. Hu Tao-ching has supplemented this source by gathering Shen's poetry from a wide range of sources in *Shen Kua shih tz'u chi ts'un* 沈括诗词辑存 (Collection of Shen Kua's extant poetry; Shanghai: Shang-hai Shu-tien, 1985).

Translations

A high standard for translation has been set in the complete, annotated Japanese translation by the History of Science Seminar, Institute for Research in Humanities (Jimbun Kagaku Kenkyûsho), Kyoto University: Umehara Kaoru 梅原郁 (ed.), *Bôkei hitsudan* 夢溪筆談 (Brush Talks, 3 vols.; Tôyô bunko 東洋文庫, 344, 362, 403; Tokyo: Heibonsha, 1978–1981). Not the least of this publication's many virtues is a detailed index. Translations into modern Chinese are sprinkled through Chang Chia-chü 张家驹, *Shen Kua* (Shanghai: Shang-hai Jen-min Ch'u-pan-she, 1962). The only book in any Western language that translates more than a few examples of Shen's writings is Joseph Needham et al., *Science and Civilisation in China,* 7 vols. projected (Cambridge University Press, 1954–), esp. in vols. III–IV. The translations are generally accompanied by more contextual information than is usual in Chinese publications. Occasionally the English version is extremely free, as when "Meng ch'i" is translated "Dream Pool." A representative selection of English translations will be included in a sourcebook of Chinese science that I am compiling.

Secondary Literature

Bibliography and biography. *Shen Kua yen-chiu* 沈括研究, 322–336, provides a bibliography of studies from 1926 to 1983. Earlier sources in Chinese are cited in Hu's *Chiao cheng* or in the footnotes to the biographies.

Chang's biography is the fullest and most accurate account of Shen's life, and pays attention to the whole range of his work. It is generally critical in method, but sometimes careless. Like other Chinese accounts of its period, it is severely positivistic, patronizing toward "feudal" aspects of Shen's mentality, and inclined to exaggerate his empathy with the common people. Another work of interest by Hu Tao-ching, overlapping to some extent the preface to his ed. of Brush Talks, is "Shen Kua ti cheng-chih ch'ing-hsiang ho t'a tsai k'o-hsueh ch'eng-chiu-shang ti li-shih t'iao-chien" (Shen Kua's political tendencies and the historical conditions bearing on his scientific accomplishments), in Li Kuang-pi 李广壁 and Ch'ien Chün-yeh 钱君晔, eds., *Chung-kuo li-shih jen-wu lun-chi* 中国历史人物论集 (Essays on Chinese historical figures; Peking: San Lien Shu-tien, 1957), 330–347. Its summary of scientific and technical achievements in the Northern Sung period from 960 to ca. 1100 is especially useful. Li Kuang-yü 李光羽, *Shen Kua* (Nanking: Chiang-su Ku-chi Ch'u-pan-she, 1983), is a brief popular biography without references.

In addition to discursive biographical studies, Shen's life has been the subject of four chronologies *(nien-p'u* 年譜), an old form that lists individual events year by year along with related data. The fullest in print (although obsolete in some respects) is Chang Yin-lin 張蔭麟, "Shen Kua pien nien shih chi 沈括編年事輯" (A chronicle of Shen Kua), in *Ch'ing-hua hsueh-pao* 清華學報, 1936, *11:* 323–358. The one appended to Hu Tao-ching's *Chiao cheng*, 1141–1156, is especially handy because of its references to jottings and to sources cited in the book's notes. The most up-to-date and accurate chronology is the concise one at the end of the Umehara translation. Hu Tao-ching, in his colophon to the 1960 ed. of Brush Talks, remarked that his own book-length chronology was in the press, but it has not yet appeared.

Yabuuchi Kiyoshi, Japan's leading historian of science, has provided a characteristically reflective discussion of the historic circumstances of Shen's career in "Shin Katsu to sono gyôseki 沈括とその業績" (Shen Kua and his achievements), in *Kagakushi kenkyû* 科學史研究, 1958, *48:* 1–6. Rejecting the prevalent tendency to prove Shen's greatness by citing anticipations of European science and technology, Teraji Jun has made a fruitful and original effort to grasp the inner coherence of his thought and work. The essay cited above influenced the present article's conclusion.

The first European introduction to Shen's life was Donald Holzman, "Shen Kua and his *Meng-ch'i pi-t'an,*" in *T'oung Pao* (Leiden), 1958, *46:* 260–292, occasioned by the Hu's 1956 ed. of Brush Talks. In addition to providing a critical and well-proportioned biographical sketch, Holzman paid more attention to Shen's humanistic scholarship than did any other author discussed in this section. He also considered some of the evidence for Shen's position in the history of science, but reached no conclusion. He tended to ask whether Shen's ideas are correct from today's point of view rather than what they contributed to better understanding of Nature in the Sung. I am working on a full-length intellectual biography. A reliable and compendious introduction to the New Policies is James T. C. Liu, *Reform in Sung China. Wang An-shih (1021–1086) and his New Policies* (Cambridge, MA: Harvard University Press, 1959).

Studies of work. The first modern study of any aspect of Shen's interests, largely responsible for the attention paid him by Chinese educated in modern science, is Chu K'o-chen 竺可楨, "Pei Sung Shen Kua tui-yü ti-hsueh chih kung-hsien yü chi-shu 北宋沈括對於地學之貢獻與紀述" (Contributions to and records concerning the earth sciences by Shen Kua of the Northern Sung period), in *K'o-hsueh* 科學, 1926, *11:* 792–807. Chu's erudite and broadly conceived article has influenced much of the later writing on the subject. Many observations on Shen's scientific and technical ideas are distributed through Needham et al., *Science and Civilisation in China,* as well as through the topical studies by leading Japanese specialists in Yabuuchi, *Sô-Gen jidai no kagaku gijutsu shi. Shen Kua yen-chiu* contains 17 articles on diverse aspects of his work.

A concise survey of Shen's positive contributions by a great historian of mathematics is Ch'ien Pao-ts'ung 錢宝琮, "Shen Kua," in Seminar in the History of the Natural Sciences (ed.), *Chung-kuo ku-tai k'o-hsueh-chia* 中国古代科学家 (Ancient Chinese scientists; Peking: K'o-hsueh Ch'u-pan-she, 1959), 111–121. The articles in Ch'ien (ed.), *Sung Yuan shu-hsueh-shih lun-wen-chi* 宋元数学史论文集 (Essays in the history of mathematics in the Sung and Yuan periods; Peking: idem, 1966) study Shen's most noteworthy problems.

There is no recent investigation in depth of Shen's astronomical activities, but Juan Yuan 阮元, *Ch'ou jen chuan* 疇人傳 (Biographies of Mathematical Astronomers, 1799; reprint, Shanghai: Commercial Press, 1935), *20:* 238–243, provides a good technical description of what were traditionally considered Shen's most important contributions.

Although physics was not a discrete body of knowledge in traditional China, it is an obvious theme for positivistic studies. Notable are Wang Chin-kuang 王锦光, "*Meng ch'i pi t'an* chung kuan-yü tz'u-hsueh yü kuang-hsueh ti chih-shih 梦溪笔谈中关於磁学与光学的知识" (Knowledge of magnetism and optics in Brush Talks), *Che-chiang Shih-fan Hsueh-yuan hsueh-pao* 浙江师范学院学报, 1956, 2: 57–67, and Li Ti 李迪, "Shen Kua tsai wu-li-hsueh-shang ti kung-hsien 沈括在物理学上的贡献" (Shen Kua's contributions to physics), *Wu-li hsueh-pao* 物理学报, 1975, *24:* 231–236. Li has also explored Shen's meteorological interests in "Shen Kua tui ch'i-hsiang hsien-hsiang ti chi-shu yü yen-chiu 沈括对气象现象的记述与研究" (Shen Kua's records and studies of meteorological phenomena), *Ta ch'i k'o-hsueh* 大气科学, 1977, 2: 159–161.

For a general view of Shen's medical knowledge see Juan Fang-fu 阮芳赋, "Shen Kua yü i-yao-hsueh 沈括与医药学" (Shen Kua and medicine), in *I-hsueh hsin lun* 医学新论 (New discussions of medicine; Harbin: Hei-lung-chiang K'o-hsueh Chi-shu Ch'u-pan-she, 1984), 230–240. Hsia Nai 夏鼐, "Shen Kua ho k'ao-ku-hsueh 沈括和考古学" (Shen Kua and archaeology), *K'ao-ku* 考古, 1974, 5: 277–289, reprinted in Hsia, *K'ao-ku-hsueh ho k'o-chi shih* 考古学和科技史 (Archaeology and the history of science and technology; Peking: K'o-hsueh Ch'u-pan-she, 1979), 15–28, 139–141, and Synnøve Vinns-rygg, "Time in Archaeological Thought: China and the West," in *Time, Sci-*

ence, and Society in China and the West (ed. J. T. Fraser, N. Lawrence, & F. Haber; The Study of Time, 5; Amherst: University of Massachusetts Press, 1986), 225–240, elucidate Shen's remarks on ancient techniques.

Rulan C. Pian, *Song Dynasty Musical Sources and Their Interpretation* (Cambridge, MA: Harvard University Press, 1967), esp. 30–32, uses and evaluates the substantial portion of Brush Talks devoted to music. Ts'ai Bin-mou 蔡宾牟 & Yuan Yun-k'ai 袁云开, *Wu-li-hsueh shih chiang-i. Chung-kuo ku-tai pu-fen* 物理学史讲义。中国古代部分 (Lectures on the history of physics. Ancient China; Peking: Kao-teng Chiao-yü Ch'u-pan-she, 1985), 130–134, provides a more technical discussion typical of Chinese textbooks. Useful for Shen's aesthetic ideas is Liu Ch'i-lin 刘启林, *Meng ch'i pi t'an i-wen pu chiao chu* 梦溪笔谈艺文部校注 (A critical, annotated study of the parts of Brush Talks on arts and letters; Harbin: Hei-lung-chiang Jen-min Ch'u-pan-she, 1986), which elucidates ch. 14 and 16–18.

Robert M. Hartwell, in an unpublished study, relates Shen's ideas concerning economic theory, the circulation of money, and similar topics, to traditions of thought on these subjects. Hu Tao-ching has published a most interesting study of Shen's military ideas: "Shen Kua chün-shih ssu-hsiang t'an yuan. Lun Shen Kua yü ch'i chiu-fu Hsu Tung ti shih-ch'eng kuan-hsi 沈括军事思想探源。论沈括与其舅父许洞的师承关系" (The sources of Shen Kua's military thought. The master-disciple relation of Shen Kua and his uncle Hsu Tung), *She-hui k'o-hsueh* 社会科学, 1980, 6: 120–124.

Retrospect

This essay, like that on Wang Hsi-shan in Chapter V and another on Li Shih-chen, was commissioned by Charles C. Gillispie for the *Dictionary of Scientific Biography.* This is not, as the length of this article makes clear, a conventional dictionary, but rather an attempt to sum up the state of understanding of its field ca. 1970 in the form of extended biographies. The essay was printed in vol. XII (New York: Charles Scribner's Sons, 1975), and reprinted for a Sinological readership as "Shen Kua: A Preliminary Assessment of his Scientific Thought and Achievements" *Sung Studies Newsletter,* 1977 (published 1978), *13:* 31–56.

The relations of science to other aspects of Chinese culture had long intrigued me; Shen's polymathic activities and interests offered an ideal focus. I was surprised to find no intellectual center uniting them. I recognized—slowly, because I did not expect it—that his technical interests were consistently shaped by his experience as a civil servant. His frequent breakthroughs in understanding owe a great deal, of course, to his brilliance and curiosity. But, like a large proportion of historical innovations, they frequently came from juxtaposing insights that did not conventionally fit together. He was able to do this because of his remarkable breadth of experience. What nurtured such connections, I believe, was his responses to a dizzying sequence of challenges, at the same time professional and personal. They arise from his piece of the action, as Vannevar Bush put it, during the tumultuous political transition that I have sketched above.

There have been a good many publications on Shen Kua since 1975. Some of this work has suggested fresh insights into one aspect or another of his writing, summarized in this revision. Most research on Shen has remained textually oriented. None has prompted a fundamental reevaluation of his life or work. I have added to the bibliography a number of recent monographs, and several books on wider topics that facilitate studying Shen and his time.

IV

COPERNICUS IN CHINA

Dedicated to my beloved teacher Giorgio de Santillana, from whom I learned the difference between mathematics and physics. As an earlier Tuscan put it, "We are concerned with a real universe, not with a paper one."

Abstract

Historians generally claim that Chinese rejected the early fruits of modern science because of some intellectual or linguistic failing, or a metaphysical indisposition. To the contrary, those best prepared to judge were quite receptive. What matters more, certain key issues were so garbled in the process of transmission that no Chinese could have comprehended them.

Jesuit missionaries, who alone were in a position to introduce contemporary scientific ideas into China before the nineteenth century, were not permitted to discuss the concept of a sun-centered planetary system after 1616. *Because* they wanted to honor Copernicus they characterized his world system in misleading ways. When a Jesuit was free to correctly describe it in 1760, Chinese scientists rejected the heliocentric system because it contradicted earlier statements about Copernicus. No European writer resolved their doubts by admitting that some of the earlier assertions about Copernicus had been untrue.

The Jesuits were also unable to discuss the wider repercussions of the Scientific Revolution, in particular Galileo's central idea that the only firm basis for knowledge of nature was the work of scientists themselves. The Church's injunction of 1616 against the teaching of heliocentricism was meant to reject this notion. To the very end of the Jesuit scientific effort in China, the rivalry between cosmologies was represented as between one astronomical innovator and another for the most convenient and accurate methods of calculation, rather than between the Scholastic philosopher and the mathematical and experimental scientist for the most fruitful approach to physical reality. The character of early modern science was concealed from Chinese scientists, who depended on the Jesuit writings. Many were brilliant by any standard. As is easily seen from their responses to the European science they knew, they would have been quite capable of comprehending modern science if their introduction to it had not been both contradictory and trivial.

IV

2

Introduction

The singular career of Copernicanism in China deserves attention because astronomy and mathematics played such an important part in the early phase of the European conquest of China. In addition, the Copernican episode emphasizes the need for care when thinking about how one culture accepts and rejects the scientific ideas of another. Any generally applicable model has to reflect the interaction of two systems of science, the one traditional and the one newly introduced. A linear theory which sees Western science imposed on a passive recipient culture, perhaps by some inevitable sequence of steps, cannot do justice to the vast historical differences in receptivity. An obvious case in point is the feverish Japanese enthusiasm for European medicine, especially anatomy and surgery, from the seventeenth century on, whereas the Chinese could hardly have been less interested for two centuries longer.

Recent generalizations about the spread of science in the past four centuries have tended to overlook this complexity. It is true that in many cultures the theory and praxis of indigenous natural science lacked conceptual strength, was unable to generate technological benefits, or was not differentiated enough from everyday knowledge, to compete for survival with Western ideas. To a large extent, too, the effective grounds for choice between old and new sciences had little or nothing to do with their content. Decisions about what aspects of Western science to accept (that is, to teach) often ignored the character and implications of the latter. They were settled by the outcome of competition between two political, economic, and cultural systems. The dynamics of interaction between the scientific traditions of two societies have a great deal in common with the dynamics of scientific revolution within a single society.

Only habit prevents us from recognizing that in the importation of modern cosmology or mechanics into France and their introduction into China, if the historical processes were not precisely equivalent, they share many dimensions. It is as much beside the point to ask why France was incapable of developing modern science solely out of her own inner resources as to pose the same question about China.[1]

1. See Chapter VII. I am grateful to Nakayama Shigeru, the late Robert Somers, Donald Wagner, Yabuuti Kiyosi (Yabuuchi Kiyoshi), and James Zimmerman for having thoroughly read and criticized an earlier draft, and to Owen Gingerich, Harold Kahn, David McMullen, Victor Thoren, Laszlo Tisza, and James R. Voelkel for helpful consultations. I thank Wang Aihe and Nakayama for help in procuring new copies of the illustrations for this revision, and to Asaf Goldschmidt for scanning them.

As Thomas Kuhn and others have shown, our comprehension of scientific change depends only to a limited extent on the comparative merits of old and new ideas, estimated abstractly with the indispensable aid of hindsight. It is essential to know who holds them, and how this scientific public envisions the cost, necessity, and benefits of revolutionary innovation.

In a transitional society, of course, the old scientific public and the enthusiasts of modern science may have little in common. In fact the latter may be quite marginal to the old social order. They are often chosen, educated, and protected from social pressure by privileged foreigners, and are often unconcerned with the survival of structures which serve only to hold them down. This was the case in late nineteenth-century China under the régime of the treaty ports. But in the seventeenth and eighteenth centuries in China, amateurs of modern science were on the whole members of the old educated élite, imbued with its values. Their first impulse was to supplement and strengthen the indigenous science, not to discard it, and their loyalty remained with their ancestral world view. In this sense they resembled their contemporaries, the Galenist professors in the European medical schools.

Another and perhaps less obvious factor is bound to complicate any general model of transmission between cultures. The transmitters themselves are sometimes unable to communicate the contemporary understanding of science. There are many ways in which this situation might arise—defective education or knowledge on the part of a traveller would be one, language problems another, narrowly defined goals leading to an overly partial representation of the science might be a third. But in focussing on Copernicanism in East Asia we find that the propagation of modern astronomical ideas was blocked, and scientific communication was distorted, by administrative decisions.

The Astronomical Missionaries

A great deal has been written on the introduction of Western science into China. It is common knowledge that for about a century and a half the Society of Jesus had a monopoly on this transmission. This monopoly came about because of the Jesuits' strategy as missionaries. By about 1600 Matteo Ricci and his colleagues were convinced that their most practical policy was to convert China from the top down, rather than from the bottom up. Since they were laboring under the inconvenience of being foreign barbarians, to make a long story short, their only sure means of access to the court was as technicians, and the one technical skill which they could count on to admit them was astronomy.[2]

It was an urgent matter of national security, as we would put it today, for the Imperial Directorate of Astronomy to be able to predict solar eclipses and other celestial events. This was because if not predicted they were ominous; that is to say, they could be interpreted (and exploited by prospective rebels) as warnings from Heaven that the virtue of the ruling house was failing. Avoiding this potential tool of political opposition was so important, in fact, that the Chinese had resigned themselves for many centuries to making use of first Indian and then Islamic specialists.

For the missionaries to establish themselves as the best astrologers was not very difficult. That was not good enough to bring them the respect they needed to become, eventually, spiritual advisors to the Chinese élite. Since the Chinese in 1600 had not yet been exposed to inundation by narcotics and diplomacy by gunboat, they were still convinced that they had nothing to learn from foreign barbarians. The Jesuits had to demonstrate that they were not really barbarians, and that Europe had a culture comparable with that of China. Religion itself was not accepted by most educated Chinese as a mark of high culture. That is why, although the Jesuits were first and last committed to the salvation of souls, they devoted a great deal of effort to the diffusion of science.

This won them a number of highly placed converts, and eventually a secure place in the Forbidden City. Their patrons established a Calendrical Office *(Li chü* 曆局) in which they worked. Despite imperial recognition for their technical publications, the Jesuits failed in the Ming to take over the official calendar and the Directorate of Astronomy *(Ch'in t'ien chien* 欽天監). The hereditary officials in the Directorate, some of them Muslim, were all too aware of the competition, and worked hard to minimize the missionaries' status. By the end

---

2. I have been brief concerning points well established and generally known. The major previous studies and reference works on the introduction of modern astronomy into China prior to the mid-nineteenth century are listed in the Bibliography at the end of the next chapter (V 20). Full references given there are not repeated in the footnotes.

Most of the earlier writings in European languages which dealt with the introduction of Copernicanism into China were written by historians within the Society of Jesus, and were meant primarily to demonstrate the contributions of Jesuits to the diffusion of science. Their viewpoint dictated a different weighting and treatment of sources, and a differently balanced assessment, than my own emphasis on central doctrines of the Copernican Revolution. Scattered through the footnotes of this chapter are examples of points on which we differ.

of the Ming, the government had reduced the staff of the Office. This pressure motivated the European astronomers to keep writing about the superiority of Western science.

Another reason this effort was so considerable is that European science, especially cosmology, changed drastically in the century and a half that the Jesuit monopoly lasted. Let me review summarily some pertinent characteristics of this change in the West before examining its repercussions half a world away.

Cosmological Background of the Jesuits' Astronomical Writings

The Scientific Revolution began with a recasting of cosmology. In classical times the word "cosmos" implied an order, which was in a sense humanity's order, finite and symmetrical about us, but at the same time serving as our pattern for ideal and eternal relations. Aristotle's conception of the world, which came to be predominant in late antiquity, was largely deductive, based on philosophical thinking about how celestial motions could be both eternal and knowable. It succeeded because it was integral with a truly universal philosophy, which provided consistent and commonsensical answers for any philosophical question that a reasonable person might ask. The integrity of Aristotle's model of the cosmos, that is to say, lay on the level of metaphysics. Professional astronomers could not be allowed to tinker with it. Modifications in the world picture, if they violated first principles, could throw the all-embracing structure of knowledge out of kilter. Although certain speculations about cosmic dimensions from Ptolemy's Planetary Hypotheses were widely diffused before the Renaissance, they were not interpreted as a challenge to Aristotle's basic conceptions; nor, indeed, were they so intended.

Ptolemy, in the Almagest, accepted the notion that the astronomer took his physics from the philosopher but could impose no constraints in return. He constructed a great system of computational astronomy whose authority lasted a millennium and a half. It was impossible for his system to yield accurate predications and, at the same time, to conform perfectly to Aristotle's reasoning about the structure of the cosmos. As a compromise his system is admirable, but he was stuck with shortcomings which sooner or later were bound to bother someone more than they had bothered him. It is well known that Islamic astronomers long before the Scientific Revolution mended what they saw as Ptolemy's compromises with conceptual rigor.[3]

---

3. See, for instance, George Saliba, "The Astronomical Tradition of Maragha: A Historical Survey and Prospects for Future Research," *Arabic Sciences and Philosophy*, 1991, *1*. 1: 67–99. When I refer to "the Ptolemaic system" below, I mean the several

Copernicus himself was at heart no revolutionary. He was a mediocre observer and a mathematician of limited competence, but he had a sufficient feeling for theoretical elegance to take the consequences of demanding it. He saw that if the sun were considered static, rather than the earth, certain factors that made no real sense in the old schema fell into place, and tied the new schema together to make a world that was coherent as well as conceptually simpler. At the same time he remained close enough to his work to ignore the price of this increase in simplicity and elegance, namely the wrecking of the traditional world view of Western man. Once the earth was moved out of the center of the universe to become a mere planet, we have gradually discovered that it goes round a second-rate sun which occupies an undistinguished location in what seems to be a commonplace if somewhat larger than average galaxy. The universe this galaxy moves in is hardly tailored to man's measure.

The philosophical consequences of Copernicus' innovations were not pressed home for more than half a century after his death in 1543. It was Galileo who saw the Aristotelian philosophy of his time as worn out to the point that it was no longer worth patching up,[4] and who insisted that the structure of natural knowledge be defined by astronomers and other scientists who could apply mathematics to the phenomena. It was precisely in that sense that he was a revolutionary. Although in many ways he remained a man of his time, he had no sympathy for the goal that had animated Copernicus to design an essentially Aristotelian universe around a static sun and moving earth.

Galileo's ambitious bid to launch a new dispensation for astronomy was backed first by the Medici court and then by the Roman papacy.[5] He was armed not only with a first-rate instinct for the patronage game, but with the new astronomical telescope. What he saw through it provided him with lethal ammunition against his philosophical target. If the sun was spotted and the moon covered with mountains, how could the earth be uniquely earthy? If Jupiter and Saturn had what appeared to be planetary systems of their own, how could the earth be the unique center of symmetry? Only the phases of Venus could serve as a direct challenge to Ptolemaic mathematical astronomy; the

---

combinations of models and methods, derived from Aristotle, Ptolemy, and diverse early modern philosophers and astronomers, that were taught in Europe ca. 1600.

4. Charles Schmitt and others have documented the flexibility and diversity of early modern Aristotelianism, but it would distort the tenor of their findings to argue that Counter-Reformation scholasticism ca. 1600 encouraged the fundamental changes that took place in the ensuing century.

5. Mario Biagioli, *Galileo, Courtier. The Practice of Science in the Culture of Absolutism* (Science and its Conceptual Foundations; University of Chicago Press, 1993).

other discoveries warned that responsibility for defining the texture of physical reality could no longer remain in the hands of schoolmen. At about the same time, Johannes Kepler, who maintained even higher standards of rigor and who had the unprecedented observations of Tycho Brahe to apply them to, reluctantly brought himself to reject the idea of eternal motion as necessarily circular. This essentially demolished what was left of classical cosmology, opening the way for modern celestial mechanics.

The Catholic Church had benefitted greatly for centuries from its sponsorship and patronage of the latest developments in natural philosophy—in the first instance, medieval Aristotelianism, on which its emerging universities were based. Galileo's proposition was that it continue to reap this benefit by actively supporting the new science. But his suggestion came at a stage of the Counter-Reformation when makers of policy could view a new cosmos only as a dangerous distraction.

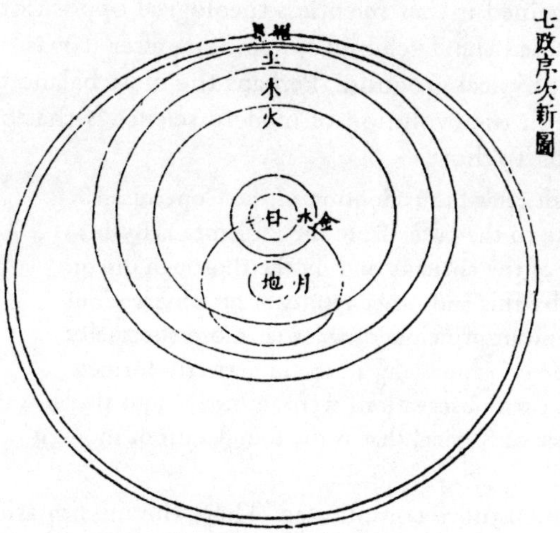

Figure 1. The Tychonic world model (from Rho). The lower of the two central characters stands for the earth; the upper is the sun, rotating about it and carrying the orbits of the planets.

The Decree of the Congregation of the Index in 1616 braked the Copernican Revolution in the Catholic countries. This ruling made it clear that Galileo had gone too far, that the doctrine of the earth's motion was not to be defended, and that *De revolutionibus* was not to be read as written. For our purpose Galileo's trial and capitulation of 1633 were anticlimactic. The Church removed *De revolutionibus* from the *Index* of prohibited books only in 1757, allowing Copernicus' world view—by that time Newtonian—to be taught in the Catholic world. Still, Galileo's name could not be mentioned in connection with it.

In 1588 the great Danish astronomer Tycho Brahe had published in rather summary form a model that made the planets revolve about the sun, which in turn revolved about the static earth (Figure 1). Tycho's world model offered many of the same systematic advantages as the Copernican system. Although he did not originate it, as the best naked-eye observer in Europe he made it influential. Besides not threatening to upset theological applecarts, this system was extremely attractive to working astronomers. They respected Tycho's attitude toward observational data, which was much more demanding than that of his predecessors had been. When the flurry over Copernicanism began, Tycho's mathematically and theologically sound replacement for Ptolemy's system generated more and more enthusiasm. By the 1620's it had become the "third system of the world," and still had supporters in the Catholic countries well into the 1680's. A critical survey of Tycho's and other similar geo-heliocentric systems concludes: "It is noticeable . . . that its supporters were particularly numerous and vociferous where the influence of the Catholic Church was most strongly felt, and the impression gained is that relentless theological opposition to the Copernican system perpetuated the Tychonic system long after it was seen to have become obsolete on physical grounds." Perhaps the most balanced evaluation of Tycho Brahe's place in the evolution of modern science is that of J. L. E. Dreyer in his biography of Tycho:

> The Tychonic system did not retard the adoption of the Copernican one, but acted as a steppingstone to the latter from the Ptolemean. By his destruction of the solid spheres of the ancients and by the thorough discomfiture of the Scholastics caused by this and other results of his observations of comets, he helped the Copernican principle onward far more effectually than he could have done by merely acquiescing in the imperfectly formed system, which the results of his own observations were to mould into the beautiful and simple system [that of Kepler] that is the foundation of modern astronomy.[6]

The situation in China was much more complicated. There the missionaries substituted the Tychonic model for the Ptolemaic in the 1630's, but the Tychonic system itself was not superseded in training astronomers until the late nineteenth century. It was still being used as a stepping-stone long after the brook had dried up at its source. So long as the stepping-stone was still there, it was easy to avoid noticing that there was no water left to cross.

This summary would be inexcusably oversimplified in any other context, but here it is meant merely as a reminder that in Europe the import of the Co-

---

6. The quotations are from Christine Jones, "The Geoheliocentric Planetary System: Its Development and Influence in the Late Sixteenth and Seventeenth Centuries" (unpublished Ph.D. dissertation, University of Cambridge, 1964), Abstract, and J. L. E. Dreyer, *Tycho Brahe. A Picture of Scientific Life and Work in the Sixteenth Century* (Edinburgh: Black, 1890), 181. Dreyer's book has been superseded in many respects by Victor E. Thoren, *Lord of Uraniborg. A Biography of Tycho Brahe* (Cambridge University Press, 1990).

pernican Revolution was philosophical, that it implied a radical new view of the universe and of man's place in it. The opposition to heliocentricism became implacable partly because these implications were repugnant, and partly because Galileo insisted upon a dimension of intellectual freedom that the administrators of the Church saw quite pragmatically as potentially subversive.

The Jesuits in seventeenth-century China were in what was to become an untenable position. They had hardly begun the labor of introducing Western astronomy when it became clear that certain developments were not to be freely discussed. It was only at the end of their period of influence in China that the Church lifted this restriction. The point is not that all the Jesuits in China would have become Copernicans promptly if the decree of 1616 had not been issued. Far from it, as I will show. But, to the extent that they were professionally concerned with astronomy and cosmology, the need at least to take the new point of view into account became steadily more pressing as time passed. By 1700, a reassessment was hardly optional. By 1740 a "modernized" epicyclic astronomy could still be taught, as it actually was in China, only at the cost of demanding that readers tolerate patent inconsistency. Although the Jesuits' Chinese writings at first reflected conservative but open-minded current thinking,[7] they gradually became hopelessly obsolete, out of touch with practice as well as theory. But the constraints under which they wrote, and the lack of competition from lay authors, meant that no one acknowledged or corrected crucial misstatements before the mid-nineteenth century.

As often happens with censorship, the injunction against Galileo did not prompt the missionaries to be silent about the new astronomy. They could speak freely about some aspects, for instance the telescopic discoveries. Taken out of context, they were mere curiosities, evidence of technical virtuosity. These and other novelties had implications for astronomical practice rather than physical reality. The Jesuit missionaries could and did give credit to Copernicus for computational improvements.[8] They could speak with caution and prudence

---

7. It is well known, for instance, that Matteo Ricci's *Ch'ien k'un t'i i* (ca. 1608; see p. 13 below) was based on his master Christopher Clavius' *In sphaeram Joannis de Sacro Bosco commentarius* (1585), which explicated the standard thirteenth-century textbook. Ricci did not, however, follow Clavius so closely as to acknowledge that the Ptolemaic system was troubled, or to write of Copernicus as a restorer of astronomy. See Li Yen 李儼, *Chung-kuo suan-hsueh shih* 中國算學史 (History of Chinese mathematics, 1937; rev. ed., Shanghai: Commercial Press, 1955), 191. Hashimoto, *Hsü Kuang-ch'i and Astronomical Reform,* traces other sources used by the Jesuit astronomical authors in China.

8. The mathematics associated with the Ptolemaic and Copernican systems differed significantly. When tracing the apparent courses of the celestial bodies, however, an

about almost every other issue. They could even bring up the possibility of the earth's daily rotation on its axis, so long as they promptly refuted it. There was, in fact, only one untouchable aspect of Copernican astronomy, namely the heliostatic (or heliocentric) hypothesis, the revolutionary displacement of the center of the celestial orbits from the vicinity of the earth to the vicinity of the sun. This reorientation, mathematically so simple but physically and philosophically so momentous, could not be so much as mentioned in China in the seventeenth century.

Would Copernicanism have been Acceptable to Chinese Astronomers?

I do not wish to imply that traditional Chinese astronomers and amateurs of astronomy would have found the conceptual leap to Copernicanism trivial. But the gap that only careful teaching could have helped them across was not that between Aristotle, Ptolemy, and Tycho on the one hand and Copernicus, Kepler, and Newton on the other. This they would have found much easier than the great majority of their European contemporaries, whose commitments to Aristotelian and Christian assumptions they did not share. Discoveries that Galileo announced triumphantly could hardly have come as a shock to the Chinese, who did not know the ramifications of the philosophy behind Ptolemy and had never felt its authority.

The early Chinese astronomer had been fairly successful at predicting the courses of the heavenly bodies algebraically, without depending on a physical or geometric model.[9] If he happened to believe that the planets went round the earth, his data came from common sense; it made no difference to his computations. His seventeenth-century successors had to learn the utility of astronomical schemata before the basic conflicts of the Copernican Revolution could make sense.

By a millennium before the Jesuits arrived, Chinese cosmological speculation and astronomical practice were no longer linked. This gap was of a different order than the discrepancies between Aristotle's Metaphysics and Ptolemy's Almagest. Far Eastern philosophers were as willing as their European scholastic counterparts to define the physics of the universe without drawing on the best

---

astronomer could use most techniques regardless of whether the cosmological frame of reference in which they originated was geostatic or heliostatic. See, for instance, Needham, *Science and Civilisation in China,* III, 446, and Shigeru Nakayama, "On the Introduction of Heliocentric System into Japan," *Scientific Papers of the College of General Education,* University of Tokyo, 1961, *11:* 166.

9. For the reasons, see Chapter II.

astronomical knowledge of their time. Chinese astronomers, however, were quite free to reciprocate by ignoring the philosophers.

Chu Hsi 朱熹 (1130–1200) was perhaps the most influential philosopher in late Imperial China. His synthesis was comprehensive and his philosophy, aimed at individual sagehood, was adapted from the Yuan on as a base for the State's conformist ideology. He was more concerned with the sky as a cosmos than as the simple collection of stars and moving planets that the technicians dealt with. Astronomical opinion to the contrary, Chu Hsi did not hesitate to argue (with recourse to yin-yang physics) that, for instance, the real motions of the sun, moon, and planets are directed in the same sense as those of the fixed stars, only slower; that the retrograde motions of the planets are in absolute terms progressive too. He declared that in positing annual rotations in the opposite sense contemporary astronomers were misguided.[10] Considering Chu's great authority, it is remarkable that astronomers so consistently ignored his opinions. The astronomer Wang Hsi-shan 王錫闡 (1628-1682) argued, in fact, that there had been a permanent split: "When it came to the Sung period, astronomy (*li* 曆) bifurcated; there was one astronomy for scholars and another for astronomers. The scholars did not know astronomical mathematics, and so they depended upon empty principles in establishing their theories. The technicians did not apprehend the principles of astronomy, but merely settled upon techniques that managed to predict the phenomena. Apparently no one mastered the measures of sky and earth and the foundations of the celestial phenomena."[11]

The Jesuits' early writings prompted many discussions by native astronomers. They make it quite clear that Chinese were capable of learning and applying—but applying critically—anything the missionaries cared to teach them about the sky. On the whole the greatest hindrance was the limited depth of the Western writings, an inevitable result of the Europeans' limited objectives. The best of the Chinese astronomers eagerly accepted Tycho's world system in the seventeenth century, and Kepler's elliptical orbits when they were introduced in a partial way in the eighteenth. There was no effective metaphysical

10. *Chu-tzu ch'üan shu* 朱子全書 (Works of Chu Hsi), *ch.* 50; Yamada Keiji, "Shushi no uchûron 朱子の宇宙論" (Chu Hsi's cosmology), *Tôhô gakuhô* (Kyoto), 1966, *37:* 41–151. See also Needham, *Science and Civilisation,* III, 400, 474–478.

11. *Wang Hsiao-an hsien-sheng i shu,* 1a. On Wang and this source, see the next chapter.

or religious orthodoxy to hold these scholars back, and on the whole they quite ignored the Jesuit digests of scholastic natural philosophy.[12]

Probably the most important immediate consequence of the Jesuit educational effort was to enable the Chinese to rediscover and revive their own tradition, which had been neglected for three centuries. By 1700 the best mathematical talents in China were getting their basic training in the more immediately accessible Western methods, and going on to devote their mature careers to the reconstruction and enrichment of the indigenous exact sciences. It was not until well into the nineteenth century that this order of study was reversed.[13]

Seventeenth-century Chinese astronomers assert repeatedly that they learned gradually to value Occidental astronomy most for the explanatory strength of its models. Typical is a 1637 statement of Chou Yin 周胤 about the series of Jesuit astronomical treatises of the 1620's and 1630's that will be discussed anon. Because writings available earlier "did not fully set out the ideas on which the new methods were based, we still did not apprehend their scope, but were aware only of the fineness of design of the Europeans' instruments and the ease with which they worked out their predictions." Chou goes on to say that in

---

12. Among the main early sources were Francisco Furtado's *Huan yu ch'üan* 寰有銓 (1628), based on a textbook version of Aristotle's *De caelo et mundo,* and Alfonso Vagnone's *K'ung chi ko chih* 空際格致, a synthesis of scholastic meteorology and theory of the earth derived indirectly from Aristotle's *Meteorologica* and printed in 1633. In contrast to the Western writings on astronomy, these books are in many respects not as well informed and or as well reasoned as the better Chinese works of their kind. The missionaries exerted themselves sedulously to prove, among other propositions, that the Empedoclean four elements were correct and the traditional Chinese five were wrong (the *wu hsing* were not in fact elements). Willard Peterson studied the content of and response to this genre of missionary writings in "Fang I-chih's Response to Western Knowledge."

13. Attempts to show that Western science originated in China have been noted by many scholars. Mikami Yoshio 三上義夫 long ago concluded that the net effect of the Jesuits' scientific propaganda, after its original impact had worn off, "was the encouragement of the study of the ancient classics rather than that of the introduction of new ideas from European sources." The two kinds of effort were not actually in competition, but it is true that the new ideas were seen as a means toward this rediscovery. See Mikami, "Chinese Mathematics," 125, and "Chûjin den ron," 185–222, 287–333. The recovery of the Chinese tradition is traced in detail in Wang P'ing, *Hsi-fang li-suan-hsueh chih shu-ju*, chaps. IV and VII. For an English summary, see the review of her book in *Journal of Asian Studies*, 1970, *29:* 914–916. On other important changes of the time, see Henderson, *The Development and Decline of Chinese Cosmology.*

sponsoring the new series of books, "the fundamental idea of Li T'ien-ching 李天經 [the official in charge of the Ming calendar reform] was that the tradition of mathematical astronomy [in China from the fourteenth century on] died out because while learning techniques we had lost sight of the principles [that determined why the phenomena were] as they were. [He felt that] the principles must be set out clearly before the techniques were taught."[14]

Influential Chinese astronomers were not only ready to rethink cosmology but ran no personal risk in doing so, as the study of Wang Hsi-shan in the next chapter shows. That this is true renders irrelevant the lists of "cultural factors" limiting the development of Chinese scientific thought that flow so readily from the pens of sinologists who have never found time to read the primary literature of science. The crucial limitation on the Chinese reception of modern cosmology, my reading suggests, was the quality and quantity of information available in China about what was going on at the other side of the world. Chinese could begin contributing to modern astronomy only after they confirmed for themselves that the achievement of the Copernican Revolution lay in its description of physical reality. That they were not able to learn until a little over a century ago.

Cosmology in the Late Ming Jesuit Writings (1608–1640)

Let us first consider the Jesuit astronomical works published in Chinese before 1616, the year of the decree by the Congregation of the Index that first defined the theological status of the heliostatic idea. These books were all written entirely from the Aristotelian point of view prevalent in Europe at the time. Typical in this respect was Ricci's Cosmological Epitome (*Ch'ien k'un t'i i* 乾坤體義, ca.1608): "These nine layers enclose each other like the layers of an onion. They are all solid, and the sun, moon, and planets are fastened into their substance like knots in a board. The motions [of the sun, moon, and planets] are entirely due to those of their proper orbs. The celestial substance is clear and colorless, and thus transparent to light, in the same way as [light] is unimpeded

14. *Hsi-yang hsin fa li shu* 西洋新法曆書 (Astronomical treatises according to the new methods of the West, printed 1646), forematter, 277b-278a.

by glass and crystal and the like."[15] He did not outline computational methods or geometric models.

Appended to Emmanuel Diaz' Catechism of the Heavens (*T'ien wen lueh* 天問略, 1615) was a short report on what he had just heard of Galileo's telescopic discoveries. Diaz did not mention Galileo's name, and did not point out to the Chinese reader the possibility of conflict with classical cosmology.[16] We can, in fact, hardly expect Diaz to have been concerned about a conflict, since his information was undoubtedly secondhand and fragmentary. We do not know whether he knew of the crucial phases of Venus. But ten years later, in 1626, we find in Johann Adam Schall von Bell's Monograph on the Telescope (*Yuan ching shuo* 遠鏡說) a full account, even including Galileo's observation that the satellites of Jupiter can be eclipsed. That Schall did not give Galileo's name is not surprising. By that time what the Jesuit historian Pasquale D'Elia called "the judgment of higher authority" had reached the missionaries in Peking.[17]

---

15. *Ch'ien k'un t'i i* (see note 7; *Ssu k'u ch'üan shu* ed.), *1:* 6a. Ricci's treatise has been studied by Imai Itaru 今井溱, "Kenkon taigi zakkô 乾坤體義雜考" (Miscellaneous researches on the Cosmological Epitome), in Yabuuchi (ed.), *Min Shin jidai no kagaku gijutsushi*, 35–47. D'Elia has collected Ricci's writings as *Fonti Ricciane* (3 vols., Roma: La Libreria dello Stato, 1942–1949). The "celestial substance" is the quintessence from which Aristotle believed the celestial spheres were formed.

16. D'Elia's translation of the pertinent passage of *T'ien wen lueh* (18–19) disproves Szcześniak's statement that Diaz was "explaining astronomy according to Galileo's observations" ("Notes," 33). Diaz merely *reported* the discovery of what were then considered the stars that attend Saturn and Jupiter, and interpreted the Milky Way as "a great quantity of small stars." This passage is in an addendum to *T'ien wen lueh,* which was not affected otherwise by the Galilean discoveries.

17. P. vii. D'Elia in his preface was frank about constraints upon the missionaries, but offered a most unpersuasive argument further on (p. 34): "Galileo is not named either here or in following texts, unless in 1640, without doubt for the simple reason that for the Chinese of that time a European name, phoneticized in Chinese, something like *Chia-li-lê-io,* would have signified little or nothing more than a barbarian name. Let one think of the reverse case of the name of a Chinese scientist in an occidental treatise on astronomy." First, Galileo *is* named (as "Chia-li-lou 加利婁) about 1637 in Rho's *Wu wei hou lun* 五緯後論 (Sequel on the planets), as D'Elia acknowledges in his note 168. Rho's work was printed as an appendix to *Wu wei li chih*; see *9:* 14b-15a. Rho named Copernicus and Tycho Brahe in the latter work by 1634 (see p. 18 below). Schall, in *Li fa hsi ch'uan* 曆法西傳 (ca. 1640; in *Hsi-yang hsin fa li shu*), 12a, referred to Galileo differently, as "Chia-li-le-a" 加利勒阿, thus making it impossible for a Chinese reader to identify him with the astronomer mentioned by Rho (I take up a similar confusion in Copernicus' name below). The current standard translit-

More to the point, Schall did not even hint at the critical implications of the Galilean observations.

The Outline of Observational Astronomy *(Ts'e t'ien yueh shuo* 測天約說, 1628) by Galileo's whilom friend Johann Schreck (or Terrentius, 1576–1630), was more concerned with observation than its predecessors. Schreck alluded to the Tychonic system, but his extensive treatment of the world's construction was still Aristotelian-Ptolemaic. He merely noted in passing that

> in modern times a celebrated mathematician in a kingdom of the Occident has constructed a telescope, with which he has observed Venus. [He thus saw that] sometimes it is dark, sometimes fully illuminated, and sometimes a crescent illuminated either in the superior or inferior quarter. It was calculated that Venus moves as a satellite of the sun. At maximum elongation, contrary to what happens with the moon, only half of its figure is illuminated. Thus one understands [the following:] Sometimes Venus is above the sun; it is then illuminated and small. (If the planet is in line with the sun and the earth, it cannot be seen. If it is only slightly distant from the sun but still above it, it resembles a nearly full moon). Sometimes it is below the sun, in which case it is dark. (Because it is in conjunction it is completely dark. When slightly distant from the sun, but still below it, it resembles the moon when it has just begun to wax; it is small but brilliant). When it is on a level with the sun, it is a crescent . . .

He said not a word about the significance of Galileo's discovery for cosmology. Nor, for that matter, did he acknowledge that his discussion of telescopic optics was mostly taken from Kepler.[18]

---

eration of Galileo's name is "Ga-li-lueh 伽利略." See, for instance, the article by Yen Tun-chieh cited in the Bibliography.

Second, as early as 1607 Matteo Ricci and Hsu Kuang-ch'i 徐光啓 (1562–1633), whose combined competence in matters of mandarin psychology could hardly be surpassed, did not hesitate to mention the names of Euclid and Christopher Clavius in their translation of the *Elements.* See the translation of Hsu's preface and Ricci's introduction by D'Elia himself: "Presentazione della prima Traduzione Cinese di Euclide," *Monumenta Serica,* 1956, *15:* 185, 187.

Third, supposing that a Chinese had invented the astronomical telescope and was the first to explore the sky with it, what indeed would we think if the author of the first European monograph on the subject refused to identify this innovator on the ground that his name would sound foreign and thus be superfluous?

18. *Ts'e t'ien yueh shuo* (in *Hsi-yang hsin fa li shu*), *1:* 16b; D'Elia, *Galileo in China,* 40. D'Elia's translation of the passage from Schreck is not very accurate, and omits without indication about half of the passage, printed in smaller type (I enclose such passages in parentheses). D'Elia's discussion of Schall's sources is also mistaken.

Giacomo Rho (1592–1638), writing his Principles of the Planetary Motions (*Wu wei li chih* 五緯曆指) in 1634 or slightly earlier,[19] was more conspicuously evasive. His accounts of both cosmology and the telescope omitted Galileo's name.[20] Rho did explain the changing aspects of the Medicean Planets (as the satellites of Jupiter were called) and dealt more briefly with the other new phenomena, but did not hint that Galileo's observations could conceivably necessitate a major revision of the system of the world.

Rho nevertheless initiated an important cosmological reorientation. He was the first of the missionaries to substitute the Tychonic for the Ptolemaic world system in his published writings. There was, of course, the danger of confusing Chinese readers. If no explanation were given, they would wonder why the Europeans were suddenly changing their story. Rho dealt with this problem by speaking of Ptolemaic astronomy as the "ancient school." Even so, he obscured the difference by stating that, among other propositions, the sun's place at the center of the planets is "also doubted neither by the ancients nor the moderns."[21] Perhaps Rho meant "the ancients" as a euphemism for the conservative

---

Hashimoto, *Hsü Kuang-ch'i*, 182–200, documents in detail Schall's use of Kepler's writings on optics. On the introduction of the telescope, see Colette Diény, "L'introduction du télescope en Chine," in *Nombres, astres, plantes et viscères. Sept essais sur l'histoire des sciences et des techniques en Asie orientale*, ed. Isabelle Ang & Pierre-Étienne Will (Memoires de l'Institut des Hautes Études Chinoises, 35; Paris: Collège de France, 1994), 177-191.

On Schreck, see G. Gabrieli, *Giovanni Schreck Linceo, Gesuita e Missionario in China e le sue Lettere dalt'Asia* (Rome: Royal National Academy of the Lincei, 1937). Needham (*Science and Civilisation*, III, 435, 447) has pointed out that Schreck described sunspots in this work in the belief that he was introducing Chinese astronomers to a novelty, but their predecessors had been recording them from the Han period on, and publishing extensive lists since the thirteenth century.

19. D'Elia gives the date of this work as 1637 (p. 97, n. 124), but it was presented to the throne in its present form on 21 January 1635. See Li Yen, *Chung suan shih lun ts'ung*, III, 37. Ch. 9, in which D'Elia noticed a date in 1636, is actually the supplement *Wu wei hou lun*, which was not submitted with the *Wu wei li chih* (see note 17 above).

20. Chap. 1 is devoted to introductory cosmology. On the telescopic discoveries see pp. 32a-35b.

21. That Rho's writing was Tychonic is obvious in the passages cited in D'Elia, 54–56. The quotation in paragraph 2, p. 55, begins with a mistranslation, since D'Elia mispunctuated his text. At the end of paragraph 2, for "solar system" read "[orb of] the sun." In lines 2–3 of the same page, for "of small orbits, and of non-concentric circles" read "of epicycles and eccentrics." The passage which D'Elia claimed "openly alludes both to Galileo and to the heliocentric system" (p. 54) mentions neither, and does not

schoolmen of his time, in which case his statement would be very optimistic. In any case, because he did not define the term, no Chinese could have understood it to mean anything but people living in ancient times.

Rho discussed a number of topics drawn from Copernicus' *De revolutionibus.* In some cases he identified them with its author. He brought up the diurnal rotation of the earth, for instance, in order to refute it, but did not express or imply any connection with Copernicus:

> Question: How about the rotation of the first moving orb[, the outermost sphere in Aristotle's cosmos, which drives all of those within]?
>
> Answer: There are two theories. Some deny that the first moving orb makes a daily revolution of the sky, rotating leftward about the earth and carrying all the orbs westward with it. When above the surface of the earth we see the celestial bodies moving leftward, [according to this view] this is not their own motion. The celestial bodies do not have a daily revolution; rather the earth along with its air and fire move as one sphere from west to east, making one rotation each day. Like a man traveling in a boat who sees trees and other things on the shore and feels that not he but the shore is moving, people on the earth, seeing the westward motion of all the celestial bodies, reason analogously. Thus by the single motion of the earth one avoids many motions in the heavens. By a terrestrial rotation of small radius one avoids a heavenly circuit of great radius [i.e. that of the first moving orb].
>
> Nevertheless all the scholars of ancient and modern times agree that this is truly not a valid explanation. It would seem that, since the earth is the center of all the orbs, and a center is like a pivot, it cannot possibly move. Further, if from the boat one sees the shore moving, why should not someone on the shore see the boat moving? The simile offered does not constitute proof [of the earth's rotation].
>
> According to the correct explanation, the body of the earth does not move. The first moving orb is a great sphere, the uppermost of the orbs, with its own poles and proper motion. The two poles of each of the orbs within, since they are contained by the first moving orb, cannot fail to move along with it. Like a man traveling in a boat, or ants moving on a millstone, although they have their proper motions they cannot fail also to follow along with the boat or millstone. Inquiries as to the thickness of the first moving orb, its substance, color, and so on, and the substance and color and so on of the other orbs, because they belong to physics [*wu li chih hsueh* 物

---

even imply that anyone ever believed the sun is at the center of the world.

理之學, literally "the study of phenomenal principles"], are not the concern of mathematical astronomy. They are taken up in detail in other books.[22]

Why did Rho want to present an explanation that no scholar of ancient or modern times believed? If no one believed that the first moving orb moves, who was disagreeing, and why? Rho did not address these points. The application of the boat simile to the question of the earth's diurnal rotation, familiar in Europe since its use by Nicole Oresme in the fourteenth century, was by no means rare in the Chinese writings of the missionaries, as we will see anon.

Principles of the Planetary Motions failed to identify Copernicus as the one who had posited the rotation of the earth about its own axis, but Rho gave him credit for a discovery he never made: "In the Middle Ages, above the orb of the fixed stars were added one orb [to account] for the east-west precession of the equinoxes and one for the north-south precession [i.e. trepidation], making an eleven-orb heaven. (This is as determined by Copernicus; since Tycho's time they are no longer used.)"

Rho explained in some detail the differences in Ptolemy's, Copernicus', and Tycho's treatment of the annual motions of the superior planets. Using the diagram reproduced here as Figure 2, he demonstrated the equivalence of the Copernican and Tychonic constructions for the equation of center, which replaced Ptolemy's combination of equant and eccentric. Here is the accompanying text: "Figure [2] is the general method of Tycho and Copernicus. The sun is taken as the center of rotation of the Five Planets. *A* is the earth. *FGH* is the deferent of the sun. The sun being located at *F, F* is the center [of rotation of the planets] . . ."[23]

Rho's diagram, taken in conjunction with these words, plainly gives the impression that, for Copernicus as well as Tycho, the earth is approximately at the center of the solar deferent, the sun rotates about it, and the planets rotate about the sun. We will see that it was understood in that way by Chinese readers.

---

22. *Wu wei li chih, 1:* 7b-8a. A note in small type at the end of this passage refers the reader to *Huan yu ch'üan* (see note 12). "First moving orb" and other terms are discussed in the Appendix, p. 51.

23. Ibid., *1:* 1b, 29b. The diagram is on the same page. I have added Latin letters equivalent to the ordinal Chinese characters to this figure and to Figure 9.

Rho used the Tychonic model heuristically throughout his work to demon strate the equivalence of computational techniques. But since he did not make it clear that two distinct models of the world were involved, he was bound to impede his readers' understanding. The policy of the Church gave him no choice.

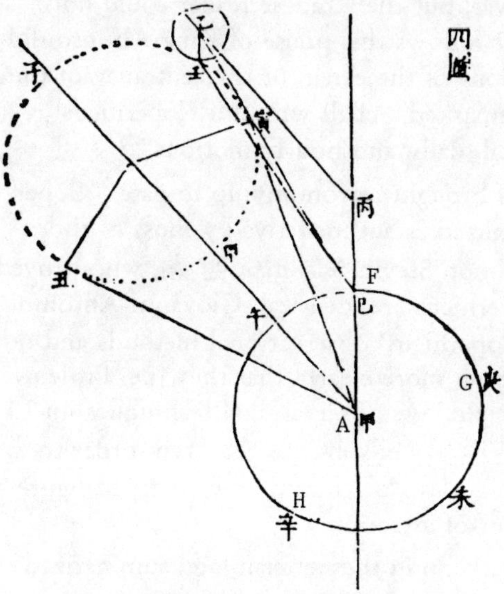

Figure 2. Rho on the General Method of Tycho and Copernicus.

Rho was prudent and willing to leave to the philosophers the physical merits of the systems he was describing. The net result was that Copernicus appeared to be a rather shadowy precursor of Tycho Brahe. That was, in fact, the way many progressive Europeans were thinking of him in 1634.

Schall, on the other hand, was direct and certainly intent on giving Copernicus and Galileo their due. But Chinese readers could only interpret his words to mean that none of the cosmological changes was fundamental.

Schall's quasi-historical treatise On the Transmission of Astronomy in the West *(Li fa hsi ch'uan* 曆法西傳, probably written about 1640) placed Copernicus and Tycho in the ancient school along with Ptolemy.[24] Oddly enough, the modern school, to which he devoted the second half of the little treatise, was represented only by Schall and Rho and their European collaborators. This kept the focus on the Jesuit writings in order to widen support for the missionaries.

Schall began with an outline of the *Almagest* chapter by chapter. He then praised Alfonso X of Castille as a king learned in astronomy who spent a vast sum on the compilation of tables: "His merit is not inferior to that of Ptolemy." He went on: "And then, four hundred years later, there was Copernicus, who determined empirically that although Ptolemy's methods were quite ex-

24. This dating is tentative; see D'Elia, *Galileo in China,* 50. *Li fa hsi ch'uan* was not among the books presented to the throne by 1635. In addition to the version printed in *Hsi-yang hsin fa li shu* (1646) there is a copy in a manuscript collection of Jesuit and Dutch material called *Ch'ung-chen lei shu* 崇禎類書 in the Tenri Library.

haustive, there was a slight lack of clarity *(wei ch'ien hsiao ming* 微欠曉明); so he created a new figure (*t'u* 圖) and wrote a book in six chapters." The word "*t'u*," which I translate "figure," can mean anything from "picture" to "map" to "diagram" to "schema." Schall generally used it elsewhere in his treatise to denote a heuristic schema that accounts for one or more of the celestial motions. We know what Copernicus' "figure" was, but the Chinese reader could not. The summary of *De revolutionibus* that follows this praise of King Alfonso did not mention the daily or annual rotations of the earth, or the centrality of the sun. In fact, as a Chinese critic later remarked, Schall wrote of Copernicus' "seeking the sun's apogee and its annual, daily and hourly motions."[25]

At the end of this summary, Schall brought astronomy up to date: "Copernicus' work, as above, has been transmitted as authoritative by most of those who came after him. There was one Simon Stevin (Hsi-man 西滿), who proved that the methods of Ptolemy and Copernicus are identical. Giovanni Antonio Magini (Ma-jih-no 麻日諾) adopted Copernicus' observational methods and revised Ptolemy's figure. We [thus] see even more clearly that they [i.e. Ptolemy and Copernicus] do not differ in principle." As observational techniques improved, progress continued. François Viète (Wei-yeh-ta 未葉大), in order to stop calculating with eccentrics and epicycles, "originated an egg-shaped figure (*t'u*) in order to explain the foundations of astronomy."[26]

Schall lauded Tycho as a great observer and theoretician, and summarized two of his works. He pointed out that Galileo's telescopic observations are important, but passed over what, aside from sheer novelty, makes them important. He also named Galileo as originator of a new "figure" that revealed what had never been revealed in all the previous study of the stars, and noted that the Tuscan wrote a book to include it. Schall had nothing to say about the charac-

---

25. See below, p. 47. D'Elia translates the passage in which the word "*t'u*" appears (p. 34; cf. note 17 above). He is, however, wrong in translating "*t'u*" as "map of the heavens," as is clear from Schall's use of the word elsewhere in the book.

26. Schall could have been hardly ignorant that the ellipse was Kepler's. Kepler, in reply to Schreck's request for information concerning the newest results of astronomical research, wrote a report and had it printed in 1629. See Max Caspar, *Johannes Kepler* (Stuttgart: W. Kohlhammer Verlag, 1950), 396 and 420. The statement about Stevin is apparently based on the fact that the author of "the first textbook destined to give a simple and easy exposition of the heliocentric theory" (*Wisconstighe Ghedachtenissen,* 1605) explained computations by both systems. It ignores the fact that in doing so Stevin "expounded the Copernican as the true beside the Ptolemaic as the untrue system." See A. Pannekoek (ed.) *The Principal Works of Simon Stevin* (Amsterdam: C. V. Swets & Zeitlinger, 1961), III, *passim,* esp. p. 6.

ter of this figure or the content of the book. He went on to discuss astronomical instruments and their applications, and what voyagers' observations had contributed to science. This remarkable sketch concludes by observing that astronomy in the West is the work of many hands over thousands of years, but "in its essentials it has not gone beyond the bounds [set] by Ptolemy."[27]

Schall introduced the same sequence of four great astronomers (Ptolemy, Alfonso X, Copernicus, Tycho), none of whom was to be considered inferior to the others in any respect, in his outline of Tychonic cosmology, Introduction to Astronomy according to the New Methods *(Hsin fa li yin* 新法曆引, 1635/1646). There he devoted only a few sentences to historical matters. He praised the four masters impartially (perhaps promiscuously in the case of Alfonso) for the proliferation of their writings, the accuracy of their observations, and the practicality of their computational methods.[28] Neither of Schall's books so much as mentioned Kepler.

Schall could also have ignored Copernicus, but he was too fair-minded and too conscientious an astronomer not to honor him in China as far as the spirit of the 1616 injunction allowed. Regardless of whether he knew of the final and generally unexpected proceeding against Galileo (1633) by the time he wrote On the Transmission of Astronomy in the West, the wording of his praise for the Tuscan astronomer was courageous. Schall's attitude toward Copernicus, Kepler, Stevin and others is representative of many astronomers of his time, who quite agreed that what physical reality underlay their observations and formulas was none of their business.

27. This summary and the citations included in it are drawn from *Li fa hsi ch'uan,* 2b-14b. There have been several biographies of Schall, none of which pay adequate attention to his astronomical work. The most useful sources for his life are Alfons Väth, *Johann Adam Schall von Bell, S. J., Missionar in China, Kaiserlicher Astronom und Ratgeber am Hofe von Peking 1592–1666. Ein Lebens- und Zeitbild* (Cologne: J. P. Bachem, 1933), and Henri Bernard, *Lettres et mémoires d'Adam Schall S. J.* (Tientsin: Hautes Études, 1942).

28. In *Hsi-yang hsin fa li shu,* 2a-2b. The dating of this book is problematic. It was not among those presented to the throne prior to 1635. Bernard, "L'Encyclopédie astronomique," 578–579, noted a separate book entitled merely *Li yin* (Introduction to astronomy), by Rho and twelve Chinese collaborators, in the National Library of Peking. Bernard thought that the *Hsin fa li yin* is a new edition of this work, although the recorded length is different and he was unable to compare the two (476). He did not notice, however, that *Li yin* is also missing from the presentation lists. For the latter reason it is probable, although far from certain, that neither work existed prior to 1635.

In Europe, however, there was a dialogue. Many sagacious and prudent men of science, convinced that the Copernican issue was unimportant and could easily lead to errors of belief, preferred being forbidden to discuss it freely. But still in Europe there were countries where a scientist could not be punished or abandoned to damnation for seeking to separate experiential and quantitative knowledge from verbal philosophy. If what was subjectively prudence and the best of intentions led to misrepresentation, some would know the truth, and would not be silent.

In China only the safe side of the argument could be presented, even long after astronomers had abandoned it in the West. Since heliocentricism was unmentionable, any discussion that attempted to do justice to Copernicus in the realm of computational astronomy would have to pay the price of distortion in that of cosmology. As the missionaries acted surreptitiously on their good intentions, the compounded outcome led Chinese readers to an erroneous conception of Copernicanism, and provoked them to reject a correct explanation when it came. The central irony of the Jesuits' situation is that they could have avoided this ultimate rejection only by not writing about Copernicus at all so long as it was a matter of conscience not to describe his work fully.

Copernicanism in Shibukawa's Synthesis

For most Chinese readers of the Jesuit astronomical treatises before the mid-eighteenth century, Copernicus' historical position was clear. He was a vague but estimable figure of the Middle Ages whose work had been made obsolete by Tycho Brahe. This consensus appears, for instance, in the Queries on Mathematical Astronomy *(Li hsueh i wen* 曆學疑問, 1693) of Mei Wen-ting 梅文鼎 (1633–1721), the most influential writer of his time on the subject. While arguing that Western astronomy was as much a result of historical development as the Chinese art was, Mei noted: "With the coming of Copernicus, there were some revisions to Ptolemy's methods. With the coming of Tycho there was a great transformation of Copernicus' methods. The methods were now on the whole complete, but the making of the telescope happened still later, and led through the accumulation of observations to enhanced precision."[29]

Ample discrepancies remained, ready to confuse those who read widely. Indeed, as I will show, confusion became inevitable later in the eighteenth century as the true importance of Copernicus began to emerge.

---

29. *Li hsueh i wen, 1:* 20a, in *Mei shih li suan ch'üan shu* 梅氏曆算全書 (Wei Nien-t'ing 魏念庭 ed., printed 1723). See also Hashimoto, "Bai Buntei no rekisangaku."

To illustrate the kind of misunderstanding that was generated by a trustful reading of the late Ming treatises, consider an important Japanese book, now rare, the Introduction to the Astronomy of the Ming Jesuit Treatises *(Sûtei reki-sho rekiin* 崇禎曆書曆引), completed about the end of 1847 by Shibukawa Kagesuke 澀川景祐 (1787–1856) and printed in 1855. Shibukawa was astronomer to the Japanese Shogun, and co-translator of Lalande's *Astronomie* from a Dutch version. He was one of the few Japanese of his time familiar with post-Newtonian astronomy, and a great compiler.[30] Since secular Dutch sources had long been available in Japan, Shibukawa's Introduction was not meant to be up to date. He was trying to reconstruct what the missionaries had said and meant. The Copernican world system his book illustrates is spectacularly muddled, but solidly based on the Jesuits' Chinese writings.

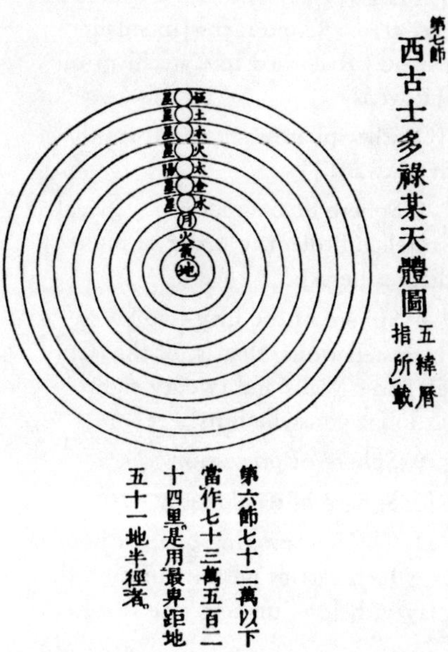

Figure 3. Shibukawa Kagesuke's Ptolemaic schema.

Shibukawa's Ptolemaic schema (Figure 3), according to its caption, comes from Rho's Principles of the Planetary Motions.[31] Shibukawa has merely added small epicycles between the planetary orbs to reflect Rho's verbal description. We see the earth at the center, the spheres of air and fire about it, and then the orbs of the moon, the planets, and the fixed stars. Shibukawa's Tychonic diagram (Figure 4) is a practically exact copy of the Rho illustration (Figure 1, page 8). The caption does not mention Tycho's name: "Figure for predicting celestial positions according to the new astronomy," and in smaller characters, "The new figure for the order of the Seven Governors [the sun, moon, and planets] as recorded in Principles of the Planetary Motions."

Figure 5 is Shibukawa's diagram for the Copernican system. Comparing it with the Ptolemaic figure, we can see two prominent differences. First, Shibukawa gave the central earth (!) features and left out its spheres of air and fire.

30. On Shibukawa, see Nakayama, *History of Japanese Astronomy*, 200–202.

31. Vol. III *(zuhen* 圖編), section "On the order of the universe." The figures reproduced here are from pp. 3a, 3b, and 4a respectively.

Second, and obviously more significant, he added precession and trepidation spheres, the first moving orb, and a "sphere of the blest," all above the sphere of the fixed stars.

The captions deserve attention:

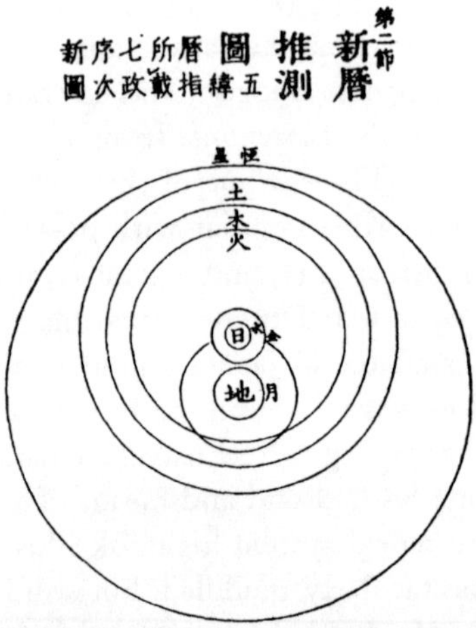

Figure 4. Shibukawa's Tychonic schema.

[In large type:] The Old Cosmos of the Westerner Copernicus [In smaller characters:] Recorded in Catechism of the Heavens[32]

天體

[On the spheres, reading from the earth outward:]

1. Sphere of the moon . . . [usual Aristotelian-Ptolemaic order of intermediate spheres] . . .

8. Sphere of the fifty-two [Western] constellations, that is, of the [Chinese] Three Walls and twenty-eight lunar lodge constellations[33]

9. Sphere of precession

10. Sphere of trepidation

11. The first moving orb, without stars, which carries with its motion the ten layers below, making one revolution per day

12. The abode of the saints, eternally immobile.

---

32. Shibukawa's accompanying text (*1:* 3b) reads: "The ancients said that from the sphere of the moon upward there is . . . a total of twelve spheres, which enclose each other like the layers of an onion. But those beyond the fixed stars are no longer amenable to observation. Since observation is the proper concern of astronomers, how can they investigate the unobservable? Thus they do not discuss [these upper spheres]."

33. I. e., the sphere of the fixed stars. The fifty-two constellations are those known before the seventeenth century. Shibukawa's text (*1:* 18b) says: "In ancient Western astronomy there were also twenty-eight houses; their significance was the same as in the ancient Chinese [art] . . . In the West [the firmament] was further divided into sixty-two *[sic]* constellations, which were also given names." The Three Walls are three chains of stars significant in Chinese astrology. See Ho Peng Yoke, *The Astronomical Chapters of the Chin Shu* (Paris: Mouton & Co, 1966), 71, 76, and 84.

It is possible to reconstruct what happened. Shibukawa found the original of his diagram in Diaz' *Catechism of the Heavens,* which does not even mention Copernicus. The Japanese scholar reproduced the captions of the twelve spheres verbatim, except that Diaz had also made the twelfth sphere the abode of God. That Name, forbidden in Tokugawa Japan, had to be removed. The Chinese

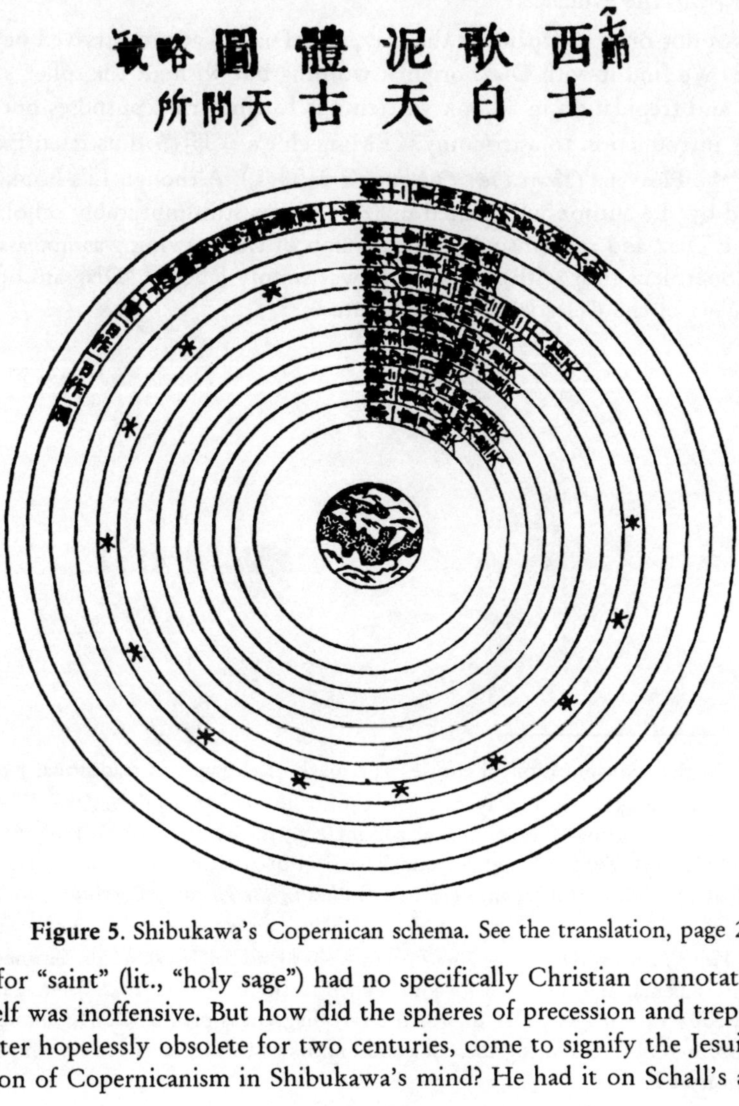

Figure 5. Shibukawa's Copernican schema. See the translation, page 24.

word for "saint" (lit., "holy sage") had no specifically Christian connotation and by itself was inoffensive. But how did the spheres of precession and trepidation, the latter hopelessly obsolete for two centuries, come to signify the Jesuits' description of Copernicanism in Shibukawa's mind? He had it on Schall's authority that Copernicus had modified Ptolemy's schema in some way that was significant, but at the same time not fundamental. In Shibukawa's search for the identity of this modification he came upon one that looked significant and had been wrongly but explicitly credited to Copernicus by Giacomo Rho in 1634, namely the addition of the two extra spheres. What, we can imagine him wondering, could merit the praise of that wise barbarian Schall more than placing two more spheres in the heavens?[34] Shibukawa can hardly be faulted for his deduction from the evidence.

It was not only in Japan, by the way, that Diaz' diagram survived beyond its time. We find it with Diaz' original wording but without the spheres of precession and trepidation in a book written not for historical purposes but as a popular introduction to astronomy, Li Ming-ch'e's 李明徹 Illustrated Explication of the Heavens *(Yuan t'ien t'u shuo* 圜天圖說). Although Li's book was prefaced by the author and printed in 1819, it was still impeccably scholastic, based on Diaz and related sources. Its version of the drawing was not associated with Copernicus (nor with Ptolemy). It was simply labelled "Diagram of the Ten Layers of the Celestial Sphere" (Figure 6).[35]

---

34. On the multiplication of entities as a mark of elegance in traditional Japanese astronomical thought, see Nakayama, "Shôchôhô no kenkyû 消長法の研究" (Studies on the cyclic variation of astronomical parameters), *Kagakushi kenkyû,* 1963, *66:* 68–84; *67:* 128–130; 1964, *69:* 8–17, and "Cyclic Variation of Astronomical Parameters and the Revival of Trepidation in Japan," *Japanese Studies in the History of Science,* 1964, *3:* 68–80.

35. *Yuan t'ien t'u shuo* (Sung Mei Hsuan 松梅軒 ed. of 1819), *1:* 5b. Li knew of the Tychonic schema, and discussed its equivalence to the Ptolemaic in *2:* 3a-4a. I have not found a copy of *T'ien wen lueh* in which Diaz' diagram is printed clearly enough to reproduce here, but taken together Li's and Shibukawa's diagrams convey an accurate impression of it.

Putative Copernicans

Historians have claimed that various Jesuits not employed in the Calendrical Office may have been Copernicans in the seventeenth century.[36] Just as we have seen regarding those in the palace, there is ample evidence to the contrary. Of the earliest, the Bohemian Wenzel Kirwitzer (1588/90–1626), there is no doubt that he accepted the heliostatic view of the universe, although his latest known statement on the matter was written (in a Latin letter) before 1616. He wrote nothing on astronomy in Chinese.

渾天十重圖

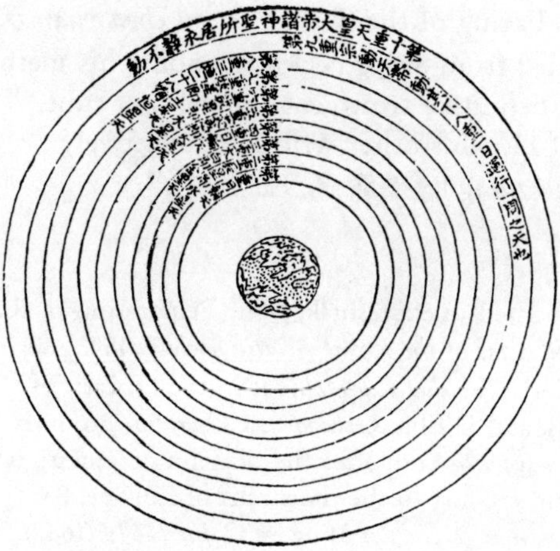

Figure 6. Li Ming-che's Ptolemaic schema.

Some are persuaded that the Pole Michael Boym (1612–1659) was a Copernican, on the evidence of a letter sent under an assumed name ("Miguel Polacco") from Macao in December 1646. In this letter Boym transmitted a copy of Kepler's *Tabulae Rudolphinae,* with a request that it be held until he or his compatriot Nikolaus Smogułęcki ("Jean Nicholas") arrived at Peking and picked it up. But the book lay in the Pei-t'ang Library until late in the twentieth century. No evidence has been presented that Boym valued it more for Kepler's heliocentricism than, as he himself put it, for its "inestimable value in calculating partial and complete solar eclipses, together with celestial movements"—for which much of the credit goes to Tycho Brahe's peerless

36. D'Elia, *Galileo in China,* 25–28 and 53, summarizes and adds to what has been written on this point. I sense that a reason for raising this issue was to refute charges that prejudice against the new science pervaded the Society of Jesus; or, as D'Elia puts it, to correct "a current opinion which would make all Jesuits think alike." Neither the charge nor the attempt to refute it by making the missionaries Copernicans has any merit.

data.[37] Whatever Boym's private beliefs may have been, they were never food for Chinese astronomical thought.

Smogułęcki (1610–1656), the last of these putative Copernicans, left an ample heritage that his disciple Hsueh Feng-tso 薛鳳祚 (ca. 1620–1680) published in Chinese from 1653 on. Hsueh, the only Chinese pupil of Smogułęcki who published anything, had studied traditional calendrical astronomy with the conservative Wei Wen-k'uei 魏文魁 beginning at the end of 1633. Hsueh says "I came to Nanking—this was twenty years later—and was further able to study trigonometry and also logarithms with Nikolaus." In another place he says: "In 1653 I wrote True Foundations of the Pacing of the Sky (*T'ien pu chen yuan* 天步眞原) according to what I had learned from Smogułęcki, emending his methods in many respects and giving them definitive treatment for the first time."[38] We also have a large printed corpus of Hsueh's other writings, collectively called Eclectic Astronomy (*Li hsueh hui t'ung* 曆學會通, ca. 1663).[39]

---

37. Boym's letter of transmittal is translated in Szcześniak, "The Penetration of the Copernican Theory into Feudal Japan," *Journal of the Royal Asiatic Society,* 1943, 60; see also the same author's "Note on Kepler's *Tabulae Rudolphinae* in the Library of Pei-t'ang in Pekin," *Isis,* 1949, *40:* 344–347, and D'Elia, *Galileo in China,* 53. Boym is chiefly remembered as the author of *Flora sinensis* and, after the Manchu conquest, as the emissary of the beleaguered Ming refugee court to the monarchs of Europe. See Robert Chabrié, *Michel Boym, jésuite polonais et la fin des Ming en Chine (1646–1662). Contribution à l'histoire des missions d'Extrême-orient* (Paris: Ed. Pierre Bossuet, 1933), and, for his contributions to European botany, Szcześniak, "The Writings of Michael Boym," *Monumenta Serica,* 1955, *14:* 481–538. There is an informative essay review of Chabrié's book by Paul Pelliot, "Michel Boym," *T'oung Pao,* 1935, *31:* 95–151, and a translation under the title *Pu Mi-ko* 卜彌格, tr. Feng Ch'eng-chün 馮承鈞 (1949; reprint, Taipei: Commercial Press, 1960).

38. Introduction to *Chung fa ssu hsien* 中法四綫 (Chinese trigonometry), I, 1b-2a; preface to the *K'ao-yen* 考驗 (Verification) section, VIII, 2a-2b, in *Li hsueh hui t'ung* 曆學會通 (Harvard-Yenching Library copy). Cf. Li Yen, *Chung suan shih lun ts'ung,* III, 46, 546. The main account of the collaboration of Smogułęcki and Hsueh is in Mei Wen-ting's bibliography *Wu-an li suan shu-mu* 勿庵曆算書目 (1702; in *Chih pu tsu chai ts'ung-shu* 知不足齋叢書, 1st ed., vol. LXXII), 33b-34b. Mei stated that Hsueh "never became a member of the Catholic Church." D'Elia claims (53) that the Polish missionary "made some proselytes among the Chinese scholars," but the source cited in D'Elia's footnote does not give pertinent data.

39. In the twelve-volume copy that I have had an opportunity to read, the first preface, 11a (in vol. I), is dated 1662, and the preface to Hsueh's table of logarithms in vol. VII, the beginning of 1663. Books by Hsueh bearing the title *Li hsueh*

Cosmology plays a negligible part in Hsueh's books, parts of which may have been dictated by Smogułęcki. These highly technical treatises, for the first time in China, applied spherical trigonometry and logarithms to calculate ephemerides. They reflect more detailed knowledge of European astronomers than could have been derived from previous publications in Chinese. For instance, Hsueh stated that an eclipse of 74 or early 73 B.C. is identically calculated according to Ptolemy or Copernicus (Ko-po-ni 歌白泥), and that another of 1509 or early 1510 "was recorded by Copernicus." Hsueh also mentioned computational advances made by Tycho, and recognized that the early Jesuit astronomical writers depended upon him: "The esoteric aspects of the contemporary Western method were explained by the Far Westerners Schall and Rho. Their principles were recondite and their computations subtle, but there remains room for deliberation. Their methods were created by the Western savant Tycho. The Westerners considered him an eminent authority on the fixed stars, but his treatment of eclipses and such was never highly regarded in Western astronomy. His work is now sixty or seventy years old, and his rules of computation are still not exhaustive. Subsequently there were the methods of Nikolaus..."[40]

What little cosmology one finds not only is not heliocentric, it is not even discernibly Tychonic. It is impossible to tell whether this is because Smogułęcki took the same agnostic position with respect to cosmology as Schall, or because Hsueh was interested only in mathematical methods. Perhaps it is significant that the great amateur and patron Juan Yuan 阮元 (1764–1849), in his biographical survey compiled at the end of the eighteenth century, as well as Mei Wen-ting a century earlier, criticized Hsueh for concentrating too narrowly on tech-

---

hui t'ung range in length between 1 and 61 *chüan.* Some citations seem to apply the synonymous title *T'ien hsueh hui t'ung* 天學會通 to the same writings, but others use it for a distinct book in one *chüan.* See, for instance Chou Yun-ch'ing 周雲青 & Ting Fu-pao 丁福保, *Ssu pu tsung lu t'ien-wen pien* 四部總錄天文編 (General bibliography, section on astronomy; Shanghai: Commercial Press, 1956); *Ch'ing shih kao* 清史稿 (Draft standard history of the Ch'ing period, 1927; Lien-ho Shu-tien ed.), 579a; Li Yen, ibid., II, 277–278, and *Chung-kuo suan hsueh shih,* 205. In some versions many chapters are devoted to European astrology. No generalization about *Li hsueh hui t'ung* will be feasible until Hsueh's printed and MS writings in Peking, Rome, Tokyo, and other places have been compared.

40. See, for instance, the pre-Tychonic diagram and its caption in *Li hsueh hui t'ung,* I, *4:* 1a. On eclipse predictions see *Chin hsi fa hsuan yao* 近西法選要 (Selected essentials of the modern Western methods), section *T'ai yin* 太陰 (The moon), XI, 21a, and, on Tycho, the first preface, I, 10a.

nique without attending to its ramifications. Juan also thought Hsueh's translation poor, and noted that, lacking experience at observation, he had accepted Smogułęcki's theories ready-made.[41]

Yabuuchi Kiyoshi has noted that, in an annotation to the section on the Earth of Fang I-chih's 方以智 (1611-1671) compendium of notes on physics and natural history, his son Fang Chung-t'ung 中通 observed that "Master Smogułęcki also had a theory of the excursions of the earth" *(ti yu chih shuo* 地游之說).[42] It is tempting to see in this annotation a reference to the earth's diurnal rotation, particularly since it follows a passage that ends "like a man sitting in a boat; the boat is moving but the man is unaware of it." This is the famous argument for the relativity of motion used in Europe both to reject and to support the idea of the earth's daily turning on its axis (see above, p. 17). But Fang I-chih was applying the boat simile to the apocryphal Chinese idea of an annual cycle of terrestrial rising and subsidence, which had been formulated in terms of the ancient *kai t'ien* 蓋天 (literally, "the sky as a cover") model of the universe.[43] This cycle he was relating in turn to seismology.

Fang Chung-t'ung's annotation outlines Smogułęcki's theory and its context, typical of late scholastic physics: "The interior of the earth is mostly hollow, with *ch'i* 氣 moving about inside. Small movements [i.e. tremors] are normal occurrences everywhere. When they are intense, the ground caves in and mountains shift. I often reflect that Chang Heng 張衡 made a bronze-dragon seismoscope in order to respond to these movements of the earth, but I do not know what this instrument was."[44]

---

41. *Ch'ou jen chuan*, 449–450.

42. Yabuuchi, "Min Shin jidai no kagaku gijutsushi," in the book of the same title (see note 15), 23. The passage discussed below is cited from *Wu li hsiao chih* 物理小識 (Little notes on the principles of the phenomena; Ning ching t'ang 寧靜堂 ed. of 1884), *2:* 22b-23a. This book contains notes written from 1631 on; Fang I-chih wrote his preface to the finished manuscript in 1643; he sent the MS to his son Chung-t'ung in 1650; and it was printed in 1664 (see note 12 above).

43. The *kai t'ien* cosmology is clearly explained in Nakayama, *History of Japanese Astronomy,* 24–35. For more detail, see the book on early cosmology by Christopher Cullen forthcoming from Cambridge University Press.

44. For a reconstruction of Chang Heng's earthquake-registering device see André Wegener Sleeswyk & Sivin, "Dragons and Toads. The Chinese Seismoscope of A.D. 132," *Chinese Science,* 1983, 6: 1–19. "I" refers to the annotator. For writings on the scholastic sciences of the earth available to Chinese thinkers, see note 9 above.

To sum up, although it is likely that individual members of the Society of Jesus in China would have accepted the idea of a solar system sooner or later if they had been allowed to do so, there is no serious evidence that Boym or Smogułęcki believed in the heliocentric theory, that they or the Copernican Kirwitzer made any contribution to the dissemination of the new world model, or that any of the three (not being members of the Calendrical Office) would have written on the subject in Chinese even if the injunction of 1616 had not intervened.

Jesuit Astronomical Writings in the Early Ch'ing (1646–1760)

The frustrations of the European astronomers ended with the end of the Ming. As the Manchus took over China, the Jesuits quickly attained the goal that had prompted them to disseminate the earlier astronomical treatises. Thanks to Schall's prompt support for the alien invaders, the Manchus retained him on a stipend within two weeks of their march into Peking in mid-1644, and gave him control over the official calendar and the Directorate of Astronomy before the end of 1645. He filled the new regime's need for a reliable calendar, traditionally one of the ritual underpinnings of dynastic legitimacy. The missionary was even able to force his old rivals, the Chinese and Islamic incumbents of the Directorate, to take an examination in Western astronomy—and soon fired the Muslims en masse. To accommodate an ephemeris of foreign origin, officials revised the ritual by which the Emperor bestowed the calendar at the New Year, asserting his control over the seasonal activities of his people. The Jesuits from that time on were in an ideal position to make converts among highly placed Chinese, in fact the only strategic position open to them.[45]

In the Ming the Jesuits had had to print their astronomical treatises privately for instructional use (the exemplars presented to the throne were manuscript copies). But under the Ch'ing these works, which usually had been collectively referred to as the Astronomical Treatises of the Reverence for Auspices Reign Period (*Ch'ung-chen li shu* 崇禎曆書), were published together for the first time by Imperial order as the Astronomical Treatises according to the New Methods of the West (*Hsi-yang hsin fa li shu* 西洋新法曆書, presented 1646).[46]

---

45. Huang I-nung 黃一農 has reconstructed this sequence of events in "T'ang Jo-wang yü Ch'ing ch'u hsi li chih cheng-t'ung-hua."

46. Bernard surveyed its history and content in "L'Encyclopédie astronomique." The Ming treatises were presented to the throne in five lots between 1631 and 1635. Some of these books were not printed before the Ch'ing, and others written after 1635 were added to the *Hsi-yang hsin fa li shu.* See, for example, notes 19 and 28 above. The

For nearly a century following the publication of the 1646 collection, no significant modern developments in world view were brought to the attention of Chinese astronomers.[47] What was taught privately to those working in the court, of course, we do not know. Only when rivals again threatened the Jesuits' position in the Directorate of Astronomy were they motivated to publish on the state of observational and computational astronomy. The writings of Ferdinand Verbiest (1623–1688), a skilled astronomer who came under strong attack from traditionalist zealots, concentrated mainly on tables, descriptions of astronomical instruments, and accounts of competitions in which the Europeans demonstrated their superiority over second-rate Chinese and Islamic astronomers by predicting eclipses and other phenomena. With respect to cosmology, Verbiest merely reprinted with minor revisions ca. 1669 some of the treatises from the 1646 collection. This last printing was generally referred to as the

---

question of whether the Jesuit treatises were printed collectively once, twice, or three times—in other words, whether *Ch'ung chen li shu* and *Hsin fa suan shu* (see below) are nothing more than conventional ways of referring to all the individual works printed separately and privately at the end of the Ming and about 1669 respectively—badly needs settling. I have found no evidence that there was ever an integral publication of either title. To give a single example, the manuscript of 1670 in 125 volumes entitled *Ch'ung chen li shu* in the Tenri Library contains post-Ming materials, and was patently copied from the *Hsi-yang hsin fa li shu.* An officially printed *Ch'ung-chen li shu* is occasionally cited (among several references I have collected is Li Yen, *Chung suan shih lun-ts'ung*, III, 37) but never with sufficient detail or precision to rule out the separately printed treatises and the 1646 collection.

Nearly four hundred pages at the beginning of the latter collection are devoted to documents concerning the Jesuits' activities in the palace in the Ming and at the beginning of the Ch'ing. Particularly informative evidence about the Ch'ing publication is given on 48b-49b and 59ff. The presentation of the completed compilation took place on 7 January 1646, and the order authorizing its use was dated 8 February. The accounts of Jesuit activities in the Standard History of the Ming (*Ming shih* 明史) and its draft (*Ming shih kao* 明史稿) were mainly based on documents in this collection. For an evaluation of the accounts in the two histories, see the long review of D'Elia's *Galileo in Cina* by J. J. L. Duyvendak in *T'oung Pao,* 1948, *38:* 321–329, especially 323–327.

47. Yu I's 游藝 *T'ien ching huo wen* 天經或問 (Queries on the astronomical classics, 1675) is a synthesis of traditional and Western knowledge, the latter derived from Hsiung Ming-yü 熊明遇 (Presented Scholar 1601), a friend of Diaz. Yu's superficial and confused book, which describes the Tychonic planetary system, had great influence in Japan but practically none in China. For a description, see Nakayama, *History of Japanese Astronomy,* 100–105.

Mathematical Treatises according to the New Methods (*Hsin fa suan shu* 新法算書), in order to deemphasize the book's foreignness.[48]

When in 1722–1724 a large editorial committee brought together the Compendium of Observational and Computational Astronomy *(Li hsiang k'ao ch'eng* 曆象考成*)*, the writings of Rho, Schall, and their colleagues remained the best available sources.[49] The Compendium was a Chinese project, meant to lay the basis for a traditional calendar reform by adapting the Jesuit writings—and necessarily reconciling the discrepancies between the views of Rho, Schall, Schreck, and their colleagues. But these sources were nearly a century old. The most active members of the large editorial committee, Mei Ku-ch'eng 梅瑴成 (1681–1763) and Ho Kuo-tsung 何國宗 (d. 1766), were among the experts of their time in Western astronomy, but they were not as up to date as Mei Wen-ting and his colleagues had been. They were unable to supplement the seventeenth-century Jesuit observations with enough new ones to meet contemporary expectations for accurate solar eclipse prediction. Their cosmology remained Tychonic.[50]

The Compendium became the basis of the annual ephemeris (*Shih hsien shu* 時憲書) from 1726 on. When an eclipse prediction failed in 1730, it became clear that the Chinese astronomers could not substantially improve their methods. The Jesuit head of the Directorate of Astronomy, Ignatius Kögler (1680–1746), and his assistant Andrew Pereira (1689–1743) were ordered to revise the Compendium. According to the memorial of the Manchu Ku-ts'ung 顧琮 proposing this revision, Kögler had earlier appended to the Compendium two substantial tables of solar and lunar motions incorporating post-Newtonian data. But he provided these tables with neither explanations nor instructions for com-

48. Because Bernard saw a book entitled *Hsin fa suan shu* in the Peking Imperial Palace, he believed that this was the collective title of Verbiest's reprints. Since Bernard was unable to examine the book carefully, it remains likely that a librarian, as librarians often do, arbitrarily chose this title for the binders of a collection of individually printed treatises. I know of no evidence that Verbiest gave his group of reprints a title. On Verbiest, see R. Josson & L. Willaert, *Correspondance de Ferdinand Verbiest de la Compagnie de Jésus (1623–1688), directeur de l'observatoire de Pékin* (Bruxelles: Palais des Académies, 1938), and Noel Golvers (ed. & trans.), *The "Astronomia Europaea" of Ferdinand Vierbiest, S.J. (Dillingen, 1687).* Monumenta Serica Monographs, 28 (Nettetal: Steyler Verlag, 1993).

49. Hashimoto, "Rekisho kosei no seiritsu." *Li hsiang k'ao ch'eng* was one of the three parts of the K'ang-hsi Emperor's (r. 1662–1722) spacious survey, Sources of Mathematical Harmonics and Astronomy (*Lü li yuan yuan* 律曆淵源).

50. See, for example, *Li hsiang k'ao ch'eng*, A, *9:* 7a.

puting. According to Ku-tsung only Kögler, Pereira, and the Mongol mathematician Minggantu 明安圖 (ca. 1692–1763) were able to use them. The tables do not seem to have been printed with the Compendium.[51]

In 1737 these three and others were ordered to revise the Compendium. In 1742 their sequel *(Li hsiang k'ao ch'eng hou pien* 曆象考成後編) was printed. It definitively treated the solar and lunar motions in order to predict eclipses accurately. The Compendium's account of the planets was allowed to stand intact. This work is an anomaly of the most remarkable sort. On the one hand, it incorporated improved constants and tables based on the observations of J. D. Cassini (1625-1725) and John Flamsteed (1646-1719), and introduced the Keplerian ellipse. It informed the Chinese reader that, since epicycles and equants were unsatisfactory for computing the solar anomaly, Kepler "posited that the deferent was an ellipse, and took equal areas of the ellipse in order to derive the daily mean motion." The compilers stated elsewhere that "the ellipse method precisely reconciles eccentrics, epicycles, and equants."[52]

Nonetheless, sixteen years after the death of Newton, thirteen years after the announcement of Bradley's discovery of stellar aberration, heliocentricism was still tabooed in China. Kepler's old master Tycho Brahe still furnished the scenery. The earth was static, with the sun revolving about it on what Kepler would have considered the earth's elliptical orbit turned inside out.

Given the necessity to relate Kepler's first and second laws to geocentric coordinates and a geostatic world system, the geometry of the ellipse could not be extended to the planetary orbits. While the sun and moon rode about the earth on ellipses, the planets were still making their rounds of the sun on the cranky epicycles of Tycho. As Hashimoto Keizô has put it, "a determination to emphasize the continuity of [ellipses and compounded circles] can be discerned throughout the book. . . . The heterogeneous circular and elliptical forms of motion were transmuted into homogeneous forms as a means toward a prag-

---

51. The memorial is given in the forematter of *Li hsiang k'ao ch'eng hou pien* 曆象考成後編 (1742; Li chih shu ya reprint of 1896), "Memorials," 4a-4b. On this compilation see Hashimoto, "Daenhô no tenkai"; Yabuuchi, *Chûgoku no temmon rekiho*, 164–168; Needham, *Science and Civilisation*, III, 448; and Hummel, *Eminent Chinese of the Ch'ing Period*, 285, a confused account. Li Ti 李迪 has written a biography of the scientific polymath Minggantu, whom that august amateur the K'ang-hsi Emperor taught astronomy: *Meng-ku-tsu k'o-hsueh-chia Ming-an-tu* 蒙古族科學家明安圖 (The Mongol scientist Minggantu; Hohhot: Nei-meng-ku Jen-min Ch'u-pan-she, 1978).

52. *Li hsiang k'ao ch'eng hou pien*, *1:* 25b-26a, *1:* 103b. On the ambiguity of these astronomical terms see the Appendix, p. 51.

matic grasp of Keplerian cosmology. One can see in the Sequel that, considered as a form of motion *compatible with classic circularity,* the elliptical method could be set out as merely a more precise technique that agreed better with observation."[53]

Just so. In 1742 the Sequel treated elliptical orbits as perfectly compatible with the Platonic eternal circles because it ignored their physical meaning. They were included merely because it was impossible to introduce important post-Newtonian developments in eclipse prediction without them. The Keplerian third law was not mentioned, and Newton figured only as a source of data.[54]

These limitations were not at all disabling. Accurately predicting planetary phenomena had always been a peripheral issue in the minds of Chinese astronomers. On the other hand, the Sequel succeeded for the first time in substantially solving the central problem of eclipse prediction. Even a century later it became the basis of a calendar reform (epoch 1834).

Tycho Brahe expired hoping in vain that his data, once in Kepler's hands, would establish the Tychonic world system. It is one of the minor ironies of the history of science that this wish came true for a time on the other side of the world after both were long dead, and in circumstances that neither could have imagined.

Introduction of the Copernican World Model

At the very end of the long chain of Jesuit scholar-missionaries, the heliostatic world model finally received a hearing in China. Its champion was Michel Benoist (1715–1774), a very competent astronomer.[55]

---

53. Hashimoto, "Daenhô no tenkai," 265, my emphasis.

54. Nakayama, *History of Japanese Astronomy,* 189.

55. The most extensive study of Benoist's writings is in Pfister, *Notices biographiques et bibliographiques,* item 377, pp. 813–826.

I believe it is to Bernard that we owe, among many other debts, that of pointing out to Europeans Benoist's importance, in his *Galilée et les Jésuites.* Presumably because by the time he wrote this article he had had no opportunity to study the Chinese treatises in detail, however, Bernard's interpretation is untenable: "As to the theories [of Galileo], they were practically useful only in the capacity of hypotheses to act as a guide in calculations, and the Chinese, who could know of geometry only what concerned straight lines, triangles, and the circle (to the exclusion of conic sections: ellipse, hyperbola and parabola), were thus incapable of understanding the laws of Kepler" (p. 380). On Benoist's introduction of the heliocentric theory Bernard asserted: "This resolution could hardly have had repercussions beyond the walls of the imperial residence,

A series of clashes between the Jesuits and members of other missionary orders in China, centering on the extent to which converts should be permitted to maintain traditional beliefs and customs—the famous Rites Controversy—came to a head at the beginning of the eighteenth century. A papal legate was sent from Rome to ask the Chinese Emperor to stop siding with the Jesuits. The arrogant conduct of the poorly prepared emissary and his party in 1706 and 1707 made all too obvious what had been decently concealed through the seventeenth century: unlike the priests of other religions in China, the missionaries were obedient agents of a foreign power. The royal patrons of their order could wield them to subvert imperial authority.

By the mid-eighteenth century, despite the missionaries' attempts to recover the good will of the Emperor, the government was taking harsh measures against them outside the court. As a Jesuit historian has put it, "because they entered the Empire illegally and stayed there against the most explicit prohibition of the Emperor, the missionaries in the provinces were considered criminals, and, in the Chinese Empire, nobody could ask for favors for such persons."[56] A small group of technicians were still allowed to reside in the capital

---

for the little old manual of Fr. Manuel Diaz, an abridged compendium of the Ptolemaic doctrine, remained almost the sole source of information for the savants of China. It was only more than a hundred years later, when the Elements of Euclid had been printed *in toto* (1865), that the principles of Galileo and Kepler could be divulged in the Far East" (p. 382).

We know, however, that Diaz' book was obsolete *as a source for Chinese* by 1635. The geometry of the ellipse was introduced in Schall's Tychonic *Chiao-shih li chih* 交食曆指 (Guide to the astronomy of eclipses, 1632). It was more fully treated in Verbiest's *Ling-t'ai i hsiang chih* 靈臺儀象志 (Treatise on Imperial Observatory instruments, 1674) and in the imperially commissioned mathematical encyclopedia *Shu li ching yün* 數理精蘊 (Treasury of mathematical principles, 1723). In 1742 *Li hsiang k'ao ch'eng hou pien* applied it to the solar and lunar orbits, with credit to Kepler. Every one of these books was printed and widely available, as were the other treatises on Tychonic astronomy and cosmology in the *Hsi-yang hsin fa li shu.*

56. Joseph Krahl, S.J., *China Missions in Crisis. Bishop Laimbeckhoven and his Times 1738–1787* (Analecta Gregoriana, 137; Roma: Gregorian University Press, 1964), 89. A useful work for the doctrinal history of the Rites Controversy is François Bontinck, *La lutte autour de la liturgie chinoise aux 17e et 18e siècles* (Louvain: Éditions Nauwelaerts, 1962). There is valuable information on the astronomical work of the mid-eighteenth century in Antoine Gaubil, S.J., *Correspondance de Pékin, 1722–1759* (Genève: Librairie Droz, 1970). On Benoist's observations in Peking, for instance, see 569, 732, 787, 832, and 843. One is left with the impression that despite skill and enthusiasm Benoist had little time for astronomical work except in 1755–1756. There is more information on

and serve the palace in such fields as astronomy, cartography, clockmaking, and optics. In the hope of moderating this persecution, Benoist, one of the group at the court, built fountains and ornamental waterworks (what George Kates has called Européenerie) to keep the Emperor amused.[57]

Benoist was able to write on Copernican cosmology because the Church's formal ban on discussion of heliocentricism ended in 1757. Considering the time required for the news to reach Benoist, his alacrity in taking advantage of his new freedom is impressive. Here is his own description, in a letter to Europe (1764), of the occasion and how he rose to it:

> I have already written you that, besides the hydraulic works with which I have been occupied for several years in the Emperor's service, I have drawn a world map of which the two hemispheres with their margins are thirteen to fourteen feet in length by seven high. In 1761,[58] when the fiftieth birthday of the reigning Emperor was being celebrated, I presented him with it, traced on [silk] gauze, which is very convenient for this sort of work here. His Majesty received my present with kindness, and kept me nearly an hour to ask me various questions on geography and physics. In the margins are drawn different mathematical diagrams, with some space left free in order to put there an explanation in Chinese of both the map and the diagrams. In my explanation I made a rather thorough exposition of the Copernican system, which was necessary because the Chinese had not yet adopted it.[59]

In another letter written three years later, Benoist offered more detail about the content and sources of his explanation. At the same time he provided a hint about the political motivation of his birthday gift to the Ch'ien-lung Emperor. The significance of this hint is clear in view of the intense rivalry between the French and Portuguese missions of the Society in Peking, for this competition reflected the national strivings of their royal patrons:

---

the period in Camille de Rochemonteix, S.J., *Joseph Amiot et les derniers survivants de la mission francaise à Pékin (1750–1795)* (Paris: Librairie Alphonse Picard et fils, 1915).

57. George N. Kates, *The Years That Were Fat. Peking: 1933–1940* (New York: Harper and Brothers Publishers, 1952), 197–200, and Maurice Adam, *Yuen ming yuen. L'Oeuvre architecturale des anciens Jésuites au XVIIIe siècle* (Pei-p'ing: Imprimerie des Lazaristes, 1936), 20–22, give information on Benoist's hydraulic works. Benoist "had at least one meal every day at the palace in the company of mandarins and nobles" (Krahl, 195).

58. This is an error for 1760. See the *Shih lu* 實錄 (Veritable records) for the Emperor Kao-tsung, *618:* 8, dated 19 September 1760.

59. Letter from Benoist at Peking to Souciet, 12 September 1764, in "Histoire ecclésiastique de l'Extrême-Orient," *Revue de l'Extrême-Orient*, 1887, *3:* 248–250.

> I have added an explanation of the terrestrial as well as the celestial globe, of the new systems for the movement of the earth and of the other planets, and of the movements of the comets, the returns of which one hopes to succeed in predicting with certainty. I summarized the great enterprises ordered by our monarch for the perfection of the arts and sciences, and especially for that of geography and astronomy, which were the subject of my writings. I described the expeditions sent to different parts of the world to observe various astronomical phenomena, to measure exactly the degrees of longitude and latitude of our globe; the men of merit whom he sent for these observations; the welcome they were given in various kingdoms . . . I cited MM. Cassini, La Caille, Le Monnier, et al., from the learned writings of whom I have taken all that I have said in mine.[60]

Benoist's explication of the heliostatic concept was short but ample, and on the whole accurate. Considerably more space in the margins of his map was devoted to astronomy, in fact, than to geography. His text contained no mathematics, but then it was meant for the Emperor rather than for the Chinese officials of the Directorate of Astronomy. Under its Jesuit Directors they had already been using the post-Newtonian constants and tables of the Sequel since 1742, and had had access to gear-driven heliocentric demonstrational instruments before Benoist's book appeared.[61] After a committee appointed to put Benoist's explanation into acceptable literary form spent nearly two years discussing the matter, Ho Kuo-tsung and Ch'ien Ta-hsin 錢大昕 (1728–1804) gave its style a final polishing and had it copied onto a redrawn map. They also made it possible to circulate the manuscript (without the illustrations) among some of China's leading astronomers.

In 1799 Juan Yuan copied the manuscript's account of Occidental cosmology into an article on Benoist in his Biographies of Mathematical Astronomers

---

60. Letter from Benoist in Peking to Papillon d'Auteroche, 16 November 1767, in *Lettres édifiantes et curieuses concernant l'Asie, l'Afrique et l'Amérique, avec quelques relations nouvelles des missions, et des notes géographiques et historiques* (Paris: Société du Panthéon Littéraire, 1843), IV, 121.

61. See Hsi Tse-tsung et al., "Heliocentric Theory in China—in Commemoration of the Quincentenary of the Birth of Nicolaus Copernicus," *Scientia sinica*, 1973, *16:* 264–376, for photographs of a heliocentric orrery and planetarium. They are also discussed in Liu Ping-sen 刘炳森 *et al.*, "Lueh t'an Ku-kung Po-wu-yuan so ts'ang ch'i cheng i ho hun-t'ien ho ch'i cheng i. Chi-nien Ko-po-ni tan-sheng wu-pai chou-nien 略談故宮博物院所藏七政儀和渾天合七政儀。紀念哥白泥誕生五百周年" (The orrery and armillary sphere in the Palace Museum. In commemoration of the 500th anniversary of the birth of Copernicus), *Wen-wu* 文物, 1973, 9: 40–44, pl. 6–7. There is a brief discussion in the original version of this chapter. See the Retrospect.

(*Ch'ou jen chuan* 疇人傳). This large book gave much more information about astronomers' technical work (often copying long excerpts from their writings) than about their careers.

Benoist's essay was published separately in 1802 or 1803 under Juan's sponsorship. For this edition, entitled World Map with Illustrated Explications (*Ti ch'iu t'u shuo* 地球圖說), Li Jui 李銳 (1765–1814) and Chiao Hsun 焦循 (1763–1820) reconstituted the illustrations on the basis of careful study of the text.[62] That they succeeded in doing so confirms that Benoist's account of heliostatic cosmology was lucid enough for Chinese outside the missionaries' ambit to understand it.

Because of the interest of its details, I translate below Benoist's account of four European cosmological systems, those of Ptolemy, Tycho Brahe, Martianus Capella (Ma-erh-hsiang and Copernicus, and his defense of the Copernican.[63] This is by no means all Benoist had to say about Copernicus, for his treatise is heliocentric from beginning to end. In this excerpt from his section on "The Order of the Seven Luminaries" there are many clues to his motives and intentions:

> From antiquity on, astronomers have developed in great detail the methods of calculating the motions of the Seven Luminaries [i.e. the sun, moon, and planets]. The scholars of the West, making full use of their intellectual acuity, have hypothesized orders of the luminaries in the universe on the basis of their own computations, and have formed different traditions *(chia*

---

62. *Ti ch'iu t'u shuo* is also referred to as *K'un yü ch'üan t'u shuo* 坤輿全圖說, a slightly more literary way of saying the same thing, and possibly Benoist's original title. On the publication date of this work, see the remarks of Chiao Hsun in his collection of jottings *I yü yueh lu* 藝餘籥錄 (Jottings stored in bamboo tubes *[or* written on odd pieces], set down in intervals between studies of the Book of Changes, printed 1886), *6:* 2a. Chiao provided what became Figure 10 of Benoist's book. My account of the vicissitudes of Benoist's manuscript is based upon Chiao's book, the prefaces to *Ti ch'iu t'u shuo*, and the letters of Benoist cited above and further on. For the biography of Juan see Wolfgang Franke, "Juan Yüan (1764–1849)," *Monumenta Serica*, 1944, *9:* 53–80. Another good source of information and references concerning Juan and other Ch'ing astronomers is Hummel, *Eminent Chinese*, s.v.

63. This translation (of about an eighth of Benoist's composition) is based on the *Ti ch'iu t'u shuo* version (in *Wen hsuan lou ts'ung-shu* 文選樓叢書), 8a-11a. The *Ch'ou jen chuan* version (see note 41 above, *46:* 601–610) does not contain significant variants. It differs only in that Example 2, near the end of the passage translated here, is appreciably shorter. Because he omitted the diagrams, Juan excised the text that originally explained them.

家), each with its own argument.[64] Let me take up the four most important founders *(tsung* 宗), but only to the extent necessary to determine [which tradition best corresponds to?] the motions of the luminaries.

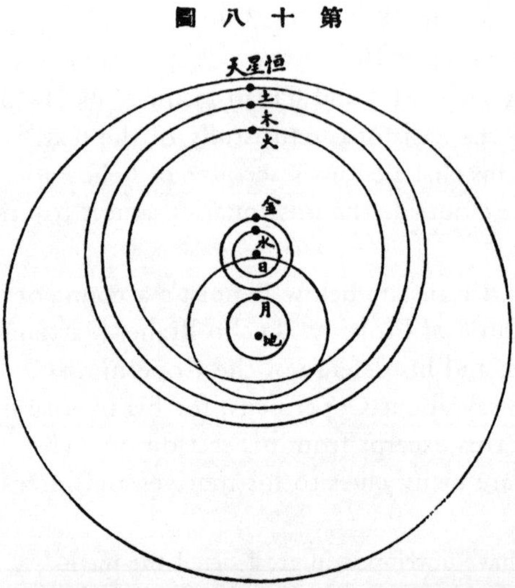

Figure 7. Benoist's scheme for Martianus Capella. The dot nearest the center is the earth. The sun travels on its second circle.

1. Ptolemy (To-lu-mou 多祿畝) claimed that the earth is the center of coordinates.[65] About the earth are the moon, Mercury, Venus, the sun, Mars, Jupiter, Saturn, and the fixed stars, each on its own deferent. All of these deferents [i.e. the spheres on which they are located] are solid; they do not join or intersect.[66] In addition to these deferents there are epicycles. Each of the Seven Luminaries travels on the periphery of an epicycle, and the center of the epicycle also travels on the periphery of the deferent. This argument, however, is inadequate to explain the principles underlying the motions of the Seven Luminaries, and today no one accepts it.

2. Tycho (Ti-ku 的谷) claimed that the earth is the center of coordinates, and that about the earth are the moon, the sun, and the fixed stars. Each has its deferent [i.e. sphere] that rotates about the earth. The deferents of the planets Mercury, Venus, Mars, Jupiter, and Saturn have the sun as their center, and on each deferent is an epicycle.

3. Martianus Capella (Ma-erh-hsiang 瑪爾象, ca. 365–440, Figure 7) held that the earth is the center of coordinates. Without leaving its place, each

---

64. Benoist naturally used the Chinese concept of *chia* to explain the relationship between the different cosmological doctrines. *Chia* were traditions of philosophy, cosmology, or for that matter any field of scholarship, often lineages based on the special teachings of a master passed down through generations of disciples. Benoist's Chinese readers would ordinarily have expected persuasions so described to coexist peaceably.

65. Literally, "the center of the Six Directions," which include the cardinal directions, up, and down.

66. Benoist's use of the term *pen lun* 本輪 is ambiguous. See the Appendix, p. 52.

day it makes one rotation about its north and south poles. About it are the moon, the sun, and the fixed stars. Rotating about the sun are the circles of Mercury, Venus, Mars, Saturn, and Jupiter.

Although in the teachings of these two traditions there are points worth

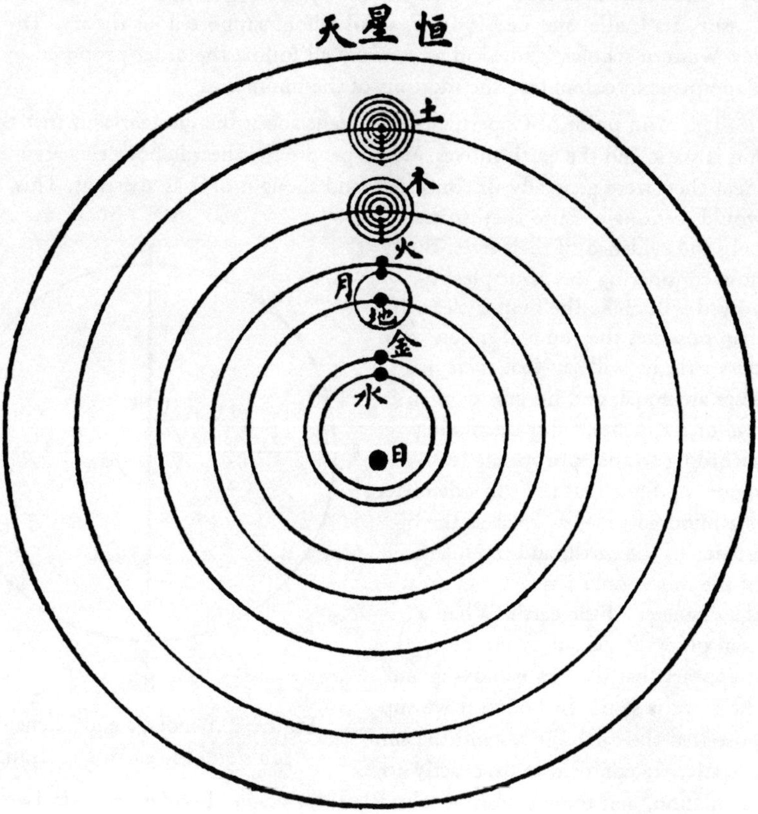

Figure 8. Benoist's schema for Copernicus.

taking up, they do not compare with the tradition of Copernicus in precision *(mi* 密).

4. Copernicus 歌白尼 (Figure 8) put the sun at the center of the universe. Closest to it is the planet Mercury, then the planet Venus, then the earth, then the planet Mars, then the planet Jupiter, and then the planet Saturn. The deferent of the moon circles the earth. Alongside the planet Saturn are five small stars that circle it. Alongside the planet Jupiter are four small

stars that circle it. Each has its own deferent that *[or* and] moves about its planet. Furthest from all these circles is the heaven of the fixed stars, eternally immobile.

The order Copernicus determined for the luminaries is, it seems, based on the argument of Nicetas, but Copernicus explained it with particular clarity. Among his successors were Kepler (K'o-po-erh 刻白爾), Newton, Cassini, La Caille, and Le Monnier, all of whom supported his theory. Today Western scholars skilled in astronomy all follow the order proposed by Copernicus in computing the motions of the luminaries.

The main point of Copernicus' argument about the luminaries is that the sun is static and the earth moves. When people first heard about this argument they were generally disconcerted, and thought of it as aberrant. This, it would seem, is because they trusted only the evidence of their eyes. But now considering the principles involved will make the matter clear. If a man observes the sun and moon from the earth, he will say that their diameters are equal, and no greater than five or six inches.[67] But calculating according to the appropriate technique, he finds that the sun's diameter is a hundred times as great as the diameter of the earth, and the diameter of the moon only a quarter as great as the diameter of the earth. When a man observes the sun from the earth, it appears that the sun is moving and the earth is static. But now, if we suppose that the earth moves and the sun is static, we conform more exactly to calculation, and there is also no obstacle in principle. Let me bring up two or three points by which the principles may be confirmed.

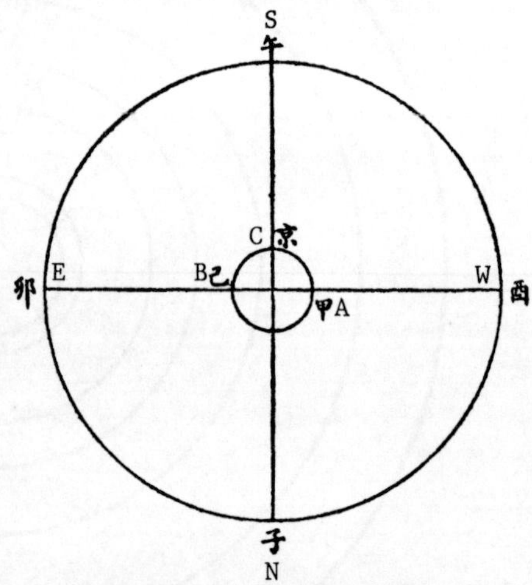

Figure 9. Benoist's equivalence of the sun's and the earth's motion.

First, when a man on the surface of the earth observes the motion of the luminaries, [they appear to] circle the earth, and the earth seems perpetually immobile. This cannot in fact be taken as evidence that the earth is still and the luminaries move. For example, consider a boat floating on the sea. When the people in it see that the relative distances of the things on the boat re-

67. Benoist or more likely his editors reverted to the ancient Chinese practice of estimating tenths of a degree of arc as *ts'un* 寸, literally "inches," decimal parts of a foot. The next sentence refers not to angular but to linear diameter.

main constant, they are not aware of the ship's motion. But when they see that the shore, mountains, islands, and things outside the boat are now nearer, now farther, now to the left, now to the right, they change their minds and suspect that it is in motion. In the present case, the earth and its surrounding air *(ch'i)*, unhindered, move equally. A man on the surface of the earth, observing the constant distances of the things around it, will be unable to sense the earth's rotation. Observing the luminaries beyond the earth, seeing that they are sometimes nearer, sometimes farther, sometimes to the left, sometimes to the right, he will say that the luminaries revolve about the earth.

Second, supposing that the earth moves and the sun is still, when one observes from the earth, it is bound to appear that the sun moves and the earth is still. But if we calculate the degree of the sun above or below the horizon according to both [hypotheses], the numerical results are bound to be the same. For instance, in Figure [9],[68] let *AB* be the horizon of the point *C* on the surface of the earth. *ESWN* is the circle described by the sun's westward motion about the earth. Supposing that the sun is at point *E*, from point *C* one will see the sun emerging from the horizon. As the sun goes from *E* to *S* it gradually rises. From *S* to *W* it gradually sets. When the sun reaches point *W*, from point *C* one will see the sun enter the horizon. The sun travels beneath the horizon, passing from *W* through *N* and again to point *E*, and again emerges on the horizon. This is according to the theory that the sun moves and the earth is static.

Now suppose that point *E* is the location of the sun, which is static, and that the earth travels rightward, revolving from west to east on its own center. Thus the points on the circle *ESWN*, located on the periphery of the sky, rotate in succession to coincide with point *C* on the earth. Thus as one observes the sun it appears to ascend, descend, rise and set on the horizon, not at all differently from the previous case. When point *C* in the figure corresponds to point *S* in the sky, one will see the sun on the horizon. As point *C* rotates toward *E*, the sun seems to ascend. When point *C* corresponds to point *E* in the sky, the sun seems to reach the meridian. When point *C* has revolved through a half-circle from *E*, [the sun's position] corresponds to point *N* in the sky, and the sun seems to set on the horizon. The remainder of the argument is similar in principle.

Third, the sun is itself a luminous body. The moon, Mercury, Venus, Mars, Saturn, and Jupiter are all dark bodies, which shine by borrowing the sun's light, as the earth does. Suppose there were men on the surface of the moon and the other planets. The earth as it appeared to them would resemble the moon seen from the surface of the earth—sometimes dark, sometimes

68. The diagram is originally Figure 19 in *Ti ch'iu t'u shuo*. See note 23 above.

> fully illuminated, sometimes an upper or lower crescent. Everyone familiar with astronomy knows this. Now since the six luminaries all resemble the earth, how could it be that the six luminaries and the sun circle the earth, which alone is still? It is best to suppose that the sun is at the center of the universe, and that the earth and the other planets circle the sun, borrowing its light. Is not this argument simpler?

Benoist then proceeded to discuss the Keplerian elliptical orbit and the proper rotations of other bodies as well as the earth, even asserting that "although the sun has no deferent, it rotates on its own center like the planets."[69]

The contrast with the prudence of the missionary astronomers writing eighteen years earlier is equally patent in what Benoist found no need to say. He did not attempt to prove the superiority of heliostatic cosmology. He simply asserted that it was the only system in current use, its ultimate authority deriving from "more precise conformity to calculation." The three arguments at the end are not meant to prove the Copernican doctrine, but merely to show that its conflict with common sense and everyday observation is only apparent. The boat metaphor for the relativity of motion was not new in China, as we have seen.

Although Benoist was free to maintain that the concept of a solar system was mathematically superior, he did not assert that it was *physically* true. As he put it in another letter: "It is not, I added, that we are sure the universe is actually arranged as we suppose it to be; we simply propose this arrangement as the one that appears most appropriate and simplest to account for the various movements of the heavenly bodies and to calculate them." In this letter, and yet another probably a bit earlier, Benoist praised (excessively, by all accounts) the astronomical aptitude and understanding of the Ch'ien-lung Emperor (r. 1736–1796), to whom he originally made the above remark.[70] The monarch greeted Benoist's explication of the earth's motion with these words: "In Europe you have your way of explaining the celestial phenomena. As for us, we have ours too, without making the earth rotate." If Ch'ien-lung was as usual smug and unimaginative, the Chinese members of the Directorate of Astronomy (who for centuries had been mostly careerists rather than expert astronomers) saw Copernicanism merely as a threat to their undemanding sinecures. As Benoist put it in 1767, "our Chinese mathematicians do not approve of all these changes. They have often heard of the movement of the earth. The tables that our missionaries have given them, and that they use in their calculations, are founded

---

69. P. 11b.

70. Undated letters from Peking shortly following one of 4 November 1773, in *Lettres édifiantes*, IV, 217–225, 209–217.

upon this system. But although they make use of the consequences, they still have not admitted the principle. Perhaps they fear that, this hypothesis once favorably received by the emperor, they might be obliged to adopt it themselves."[71]

Whether these bureaucrats dreaded the Copernican system as a novelty is hardly to the point. Serious amateurs, in particular the outstanding group about Juan Yuan, had access to the manuscript of Benoist's explanation before Juan published it at the turn of the nineteenth century. They responded not with airy dismissal or defensive rejection, but with engagement. Juan's Biographies gave full and acute coverage to Copernicus, Galileo, and Benoist, even though the Jesuit writers' failure to settle upon a single Chinese transcription for the names of Galileo and Copernicus led Juan to treat each of them as at least two different people.[72]

Juan's Biographies is not simply a historical compilation, as it appears superficially to be. It juxtaposes traditional and Western astronomy, encouraging the study of the latter in order to improve the former. Juan emphasized the old idea that the roots of modern astronomy are to be found in ancient China, but that the tradition continued to develop in the barbarian outer reaches of civilization (that is, Europe and Islam) after it was lost in the Middle Kingdom. The observation that this thesis is historically untrue, although often ponderously set down by historians of science, is beside the point. Juan was not trying, as some writers who have not found the leisure to read him suppose, to denigrate European astronomy. Exactly to the contrary, he was providing a myth that would legitimize its study—not as an exotic novelty, but as knowledge affiliated with the classical tradition despite the unfamiliarity of its expression, and notwithstanding the unacceptable non-scientific ideas with which the foreigners associated it. Such a myth was indispensable in a culture that habitually used the past to sanction innovation, and that had long since lost the curiosity that could put such ideas in vogue *because* they were foreign.

---

71. Letter of 16 November 1767, in *Lettres édifiantes,* IV, 122–123.

72. Copernicus was given separate articles as two unrelated persons: Ko-po-ni , the Ptolemaist of Schall's *Li fa hsi ch'uan,* and Ni-ku-lao 泥谷老 (literally, "the old man of mucky valley"), who made observations in 1525 and whose computational innovation was later rejected as unsatisfactory by Tycho (*Ch'ou jen chuan,* ch. 43). Rho had also cited Copernicus' methods as those of Ku-po-ni 谷白泥; see *Wu wei li chih, 3:* 10b. The *Ch'ou jen chuan's* resumé of Benoist, which occupies two-thirds of ch. 46, or about fifteen per cent of the space devoted to non-Chinese astronomers, also cites Copernicus as Ko-po-ni. See also note 17 and p. 29 above.

Juan was not even an innovator in this regard. The myth of the Chinese origin of mathematical astronomy had been used effectively before him to justify study of the Western art by Mei Wen-ting and his grandson Ku-ch'eng, who were at the center of amateur astronomy in previous generations, and even by the K'ang-hsi Emperor (r. 1662–1722), the most able and learned patron of astronomy among the Manchu emperors.[73] This myth may be compared to the European notion that ancient and sophisticated societies could be "discovered," an accomplishment that justified subjugating them and destroying their traditional religions.

It was in the last analysis Juan Yuan and the scientific amateurs about him on whom the fate of Copernicanism in China was bound to depend. They, and not the functionaries, were the experts of their time. They often performed astronomical services for the government, as we have seen, but they did so as what we would call civil-service generalists or consultants rather than as career technicians. They had absorbed all the knowledge the missionaries offered. They were systematically reconstructing the mathematical methods of their forebears, which, although quite available in Matteo Ricci's time, few had read and hardly anyone comprehended.

In these men the values of their culture were still intact. As they learned modern astronomy they naturally tried to reconcile their new knowledge with the methods that had evolved in their own society. This reluctant but inexorable urge toward integration should hardly astonish those who study the growth of science. The difficulties we witness in China are comparable to those of seventeenth-century European astronomers facing aspects of the new science that could not be reconciled with late scholastic metaphysics.

In order to comprehend the net effect of Benoist's labors, it is necessary to reconsider Juan Yuan's evaluation of Copernicus. Historians have derided state-

---

73. Wang P'ing discusses the history and significance of the belief that the origins of Western astronomy lay in China in *Hsi-fang li-suan-hsueh chih shu-ju,* especially 69–74 and 97–103; see also note 13 above. Some modern historians have been so offended by the historical error that they ignored its purpose. It misses the point to describe Mei Wen-ting, probably the greatest expert in Occidental astronomy of his time and certainly the man who did most to encourage its study, as one "who, during his long life (1633–1721), never ceased to poke fun at those who, lacking patriotism, went to take lessons from foreign barbarians"; Bernard, "L'Encyclopédie astronomique," 479. The specter of nationalism which Bernard raises is quite irrelevant to the issue. On the role of the emperor see Catherine Jami, "L'empereur Kangxi (1662-1722) et la diffusion des sciences occidentales en Chine," in Ang & Will, *Nombres, astres* (see note 18), 193-209.

ments like this one as typical xenophobia, because they had no better explanation for Chinese "failure" to adopt the concept of a solar system before the Opium War era:[74]

> Michel Benoist says that Copernicus, in his discussion of the heavenly bodies, stated that the sun is static, that the earth moves, and that the sphere of the fixed stars is forever immobile. [Benoist goes on to say that] the finest students of astronomy among Western scholars support his theory. This differs considerably from what Schall says in On the Transmission of Astronomy in the West. According to Schall, Copernicus had an explanation of the celestial motions in terms of circles, and sought to determine the sun's apogee and its angular motions. Now if the orbs move in circles, and if the *sun's* distance varies and it has an angular motion, then the firmament and the sun must be moving, not static. How can the theories of one and the same Westerner be so self-contradictory?[75]

Juan saw this contradiction in his sources, but he had no way of knowing how it had come about. Benoist had said not a word about the vicissitudes of Copernicanism in Europe, nor did he admit that his account conflicted in any way with those of his predecessors. The simplest conclusion for Chinese readers—who had no reason to suspect that any of the missionary writers had been less than candid, or subject to any ideological restraint—was that Benoist was merely amplifying and reassessing earlier accounts, and that Copernicus was too inconsistent to take seriously.

Given the confusion that had been accumulating for nearly two centuries, it was inevitable that the skepticism of China's best astronomers at the turn of the nineteenth century should not have been limited to Copernicus alone. The Jesuits had taken great care to explain the transition between Ptolemy and Tycho as a historical development. Their Chinese readers, as we can see from *their* writings, generally understood the point clearly. But Rho and Schall had denied that there was any fundamental cleavage between Tycho Brahe and his precursor Copernicus. Early assurances that the Tychonic and Copernican world pictures were equivalent, given equal credence with Benoist's sanguine

---

74. L. van Hée, "The Ch'ou-jen Chuan of Juan Yuan," *Isis*, 1926, *8:* 117–118; George H. C. Wong, "China's Opposition to Western Science during Late Ming and Early Ch'ing," idem, 1963, *54:* 29–49. On the latter cf. Sivin, "On 'China's Opposition to Western Science during Late Ming and Early Ch'ing,' idem, 1965, *56:* 201–205.

75. *Ch'ou jen chuan, 43:* 554. This translation overlaps that given in Wong, 47. The point that Wong makes there depends on a serious misunderstanding of the phrase *t'ung i hsi jen* 同一西人 (literally, "one and the same man of the West") in the final sentence. Juan was using Schall and Benoist as sources, not as symbols.

description of their differences, were bound to cast doubt upon the rigor of Tycho as well as of Copernicus. To make matters worse, Schall and others associated highly accurate constants with Ptolemy without explanation. They prompted Juan Yuan to charge that, although the computational methods of Europeans, like those of Chinese, had evolved gradually, in their discussion of Ptolemy "is it not that Schall and his ilk were boasting in order to deceive us Chinese, and that Hsu Kuang-ch'i and Li T'ien-ching were taken in?"[76]

For Benoist himself such issues were not urgent. Like his predecessors, he could draw cosmological distinctions without violating his conscience because he had been taught to think of the concept of a solar system as nothing more than a heuristic device. The limits of discussion had simply been widened by a new administrative decision.

Chinese readers who had explored the Jesuit astronomical literature thoroughly were left with no alternative but a position parallel to Benoist's own. They came to believe that Tycho was as internally contradictory as Copernicus, or else Rho and Schall would not have associated the two. If consistency was so unimportant to such eminent Europeans, their cosmology was obviously not meant seriously.

In the Ming, Chinese had quickly recognized and valued Western cosmologists' ability to explain the reasons underlying the phenomena. By the end of the eighteenth century, what had earlier asserted itself as a science, rather than a mere collection of computational techniques like late Chinese calendrical astronomy, seemed to have no more depth than the latter. The study of Western methods could be encouraged all the more because they could be adapted to the ends of the traditional art without endangering its essence. These themes emerge in Juan Yuan's evaluation at the end of his article on Benoist:

> The ancient masters of astronomical reckoning, in ordering the motions of the Seven Governors, spoke of their inequalities but not of the reasons that their motions are irregular. Convinced that the Way of Heaven is too subtle for human power to spy out, they spoke only of what was appropriate for it to be, and did not press further to extort knowledge of why it is what it is. Such was the prudence of the men of old when they established their teachings.
>
> Since the Europeans, attracted by the Emperor's civilizing virtue, came from afar and translated their techniques for pacing off the heavens, we have had the mathematics of deferents and epicycles. These, it would seem, are simply hypothetical figures, used to demonstrate deviations from mean values. Because they may be used to explain why [we observe] inequalities in

76. *Ch'ou-jen chuan, 43:* 553.

> the motions of the sun, moon, and planets, however, the undiscriminating make the mistake of believing that there actually are such circles in the blue heavens. This is truly a great delusion!
>
> And then, after not very long, there was a change. For what have all along been called circles they have substituted ellipse techniques, and they hold that the earth moves and the sun is static. This means that the Westerners were unable firmly to maintain their previous arguments. If they were simply making use of figures to illustrate computational principles, then speaking of ellipses would be fine, and why not say that the earth moves and the sun is static? But in their [new] doctrines, going so far as to reverse above and below [i.e. the positions of the heavenly bodies with respect to the center], and turn moving and static topsy-turvy, shows them to be heretical, and incapable of edifying others [on the subject of physical reality]. Never has there been a worse instance. From Tycho's time to the present has been only somewhat over a century, but how many times have they changed his methods! I cannot imagine how much further they will go. They are certain to surpass these beginnings, boasting of knowledge that only they have, inventing absurd theories.
>
> This being so, how, then, can claims that the Westerners' discourses on celestial phenomena can clarify their reasons make them comparable to the perennially flawless [Chinese approach to] analysis of irregularities in the celestial motions, which only asserts what it is appropriate for them to be, and not why they are what they are?[77]

We have seen that the late-seventeenth-century Chinese astronomers were quite willing to explore the physical patterns of the cosmos, once encouraged by European approaches. They were also able to understand why the missionaries' accounts were inconsistent. Mei Wen-ting in 1702 explained that, although all the seventeenth-century writings were referred to as the "new methods of the West, what Schall translated was mostly based on Tycho, and diverged in many ways from Ricci's accounts. There was also the Westerner Smogułęcki . . . who in turn greatly differed from them. . . . as well as subtle differences in [the writings of] Verbiest. If we consider it analytically, Ricci and

---

77. *Ch'ou jen chuan, 46:* 610. See the interesting discussion of this passage in John Henderson, *The Development and Decline of Chinese Cosmology,* 256. Henderson sees in this passage the denial of a cosmic order, a denial that he argues was peculiar to the Ch'ing. The issue, I believe, is rather the limits of empirical knowledge. Statements of the same sort can be found in astronomy from the Han on. See Sivin, "On the Limits of Empirical Knowledge in Chinese and Western Science," in Shlomo Biderman & Ben-Ami Scharfstein (ed.) *Rationality in Question. On Eastern and Western Views of Rationality* (Leiden: E. J. Brill, 1989), 165–189.

Schall, Schall and Verbiest all differ. That is why I say that [what they have described] is not a single 'Western method'; they have gradually refined their art through practice. Without reading their books in depth it is impossible to understand why this is so."[78]

This progressive view was no longer visible at the end of the eighteenth century. European cosmology had been discredited by its incoherence. Juan Yuan actively and effectively supported the study of Western mathematical techniques. But here he was voicing the definitive judgment of his era not only upon Copernicus but upon the claim of early modern scientists to be responsible for describing physical reality.

Benoist, who began the restoration of the balance, precisely marks the effective end of the Jesuits' effort in China. The failure of their hopes grew out of the Rites Controversy. The antagonism of the other missionary orders in China simply reflected a growing threat to the Society's existence in Europe. Benoist died of a stroke in October 1774, upon hearing the news that the Society of Jesus had been abolished by the Pope.[79] The remaining missionaries made no significant contribution to the dissemination of modern cosmology.[80] Not until the mid-nineteenth century, when Protestant missionaries translated modern textbooks and used them to train professional astronomers who had no stake in the old society, did Chinese have an opportunity to accept post-Newtonian cosmology as one of the foundations of science.

---

78. *Wu-an li suan shu-mu,* 3b-4a.

79. Pfister, *Notices biographiques et bibliographiques,* item 377; Krahl, *China Missions in Crisis,* 223. Krahl says simply that Benoist "died of grief."

80. The Lazarist Monteiro de Sera, the last European to serve as a member of the Bureau, resigned in 1826, after which the post reserved for foreigners was abolished. On Monteiro see the book review by Ch'en Hsiang-ch'un in *Monumental Serica,* 1938, *3:* 325. For a chronology of leading officials in the Directorate see Po Shu-jen 薄树人, "Ch'ing Ch'in-t'ien-chien jen shih nien-piao 清钦天监人事年表 (Chronology of personnel in the Ch'ing Directorate of Astronomy), *K'o-chi-shih wen chi* 科技史文集, 1978, 1: 86-101.

APPENDIX
Chinese Epicyclic Terminology

Shifts in Chinese astronomical terminology before it was standardized in the late nineteenth and twentieth centuries deserve to be studied. This appendix is concerned with a subsidiary topic, the language of epicyclic astronomy as it appears in texts cited above.

Giacomo Rho (p. 16 above) used *t'ien* 天, literally "heaven," to refer to the celestial spheres. I translate this word consistently as "orb"; it does not connote spherical shape. Rho also used *ch'iu* 球, which exactly means "sphere," for other components of the Aristotelian universe—the earth with its upper spheres of air and fire, and the "first moving orb" (*primum mobile, tsung tung t'ien* 宗動天) at the periphery of the cosmos, which imparts motion to all the orbs within. *Kuei-tao* 規道 is "orbit."

Rho's set of terms for the geometric elements of the Ptolemaic, Copernican, and Tychonic models remained standard until the mid-eighteenth century. In his Ptolemaic theories the usual general term for deferent was *pen ch'üan* 本圈, literally "proper circle." He called the planetary deferents *hsing pen hsing t'ien* 星本行天, ("orb of proper motion of the planet"), shortened to *pen t'ien* 本天 ("proper orb"). Epicycles in general are called *hsiao lun* 小輪 ("small wheel"). His word for "planetary epicycle" was *pen lun* 本輪 ("proper wheel"), and for "annual epicycle" *tz'u hsing ch'üan* 次行圈 ("circle of secondary motion").

In the Ptolemaic theory of all the planets except Mercury, the deferent is centered upon an eccentric point some distance from the earth, but the motion of the center of the planet's epicycle is constant not with respect to the eccentric point but to the equant, which lies on the other side of the eccentric point from the earth and at an equal distance. In other words, the three points Earth–eccentric point–equant lie equidistant in that order along the line of apsides. In Chinese the deferent, *chün ch'üan* 均圈 ("uniformity circle," a term derived from *chün hsing chih ch'üan* 均行之圈, "circle of uniform motion"), rotates about the eccentric point, halfway between the center of the earth, *ti hsin* 地心, and the equant point, *pen ch'üan hsin* 本圈心 ("center of the deferent"). Note the confusion inherent in the literal meanings of the terms Rho coined, all the more so because "uniformity" suggests "equant."

In the Copernican system, which abolished the equant, three-quarters of the Ptolemaic eccentricity of the equant point is assigned to an eccentric deferent and one-quarter to an epicycle. The terms for the eccentric deferent and its center were the same as above, but Rho called the epicycle *hsiao chün lun* 小均輪 ("small uniformity wheel"), perhaps to record its derivation from the old equant system. Additional compounded epicycles, as in the Copernican lunar model, were called *tz'u lun* 次輪 ("secondary wheel").

The Tychonic planetary system, which Rho recommended as modern, uses a simple geocentric deferent, *pen t'ien* 本天, and compounded epicycles. The first, whose radius corresponds to the Ptolemaic eccentricity, he called *pen hsing lun* 本行輪 ("wheel of proper motion"). This is equivalent to the Ptolemaic *pen lun* 本輪. But Rho called the second epicycle *chün lun* 均輪 ("uniformity wheel"), cognate to the Copernican *hsiao chün lun.*

Wang Hsi-shan 王錫闡 (1628–1682; see the next chapter) followed Rho generally, but in his planetary theory the centers of the planetary epicycles rotate on a figure "that resembles an ellipse" called the *li chou* 曆周, literally "perimeter of travel," a term used traditionally in other senses.

In the mid-eighteenth-century the *Li hsiang k'ao ch'eng* of Mei Ku-ch'eng and Ho Kuo-tsung (p. 33) still used the old terminology, but they distinguished more clearly *ch'üan* for both the orbit and for orbital motion, and *t'ien* for the physical orb.

Michel Benoist, writing forty years after Mei and Ho, used *pen lun* 本輪 to stand for both the geometric deferent and the physical sphere. I translate it "deferent" in the excerpt from the World Map (p. 40) in order to make this ambiguity manifest in English. This sense was a radical departure from the old sense of "epicycle," but Benoist did not point out the change. *Chün lun* 均輪 was his general term for planetary epicycles. He wrote, for instance, that in the Ptolemaic system the sun, moon, and planets travel "on the periphery of a *chün lun,* and the center of the *chün lun* travels on the periphery of the *pen lun.*"

Retrospect

This essay was commissioned by the International Union of the History and Philosophy of Science for the five hundredth anniversary of Copernicus' birth, and published by the Polish Academy of Sciences in one of the volumes of its series *Studia copernicana* devoted to this celebration. This version is revised in several respects:

1. The discussion of Wang Hsi-shan's response to the Jesuit writings overlapped the analysis in the biography of Wang (Chapter V). I have replaced it here with a cross-reference.

2. I have excised Appendix B, "Heliocentric Orreries in the Chinese Court." In 1973 this was the only study of these instruments in a European language, but two essays published in the same year offered details and actual photographs, making my discussion, based on a book of 1766, otiose (see note 61 above). I was able in 1977 to document these instruments, but the detailed discussion they deserve would be out of place here. I have also expunged Appendix C, "The World System of Ma-erh-hsiang and the Copernicanism of Mersenne". I originally identified the mysterious Ma-erh-hsiang as Marin Mersenne (1588–1648), merely mentioning the early medieval encyclopedist Martianus Capella as a less likely alternative. Several years after the essay was published, Bruce Eastwood, a scholar of early medieval science, told me he had discovered that Martianus included in his Marriage of Philology and Mercury *(De nuptiis philologiae et mercurii et de septem artibus liberalibus libri novem,* 410) a system that undoubtedly was Benoist's source. Martianus' survey of classical learning incorporated one of the few moderately sophisticated cosmologies available for some centuries after it was written. It has rarely been studied since Tycho used it as one of the sources of his model, and Kepler mentioned it rather unclearly, but a Jesuit astronomer would more likely be aware of it than of Mersenne's private thoughts.[81] Students of the Galileo affair and its aftermath, in which Mersenne played an important but enigmatic part, may find some parts of my investigation in Appendix C useful; they are welcome to consult the first edition of this study. Because the Bibliographical Notes overlapped considerably with the Bibliography of the next chapter, I have combined the two and placed the combination at end of the latter (V 20).

3. I have mended many typographical and other printers' errors that the original printers were unable to correct.

---

81. See Eastwood's two 1982 studies reprinted in *Astronomy and Optics from Pliny to Descartes* (London: Variorum, 1989). On the role of Martianus' system in Tycho's cosmology see Thoren, *The Lord of Uraniborg* (see note 6 above), 239.

V

WANG HSI-SHAN

It is important that the first substantial encounter in China between its cognitive traditions and those of Europe was in mathematical astronomy. Medicine provoked no such response, the limited appeal of religion was to faith and not theology, and confrontations of political ideas were negligible. The style of the mid-seventeenth-century response to Occidental astronomy set the style of less abstract encounters later, as may be seen in the reaction to the all too tangible Western military technology of the late nineteenth century.

The early Chinese response to the exact science of Europe is poorly understood. The usual conceptual model for the travel of Western technical concepts and methods, which assumes that their wholesale triumph is automatic and that any hesitation must be the result of ignorance and xenophobia, is not a serviceable tool for this explanation. Experts on "China's response to the West" tend to chide their subjects for not rejecting their culture immediately and wholesale in favor of the new ideas. That expectation betrays a remarkable innocence about the way cultures work, as well as about the value of Occidental ways in Oriental circumstances. The broad pattern of response that we find generally in such historical confrontations—misconception, ambiguity, reinterpretation and selective adaptation over a long period—can be seen equally in American physicians' discovery of acupuncture from the 1970's on.

If we want to understand the indirect paths by which mathematical astronomy exerted its influence, a largely unexplored range of questions becomes appropriate. The last chapter has dealt with one of them, namely, what were the limits of the foreign knowledge that reached China? This one, using a single case, will take up a number of other issues. In what circumstances of society and culture were those who responded to European astronomy? What did they find familiar in it? What did they value as new and useful? For what did they have no use, and what reasons did they give for rejecting it? They lived in a society that esteemed the unity of knowledge. They did not delude themselves that the practice of science can or should be value-free. Like pre-modern European scientists, they would not have found a science divorced from values desirable. What values, then, did they perceive as inherent in the new science, and how were their own values affected by what they adopted from it?

This meditation on the career of Wang Hsi-shan is an experiment along these lines. Wang was a remarkable astronomer, and a lonely man in a dislocated time. His predicaments, social and intellectual, resembled those that formed the careers of many contemporaries and shaped trends of thought. Even the character of his approach to astronomy has much in common with what came to be the dominant trend in Ch'ing neo-Confucianism. I suggest below that this was no coincidence, but represents one of the most enduring influences of Western astronomy.

Life

Wang Hsi-Shan 王錫闡 was born 23 July 1628, and died 18 October 1682. His place of registration was Wu-chiang 吳江, Soochow prefecture.

Wang was the son of Wang P'ei-chen 培眞 and his wife, née Chuang 莊. Wang Hsi-shan's epitapher, who would have been expected to cite distinguished ancestors in the preceding few generations, did not do so. Wang was designated to continue the family line of a childless uncle, which suggests that he was not the eldest child or that his family was poor. Nothing is known of his education except that he was self-taught in mathematics and astronomy. He probably did not have access to all the available treatises on the European sciences, for in a letter, probably of 1673, he asked a friend to look for some of them in Peking.

He had no son to arrange for posthumous publication of his writings. One of his few disciples sketched, in twenty-five Chinese characters, the impression Wang made: "emaciated face, protruding teeth, tattered clothes, and shoes burst through the heels. His character made him aloof, as though no one could suit him; but when someone inquired about a scholarly topic, he was forthcoming as a river in flood."[1]

The conventional road to social advancement for the son of an obscure gentry family was the civil service examinations, which required many years of preparation. This path was blocked by the Manchu conquest.

The invaders from the north overran Wang's district in 1645. Whether to collaborate with the alien government was an issue for all Chinese who in normal times would aspire to government service. Wang was only sixteen, but he made it clear that he did not wish to live with the new order: "In a burst of passion, wanting to die, he jumped repeatedly into the river. It always happened that someone was there to save his life. He refused to take food for seven days, but still did not die. His parents were persistent; he had no choice but to resume eating. Renouncing worldly ambition, he dedicated all of his powers to learning." As he put it, in 1649 he closed his door to all worldly pursuits,

1. This and the next quotation are from the funerary inscription by Wang Chi, *16:* 1a-1b. For full citations see the Bibliography, p. 20 below.

including the examinations. Hopes for a Ming restoration faded, but in letters and manuscripts he never acknowledged the new Ch'ing dynasty, and reproached those who did.

His friends were like himself Ming loyalists prevented by scruple from following conventional careers. Among them were the great transitional figure of neo-Confucianism, Ku Yen-wu 顧炎武 (1613–1682), as well as P'an Ch'eng-chang 潘檉章 (also of Wu-chiang, died 1663) and Chuang T'ing-lung 莊廷鑨 (died ca. 1660), both of whom planned histories of the Ming, to be compiled by coteries of loyalists. Wang took part in the first, the result of which was destroyed in manuscript. Apparently, disguised as a Buddhist monk, he visited in prison some of the seventy people condemned to death in 1663 when the publication of the second led to a charge of sedition.

Wang was not widely traveled; he never met Mei Wen-ting 梅文鼎 (1633–1721) of Anhwei or the northerner Hsueh Feng-tso 薛鳳祚 (died 1680), now considered the other two great astronomical scholars of the time. Mei acknowledged Wang's preeminence, and wrote commentaries, never published, on several of his books.[2]

Wang apparently made an indifferent living by teaching mathematics and astronomy to a few disciples. He was supported for a time by P'an Ch'eng-chang and later by Lü Liu-liang 呂留良 (1629–1683), another prominent Ming loyalist, who practiced medicine. Wang's career was impeded by isolation and illness, including partial paralysis in his later years. The year before he died, he wrote, in connection with the prediction of a solar eclipse: "Whenever there is a conjunction I have always checked the accuracy of my computations against observation, despite sickness, cold, or heat, for thirty years and more."[3]

Wang's technical writings circulated in manuscript among astronomers after his death. Their preservation was not guaranteed until a major composition was included in the enormous imperial manuscript collection of rare texts, the Complete Library in Four Repositories (*Ssu k'u ch'üan shu* 四庫全書, compiled 1773–1785). The descriptive and critical catalogue of this collection (printed 1794) brought Wang's contribution to general attention.

---

2. Described in *Wu-an li suan shu-mu,* 34b-35a. On the relations of Mei and Wang, see Hsi Tse-tsung, "Shih lun Wang Hsi-shan te t'ien-wen kung-tso." Reliable short biographies of Mei and many other figures mentioned in this essay appear in Goodrich, *Dictionary of Ming Biography,* and Hummel, *Eminent Chinese of the Ch'ing Period.*

3. Cited by Hsi (63) from "Wang Hsiao-an hsien-sheng i-shu pu-pien."

The Setting of Wang Hsi-Shan's Astronomical Career

As I have shown in Chapter IV, Chinese astronomers formed their impressions of European astronomy and cosmology from writings that, about 1630, were typical of textbooks and handbooks current in the Church's educational institutions in Europe. After 1616—in other words, not long after the missionaries began writing about astronomy in Chinese—the Inquisition's warning to Galileo prevented them from describing current changes in cosmological thought. At the same time, their competence and good intentions led them to drop hints about these changes that turned out to be contradictory and in the final analysis misleading. It was with this in mind that Hsi Tse-tsung remarked "We can imagine, if Wang Hsi-shan had only come upon [Copernicus' *De revolutionibus,* Galileo's *Dialogo,* and Kepler's *Epitome astronomiae copernicanae,* all of which the missionaries kept for their private use in Beijing], how much greater his contribution to astronomy would have been."[4]

Wang, Mei Wen-ting, and Hsueh Feng-tso were the first astronomers outside the court to respond to the exact sciences that were being introduced from Europe, and to shape the influence of these new methods and conceptions on their successors. They were, in short, responsible for the scientific revolution discussed in Chapter VII. There I argue that the social displacement of that one transitional generation opened new possibilities of response.

At the time there were no socially marginal students of astronomy alienated from traditional values and protected by association with privileged foreigners, as would be the case in late nineteenth-century China and elsewhere during the heyday of imperialism. The only astronomers who could respond to the Jesuits' writings were members of the old intellectual elite. Yabuuchi Kiyoshi 藪内清 has noticed that among those interested in foreign scientific learning "many were members of the 'pure discussion' faction, related to the Eastern Grove Party (Tung-lin Tang 東林黨) and the Revival Society (Fu She 復社, two reform groups), who lived as commoners after the fall of the Ming. They mostly came from commercially and industrially prosperous cities in Fukien, Kiangsu, Kiangsi, and Anhui."[5] Members of the upper social strata, whether descended from organized Ming reformers or not, were bound to evaluate innovations in the light of established ideals that they felt an individual responsibility to perpetuate.

The period of the Manchu conquest was crucial in more ways than one. The change of dynasty gave the Jesuits the official status in the court that they

---

4. Ibid.

5. Yabuuchi, *Min Shin jidai no kagaku gijutsushi,* 26, n. 3.

had been seeking for half a century, and encouraged the dissemination of their diverse writings. The takeover rent the lives of individuals all over China for that one generation and narrowed their career options to professions that in other circumstances, to conventional families, would not have been desirable alternatives to civil service. The children of those internal exiles did not have to face the same choice.[6]

The transition also encouraged elite scholars to seriously study Western astronomy, distinct from the traditional art in method and conception. Those who had first pursued it late in the Ming naturalized it by creating a foundation myth. They claimed that European mathematics had evolved out of certain techniques that had originated in China and had been transplanted to the extreme fringes of civilization (Europe and Islam) before being neglected and more or less dying out in their original home. This myth circulated by word of mouth until, once the Ch'ing regime was established,the K'ang-hsi Emperor (r. 1662–1722), the most powerful of amateur astronomers, took it up. He prompted his scientific associates, including Mei Wen-ting, to take it seriously. Mei began propagating it in writing, and it spread widely.[7] Wang Hsi-shan was among the many who accepted it as a warrant for considering the two traditions side by side.[8]

This foundation myth, an appeal to the Chinese tendency to see perfection in high antiquity, connected Western science with certain ambiguous references in ancient historical writing. Since it was not intrinsically foreign, it could be taken seriously. Mei and his successors used this notion to legitimate the institutions—small private groups of masters and disciples—that organized themselves around the new astronomy and taught it. There is nothing inherently

---

6. This norm generally affected only the transitional generation in families that had served the previous dynasty. Refusing to take the examinations or accept office was in any case a matter of conscience, seldom of social coercion. Zurndorfer 1988, esp. 22–29, has some interesting things to say about responses to Western astronomy as a generational matter.

7. Jacques Gernet has suggested (letter, 5 Oct. 1978) that Mei might have been influenced by his teacher Matteo Ricci's habit of claiming Western origins for things he saw in China. Chiang Hsiao-yuan, "Shih lun Ch'ing-tai 'Hsi-hsueh Chung yuan' shuo," believes that the notion of Western origins can be traced back to unpublished remarks by Huang Tsung-hsi 黃宗羲 (1610–1695). For the emperor's involvement see Wang P'ing, *Hsi-fang li-suan-hsueh chih shu-ju,* 77–79, 97–103. Martzloff, "Space and Time," argues that Mei, Wang, and others sincerely believed it.

8. See Wang's *Tsa chu* 雜箸 (Miscellaneous Essays), in *Hsiao-an i shu*, XXXV, 1a-2a, 10b-11a.

Chinese about the use of such myths. An analogy that comes to mind is the equally remarkable European myth that non-European societies could be "discovered," a change of status that authorized their economic despoliation and the systematic destruction of their ways of life during the Age of Discovery.

Mei, Wang, and others were aware that, whatever the lost grandeur of archaic times may have been, from about 100 B.C. through the thirteenth century, computational astronomy in China had actually continued, within its stylistic limits, to grow in power and range. From the Yuan period (1279–1368) on, it came very gradually to be little practiced outside the Astronomical Bureau, which was dominated by foreign technicians. In some periods, laws prohibiting unofficial astronomy were actually enforced.

By 1600 it is not clear that anyone was able fully to comprehend the old numerical equations of higher order, proto-trigonometric approximations, applications of the method of finite differences, and other sophisticated techniques. At the same time, the official system used to compute the Ming calendars was regularly failing. This was not surprising, for it had been in use since 1384, and was only a modified version of the Yuan system of a century earlier.

The major activity of the Directorate of Astronomy, as the Ming drew to a close, was repelling attempts of outsiders to revise the computational methods. A famous example was that of Hsing Yun-lu 邢雲路 (Presented Scholar 1580), who had a remarkably comprehensive knowledge of methods used in the past. His proposal for reforms, made in 1596 when he was a middle-level provincial official, was criticized not on technical grounds, but as seditious. The officials of the Directorate countered that publicly admitting official predictions had been failing would simply encourage insurrection. Given the general paralysis of government from ca. 1580 on, with attempts at reform savagely punished, Hsing was fortunate not to be prosecuted.[9]

A generation later, as the Ming slid deeper into crisis, the Jesuits, despite their record of accuracy and their support by several high officials, were unable to get permission to carry out an official reform.

With this debacle, unrelieved until the Manchu conquest, in mind, the greatest astronomical figures of the early Ch'ing era were convinced that the time

9. Hsing's proposals are recorded in *Ming shih* 明史, *31:* 527, and *Ch'ou jen chuan*, *31:* 378–382. On the response see *Ming Biographical Dictionary*, I, 369. Hsing's *Ku chin lü li k'ao* 古今律曆考 (Studies of mathematical astronomy and harmonics, ancient and modern, ca. 1600), is the largest and, in the parts I have used, most percipient survey of its kind. A particularly vivid picture of the setting is Ray Huang, *1587. A Year of No Significance* (New Haven: Yale University Press, 1980).

was overripe for a renascence. That priority defined the proper field for application of Western knowledge. It was to be part of a larger rebirth, which I will take up below (p. 12).

The Chinese scholars' use of Western cosmology and computation, naturally enough, was highly selective and, in Wang's case, astutely critical. Logic and geometric rigor interested them a great deal less than tangible numerical methods. J. C. Martzloff has suggested that Mei's and Wang's appreciation of Western (Ptolemaic and Tychonic) cosmology was "global, concrete, qualitative, and topological," freely drawing for explanation on metaphors from anatomy, zoology, and so on. Martzloff has traced this selectivity to the difference in the two cultures' conceptions of space and time, and in their assumptions about the character of astronomical knowledge.[10]

Astronomical Work

Wang and his contemporaries were motivated by two central questions. The first was how the new knowledge from abroad might be useful in reviving the exact sciences. Traditional knowledge was recorded, and the perennial problems of prediction were set, in the easily accessible standard histories of the various dynasties. Each incorporated technical treatises, which, among other matters, recorded in detail many of the complete systems for computing ephemerides that had been proposed or accepted for official use since about 100 B.C. Wang was familiar with the chief writings of this sort and with many of the Jesuit treatises of the early 1630's.[11] There is no reason to believe that he met any foreigner,

The second problem was how to resolve the internal contradictions of European astronomy. Since circumstances had ruled out a unified set of treatises, some discrepancies were due to divergences of approach and varying choices of constants, and some to limitations of the missionaries' skill. The most important source of inconsistency was the different cosmological viewpoints through which European writers tried to convey the best knowledge of their time, before and after the decree against the teaching of heliocentricism limited debate about the system of the world (see Chapter IV for details).

Historians have occasionally asserted that Chinese were incapable of responding creatively to geometrical models, or that a bias against abstraction would have prevented them from taking up Copernican cosmology had it been available. Wang's response to the Jesuits' astronomical writings belies these

10. Martzloff, "Space and Time," 73.

11. On his efforts to buy these treatises see *Dictionary of Ming Biography,* 1381a.

opinions. In adapting the missionaries' version of the Tychonic cosmos to his own uses, Wang was alert to contradictions in its presentation. He noted, for instance, that a secular diminution in the length of the tropical year had been mentioned, but was ignored in a discussion of the precessional constant (which, by implication, should increase). This was not the modern variation in the length of the year, but one of much greater magnitude, introduced by the missionaries, that was obsolete in Europe by the time Wang wrote.

Wang's New Method *(Hsiao-an hsin fa* 曉菴新法, completed 1663) presented, in conventional form, a complete system of ephemerides computation centered on eclipse prediction, with tables that made only simple arithmetic necessary for calculating the ephemerides. It provided, for the first time in China, methods for predicting planetary occultations and solar transits. The missionaries had dealt with these problems only in principle. The Compendium of Observational and Computational Astronomy *(Li hsiang k'ao ch'eng* 曆象考成, 1713, printed 1724), part of a great survey of the mathematical arts sponsored by the K'ang-hsi emperor, included some of Wang's techniques along with post-Newtonian data.

Wang's On the Angular Motions of the Five Planets *(Wu hsing hsing tu chieh* 五星行度解, completed by the autumn of 1673) was a geometrical description of his own world model. The work reflected familiarity with European trigonometry. It drew on Tycho Brahe's scheme, but Wang was critically testing every assertion as he studied it.[12] His own system differed from Tycho's unfinished planetary theory primarily in substituting eccentrics for major epicycles, and making the superior and inferior planets rotate in opposite directions. His description, unlike those of most of his Chinese predecessors, was as unambiguously physical and spatial as Tycho's description of his own had been. Here is an example:

> Of the Five Planets, Saturn, Jupiter and Mars revolve leftward, carried toward the east by the sun. Venus and Mercury rotate toward the right on their own orbs, each with its own angular speed. They also accompany the orb of the sun in moving one degree daily. But the orbs of these two planets are too small to enclose the earth, so their mean motion must be that of the sun. According to Western [i.e. Tychonic] astronomy, all five planets rotate toward the right. This does not agree with the celestial motions [i.e. the phenomena], and will be corrected below.
>
> In Western astronomy Mars, Venus, and Mercury are sometimes above the sun, so that their orbs should be outside that of the sun, and are some-

---

12. The diagram from Wang's book reproduced by Needham *(Science and Civilisation,* III, 455) was not meant to explain the Tychonic theory, but Wang's own.

> times below the sun, so that their orbs should also be within that of the sun. So it is said that the spheres are able to interpenetrate, and cannot be substantial. They do not know that the orbs of the Five Planets are all within that of the sun. But while the Five Planets are located on the periphery of their orbs, the sun occupies the center of its orb (this is also the center of the spheres of the Five Planets), except that it is displaced slightly upward (in the direction of the *primum mobile).*[13] The sun rotates with its sphere, [the points through which it passes] forming the "track of the sun's motion." This track is actually without solid substance, so the three planets pass through it without hindrance. As to the orbs of the Five Planets, they are all substantial.[14]

Wang Hsi-shan's feeling for rigor—necessarily exercised within the limits of what Tychonic cosmology had been described in Chinese—led him to criticize fundamentally the clarity, consistency, and accuracy of the Jesuit writings.

The most original idea in On the Angular Motions was a physical one. To explain the eccentricity of the sun-centered figure on which the centers of the planetary epicycles rotate, Wang posited a physical force radiating from the outermost moving sphere *(tsung tung t'ien* 宗動天, the Aristotelian *primum mobile)* and attracting each planet to an extent maximal at apogee. This theory used the old Chinese concept of *ch'i* 氣, which in modern terms would be to some extent pneumatic and to some extent energetic:

> The cause of the apogee and perigee of the orbit would seem to be that, as the *primum mobile* . . . carries the luminaries around . . . , its *ch'i* exerts an attraction upon the sun, moon, and planets like that of a lodestone for a needle. When one of the celestial bodies reaches a certain point, it rises toward the *primum mobile.* As it departs from a certain point, it sinks away from [?] the *primum mobile.* The mode of its rising and sinking is not a rectilinear but a circular motion. (All celestial motions are circular motions).[15]

---

13. More precisely, it develops later, the earth is at the center of the sun's sphere, although it too is displaced slightly to account for the first inequality. The sun's displacement is of course approximately equal to the distance from the earth to the sun. The *primum mobile* is the outermost moving sphere of the cosmos, which drives those within. The remarks enclosed in parentheses in this and later translations are notes in smaller type in the originals.

14. *Wu hsing hsing tu chieh,* 1a. On the Chinese term for "orb" in this quotation, see Chap. IV, Appendix A. The point about insubstantiality of the orbs was Tycho's.

15. Ibid., 7b. My translation of the next to last sentence is tentative, since the character *ta* 達, literally "to arrive at," is probably in error; a roughly opposite meaning seems to be called for.

Wang's force, although not universal like Newtonian gravitation, applied to all the planets known to him. Wang went on to work this peripheral, oscillatory force, emanating toward what seems to be one point in each orbit, into his geometrically conceived model.

In Ptolemaic astronomy there was no need for forces to explain celestial motions, for circular motion, steady and closed on itself, was eternal by nature. Only when Kepler reluctantly did away with the millennial notion that orbits are compounded of circles did forces become necessary to explain why the heavenly bodies keep moving. Kepler's force, like Wang's only in its vagueness and lack of quantitative significance, radiated from the central sun rather than emanating from the periphery of a system.[16]

Kepler's new celestial dynamics had not been explained in China. The closest thing to an allusion was a vague statement in Giacomo Rho's Principles of the Planetary Motions (*Wu wei li chih* 五緯曆指, 1634 or slightly earlier). Rho's force was central and, unlike that of Kepler, implied no magnetic analogy. In the course of arguing for the *Tychonic* arrangement of the planets Rho simply asserted "As for Venus and Mercury, in ancient and modern times no one has failed to agree that the mean motion of the sun is the mean motion of their orbs. Thus the motions of the three orbs (sun, Mercury, Venus) are all due to one potential moving force. This potential force is located in the body of the sun" and, again, "The sun is to the planets what the magnet is to iron. The planets are compelled to revolve according to the revolution of the sun."[17]

In other words, the equal mean sidereal periods of the sun and the two inner planets point to a common motive force for the three. An obvious corollary would be that this force does not also affect the other planets, whose mean sidereal periods are not a year. Rho did not connect this assertion with the remainder of his planetary theory, and did not suggest that it could be extended to the superior planets. In context Rho's remark was too pointless to carry any Chinese reader far toward the concept of a universal force. After all,

---

16. Johannes Kepler, *New Astronomy* (trans. William H. Donahue; Cambridge University Press, 1993), 379–380 and elsewhere.

17. *Wu wei li chih* (in *Hsi-yang hsin fa li shu* 西洋新法曆書), *1*: 6a, 33a. The first text reads "moon" for "Mercury," an obvious error. "Potential" tentatively translates *neng* 能; if the word is not being used in this technical Aristotelian sense, the rather clumsy phrase would mean merely "one force capable of causing motion." Hashimoto Keizô 橋本敬造 first noted this theme in Rho; "*Rekisho kôsei no seiritsu*," 86 and n. 122. On another anomaly in this treatise that caused Chinese astronomers some difficulty, see Martzloff, "Space and Time," 88.

the Tychonic framework Rho was presenting needs no explanation for the periods of the lower or upper planets. Unlike Ptolemy, Brahe resolved the mean sidereal periods of the inner planets into the sidereal period of the sun about which they are rotating.

Wang Hsi-shan certainly knew of Rho's passage, although he did not cite it. He apparently made the leap on his own to a force that moves the sun, moon, and all the planets.

Wang's notion of synthesis went deeper than reconciling ancient schemes of calculation with foreign techniques. The power of Western models not only to predict phenomena but also to exhibit their inherent patterns attracted many Chinese. Wang sought to establish metaphysical links for further exploration into celestial reality. This motive lay behind his suggestion that the circle be divided into 384 degrees. The traditional division made each degree *(tu* 度) equal to one tropical day's mean solar travel (so that in Wang's system there would be 365.2422 degrees). He was aware that the European division of a circle into 360 ($2^3 \times 3^2 \times 5$) parts was convenient, especially in manufacturing graduated instruments. He did not copy this constant, but analogously chose the number 384 (3×2^7). As the number of lines (6×64) in the sixty-four hexagrams of the Book of Changes, it also related astronomical quantities to the fountainhead of conventional speculation about cosmic change, uncovering another layer of significance.[18]

Despite his dedication and critical intelligence, Wang could not hope, any more than his contemporaries, to succeed in a mature synthesis of traditional and modern science. Information from the West was inadequate in many respects, and several generations more were needed to reclaim the traditional corpus of Chinese mathematics and astronomy as part of the astronomer's repertoire. But the astronomers of Wang's generation provided tools and methods as well as a goal.

For many decades students began with the Western writings and went on to study the Chinese technical classics. The latter, once mastered, increasingly defined the style of research in the exact sciences, even as researchers concerned themselves with new problems. By the early nineteenth century, Western mathematics and astronomy were no longer novelties. They had been studied continuously for two hundred years. The basic training in the decades before the Opium Wars (ca. 1840) was in the native writings.[19] They served as excel-

18. *Hsiao-an hsin fa*, 2a.

19. This change in priority of learning is a main theme of Wang P'ing, *Hsi-fang li-suan-hsueh chih shu-ju*.

lent preparation for up-to-date Western treatises that gradually began to appear as part of a new confrontation—this time a total confrontation—between China and the West.

General Significance for Chinese Thought

Wang Hsi-shan's lifetime was a critical epoch in the evolution of Chinese philosophy. What Western historians call neo-Confucianism, like earlier Confucianisms, was a sustained quest for doctrines of education, self-cultivation, and moral conduct. Its successive new departures depended upon scrutinizing antiquity to identify and interpret (differently for each age) the authentic core of Confucian teachings. Well before the late Ming period, expanded scope for self-consciousness, increased blurring of social barriers, and the more penetrating influence of Buddhist and Taoist practices had deepened religious and moral awareness. This trend affected both the tradition of Chu Hsi 朱熹 (1130–1200), which explored the phenomenal world (including the mind and experience recorded in books) to grasp the single coherent pattern inherent in all change, and that of Wang Shou-jen 王守仁 *(or* Yang-ming 陽明, 1472–1529), which emphasized enlightenment through self-awareness, particularly of the mind engaged in conscientious social activity.

The great intellectuals of the dynastic transition were, on the whole, Ming loyalists. They were among the minority who did resist, in the main passively, and after the transfer of power had taken place. They were convinced that assessing the failure of their intellectual predecessors would guide them toward philosophical and spiritual reinvigoration, and responsible engagement in the world of affairs. Among the most influential was Wang Hsi-shan's friend Ku Yen-wu. He charged that his late Ming predecessors let themselves be distracted from moral commitment and public responsibility by sectarianism, by pedantry and triviality in the Chu tradition, and, in the tradition of Wang, by a subjectivity and individualism ignorant of the authoritarian and hierarchic requirements of social order. Above all, Ku and other Ch'ing survivors were convinced that the rivalry of schools, blinded by selfishness and pride, had corrupted Confucian doctrines, leaving them unable to rise above the political futility that led to the disintegration of the Ming.[20]

---

20. I am grateful for this formulation, and for a number of helpful criticisms, to Lynn Struve. I am also thankful for suggestions by Judy Berman, Jacques Gernet, Dianna Gregory, and Yü Ying-shih. The outline that follows draws on Elman, *From Philosophy to Philology.*

Ku's prescription for the ills of thought was to purge postclassical influences that hid the original principles of Confucius and his orthodox followers. Those who agreed with him needed a method for critically examining texts in order to determine what was authentic. The thorough research that widespread printing and large libraries had made feasible revealed to Wang's contemporaries how easily the understanding of their predecessors had been led astray. Some now endeavored to recover the earliest—the least corrupted—versions and interpretations of the classics. Others studied the working out of canonical moral patterns in the events of history. These reformers saw their work as the surest way to a living philosophy. The fortunes of the empire had given this search a new poignancy.

By the mid-eighteenth century, narrowly defined research methodology had become an end in itself, narrow in interpretation and intolerant of the urge to generalize. The call for "social utility, concrete practicality, and tangible evidence," which had promised philosophic regeneration at the beginning of the Ch'ing, outlived the openness to the unexpected that was implied in its original motivation.[21] Classicism flourished, despite the atrophy of metaphysics, because it yielded a succession of critical breakthroughs, and because it posed no threat to a state which insisted that collective intellectual activity be apolitical.

This final evolution of disciplinary specialization out of a philosophic renaissance is not of further concern here, but how the Ch'ing style of critical neo-Confucian scholarship began to take shape in Wang Hsi-shan's lifetime bears examination.

Many important neo-Confucians of the late Ming and early Ch'ing, especially among those close to the Chu tradition, wrote on mathematics and astronomy. A number of their treatises reconstituted early computational techniques, and about the same number used mathematics in chronological or other studies of canonical writings before about 200 B.C.[22] In the mid-seventeenth century the people who innovated in both philosophy and the sciences eschewed politics and public service. This is perhaps not remarkable, since re-evaluation and the eclecticism that yields new insights are likely to begin with talented and ambitious people subsisting on the margins of the elite. But in addition to obvious consequences of this social overlap, philosophers, scientists,

21. W. T. de Bary, "Neo-Confucian Cultivation and the Seventeenth-Century 'Enlightenment,'" in *The Unfolding of Neo-Confucianism*, 193.

22. See the list in the Appendix, p. 17. The scientific writings of orthodox scholars are an important source for Elman and for Henderson, *The Development and Decline of Chinese Cosmology.*

and mathematicians shared important convictions about the means and ends of research.

To sum up the argument so far, certain critical motifs recurred in neo-Confucianism just after the Manchu conquest, pointing the way toward new departures. Intellectuals emphatically rejected what they saw as decadent and destructive tendencies at the end of the Ming. They believed that those tendencies arose partly because of inadequate study, misunderstandings and textual corruptions, and partly because heterodox ideas had insinuated themselves into texts and undisciplined scholarly writings. They were convinced that, to understand the inherent patterns of cosmic and human activity *(li* 理) and the moral imperatives they imply, they must critically reexamine classical literature and history.

All of these ideas motivated Wang Hsi-shan. We are told by his biographer that after he renounced worldly ambition, "he excoriated heterodoxy [this usually refers to Buddhism, sometimes to Christianity as well], attacked 'innate moral consciousness' *[liang chih* 良知, the characteristic doctrine of the Wang Yang-ming school], and accepted the orthodox Confucian tradition of the Chu Hsi line as his personal mission." The preface to his "New Method," instead of conventionally affirming the high antiquity of astronomy, began by taking up questions that had been raised about the authenticity of seven calendars that the historians dated prior to the Han period (206 B.C.). Wang stated flatly, "There is no doubt that they were forgeries of the Han." He declared that astronomers of the recent past comprehended less than their predecessors. The remedy was rediscovering the lost meaning of the technical classics.

Earlier scientists had argued that, although mathematical astronomy could provide useful knowledge and advance understanding, the subtle texture of the natural order could ultimately be penetrated only by illumination.[23] Wang did not reject this view, but he saw number as a means toward that penetration: "One who seeks rigor must reach it through computation. Numbers are not themselves the inherent pattern *(li);* but because the pattern gives rise to number, through number one may reach enlightenment as to the pattern."[24]

23. See Sivin, "On the Limits of Empirical Knowledge in Chinese and Western Science," in *Rationality in Question. On Eastern and Western Views of Rationality* (ed. Shlomo Biederman & Ben-Ami Scharfstein; Leiden: E. J. Brill, 1989), 165–189, reprinted below as Vol. II, Ch. IV.

24. *Tsa chu*, 4a. This conviction was almost certainly influenced by the argument of Matteo Ricci, in his preface to the Chinese translation of Euclid's Elements *(Chi-ho yuan pen* 幾何原本, 1607), that geometry is a unique means to knowledge of *li.* Ricci

These parallels, and others in the writings of Wang's scientific contemporaries, suggest a close connection between the scientific revolution of seventeenth-century China and the evolution from philosophy to exact scholarship that took much longer to run its course. In particular, they suggest that Western influence on main currents of early Ch'ing philosophy—on the frontiers of Chinese self-awareness—should not casually be ruled out. Historians have usually ruled such influence out (or seen native thought as a reaction against it) because they have not studied the scientific literature and because they have relied on crude and narrowly defined tests for intellectual influence that ignore the mathematical dimension of human thought. A generation earlier Sinologists just as grossly overestimated Western influence. Elman has laid both fallacies to rest, showing in how subtle a way the influence exerted itself.[25]

This mathematical challenge to values coincided with an even more traumatic challenge, the Manchu invasion. Some shade of ambiguity toward Western science must have come from the Jesuits' prompt tender of services to the Manchus and the immediate official adoption of their astronomical system (although some missionaries accompanied the refugee Ming court southward to cover the eventuality of a restoration). It was not, however, characteristic of Chinese to reject what was foreign simply because it was foreign; by 1650 nationalism had barely been conceived. A willingness to adopt the behavior patterns and rituals of well-bred Chinese gave Manchus and Jesuits a right to be where they were—although nothing compelled loyalists to collaborate with either. But Wang Hsi-shan's type of loyalism had become, a generation later, only a memory, and no longer deterred the study of new ideas.

It is merely reasonable to suggest that the early success of astronomers in applying the foreign tools (in eclectic combination with traditional ones) influenced philosophers to reexamine classical learning. Scrutiny of ancient observations and predictions was an established part of the astronomer's work. Not long after Wang Hsi-shan's lifetime, technical examination of the philosophic classics to fix dates and test authenticity became the explicit end of most astronomical exploration, with the revival of traditional science as an intermediate means that fully occupied many scholars.

Even more important in assessing this channel of European influence is the fact that as time passed, leading teachers also became mathematicians used to

---

consistently used *li* for Latin *ratio,* knowledge that does not depend upon individual belief and thus can overcome individual doubts. His Asian readers would not have understood him that way.

25. *From Philosophy to Philology,* 79ff and elsewhere.

V

working with Western techniques and concepts. Elman has shown that these scholars congregated in private schools and maintained close relations. Even those who never applied European science in their writings were aware of it through discussing it with their associates and through reading their monographs.

None of this suggests a simple causal relation between European astronomy as described in the Chinese language in the early seventeenth century and the forms of Chinese thought that gestated then and became dominant by the middle of the eighteenth. Philosophy's relation with astronomy remained a dynamic one, despite the conventional moral and social focus of self-cultivation. At the beginning the new science offered what were seen as powerful tools toward a reformation of thought. Wang Hsi-shan's belief that number could bring ultimate insight into the universal pattern was pregnant in precisely this way. Be that as it may, the consensus ultimately formed in more conventional quarters, and was barren for a new synthesis of nature, society, and man.

Still, the new techniques could never be mere tools. To use them, as so many thinkers did, was to form habits that reinforced long-held convictions about the usefulness of scholarship in exploring reality. Seventeenth-century European science was not, after all, modern science. In China the lack of sources alternative to the missionaries' writings artificially perpetuated it for two centuries. Orthodox thinkers, whose sense of man and the cosmos was in part formed by study of canonic books, responded to the universal explanatory character that this foreign science derived from its Scholastic framework much more than to the grip on direct experience of nature that Wang valued.[26] Astronomy, as Confucian scholars understood and used it, converged with philology, gave it added weight, and obviously played a part in tipping the scale.

It would be premature to suggest any particular line of development between the recourse to Western astronomy among philosophers at the beginning of the Ch'ing period and the eventual swamping of earlier philosophical concerns by exact scholarship—scholarship of a kind that finally integrated mathematical astronomy as one specialism among many. The career of Wang Hsi-shan suggests a range of possible patterns that, tested against many other careers, can throw light on the central enigmas that shroud the collapse of imperial China.

---

26. Willard J. Peterson, in his perceptive "Fang I-chih: Western Learning and the 'Investigation of Things,'" has shown how Fang 方以智 (1611–1671) used his knowledge of Western sciences to argue for greater emphasis in philosophy upon accumulating knowledge of "physical objects, technology, and natural phenomena."

Appendix
Writings in Mathematics and Astronomy by Important Figures in Late Ming and Early Ch'ing Neo-Confucianism

The list that follows is meant to demonstrate the ubiquity of mathematical studies among late Ming and early Ch'ing orthodox scholars. It will perhaps set to rest the hoary but far from dead generalization that the Chinese literati had only literary interests and were opposed to taking up technical studies. It also shows how pervasive the influence of European scientific learning was among leading intellectuals.

This list was compiled by comparing standard biographic sources on the history of Ch'ing philosophy with those on the exact sciences. It includes a representative assortment of the best-known authors, teachers, and patrons of the various schools from the Ming-Ch'ing transition to about 1800, as recognized by Ch'ing authorities. Their recognition is, in diverse ways, tendentious. There is room for disagreement about principles for compiling such a list, or about individual additions or deletions. Still, various trials suggest that so long as only major figures are considered, the general shape of the outcome will not vary significantly. I have excluded people such as Mei Wen-ting whose achievements lay *primarily* in the exact sciences. I have also excluded the subject of mathematical harmonics. Although traditionally linked to astronomy, it was little affected by European techniques. The list also considers cosmology only when practiced in conjunction with astronomy, natural history, geography, medicine, studies of the Book of Changes, and other pursuits that use number and measure in various ways but are not primarily mathematical.[27]

These limitations should not obscure the fact that many not included were strongly rooted in the traditional qualitative sciences. Chiao Hsun 焦循 (1763-1820), for instance, was a polymath of the old type, with writings on medicine, agriculture, natural history, and applications of mathematics to the Book of Changes (a topic that interested several other renowned orthodox scholars of the time). Yen Yuan 顏元 (1635-1704) earned a living as a physician for part of his life, and Ch'eng Yao-t'ien and Chiang Fan are remembered for their contributions to natural history.

---

27. Cf. the list of early Ch'ing cosmologists in Henderson, *Development and Decline,* 142-147.

V

18

I list below eighteen names, in order of birth, along with a brief description of each individual's writings on mathematics and astronomy. I do not reproduce lists of the books, since the specialists who have use for them will find the particulars readily in the sources cited and in standard biographies in Chinese (they are often omitted from biographies in English). In order to provide ready access to traditional evaluations of these scholars' work in the sciences, each name is followed where possible with a page reference to *Ch'ou jen chuan* 疇人傳 (Biographies of astronomers and mathematicians) and its supplements, as follows:

- Juan Yuan 阮元, *Ch'ou jen chuan* 疇人傳, 1799 (*ch.* 1-46 in *Wan yu wen k'u* 萬有文庫 ed.; reprint, Shanghai: Commercial Press, 1955; page number prefixed A in references below),

- Lo Shih-lin 羅士林, *Hsu Ch'ou jen chuan* 續疇人傳, 1840 (*ch.* 47-52 in idem; references prefixed B),

- Chu K'o-pao 諸可寶, *Ch'ou jen chuan san pien* 疇人傳三編, 1886 (separately paginated in idem; references prefixed C), or

- Huang Chung-chün 黃鍾駿, *Ch'ou jen chuan ssu pien* 疇人傳四編, 1898 (Shanghai: Commercial Press, 1955; references prefixed D).

In order to facilitate access to typical information about the neo-Confucian school with which historians have associated each of those listed, I also provide, following the reference to *Ch'ou jen chuan*, a citation in Liang Ch'i-ch'ao's 梁啓超 concise and accessible *Ch'ing-tai hsueh-shu kai lun* 清代學術概論 (1921), translated by Immanuel C. Y. Hsu as *Intellectual Trends in the Ch'ing Period* (Cambridge, MA: Harvard University Press, 1959; references prefixed by L).

1. Huang Tsung-hsi 黃宗羲 (1610-1695). Wrote at least a dozen books on such topics as trigonometry, Islamic and European astronomical methods, and chronological notation, and two textbooks for teaching the Yuan methods of calendrical computation (A, III, 455; L, 36).

2. Yen Jo-chü 閻若璩 (1636-1704). Applied Chinese and European astronomical techniques to test the authenticity of the Ancient Script version of the Book of Documents in *Shang shu ku wen shu cheng* 尚書古文疏證; A, III, 502; L, 32).

3. Liu Hsien-t'ing 劉獻廷 (1648-1695). Jottings of scientific interest in his *Kuang-yang tsa chi* 廣陽雜記. Argued for including astronomical and meteorological information in local gazetteers (D, 84; L, 44).

4. Hsuan-yeh 玄曄, the K'ang-hsi Emperor (1654-1722, r. 1661-1722). He commanded neo-Confucian status in his capacity as, so to speak, pope of its

established church. He was taught mathematics by Jesuit missionaries, and took a part in educating a generation of leading Chinese practitioners of the exact sciences. He write a book on geometry, *San-chiao-hsing t'ui-suan fa* 三角形推算法 (On computation involving triangles). It is appropriate to mention in passing one of the most interesting collections of jottings on natural history and physical studies of the early Ch'ing, his *K'ang-hsi chi hsia ko-wu pien* 康熙幾暇格物編 (K'ang-hsi's notes on the investigation of things in moments of respite from the minutiae of office).[28]

5. Hui Shih-ch'i 惠士奇 (1671-1741). Author of *Chiao-shih chü yü* 交食舉隅 (Eclipses: hints for the perceptive; A, III, 512; L, 23).

6. Chiang Yung 江永 (1681-1762). Author of half a dozen mathematical books, including the influential *Shu-hsueh* 數學 (The study of mathematics), and a book on mathematical astronomy (A: III, 527; L, 23).

7. Hui Tung 惠棟 (1697-1758), son of Hui Shih-ch'i. Author of supplementary commentaries to the treatises on astronomy, astrology, and Five-Phases phenomena of the standard history of the Later Han period (L, 51).

8. Chiang Sheng 江聲 (1721-1799). Author of *Heng hsing shuo* 恆星說 (On the fixed stars; D, 91; L, 23).

9. Wang Ming-sheng 王鳴盛 (1722-1798). Included mathematical jottings in *I shu pien* 蛾術編 (Antlike learning; L, 23).

10. Tai Chen 戴震 (1724-1777). Wrote half a dozen books on astronomy and mathematics, including one on Chinese Napier's Bones and a historic treatise on astronomy. Restored the Sung government's anthology of ancient mathematical texts, *Suan ching shih shu* 算經十書. Constructed models of ancient mechanisms, including an armillary sphere (A, III, 529; L, 54).[29]

---

28. For a variorum text and translation into modern Chinese see Li Ti 李迪, *Kang-hsi chi hsia ko-wu pien i chu* 康熙幾暇格物編譯注 (Shanghai: Shang-hai Ku-chi Ch'u-pan-she, 1993). Use of this notebook would have enriched an otherwise outstanding biography, Jonathan Spence, *Emperor of China. Self Portrait of K'ang Hsi* (New York: Alfred A. Knopf, 1974).

29. Tai wrote *Hsu t'ien-wen lueh* 續天文略 (Sequel to the Treatise on astronomy) for a sequel to the twelfth-century *T'ung chih* 通志 (Comprehensive treatises). It has been translated into Japanese in Yasuda Jirô 安田二郎 & Kondô Mitsuo 近藤光男, *Tai Shin shû* 戴震集 (Tokyo: Asahi Shimbunsha, 1971), 399-499. On the models, see the preface of K'ung Kuang-sen 孔廣森 to Tai's *I shu* 遺書.

11. Ch'eng Yao-t'ien 程瑤田 (1725-1814). Used mathematics generally in study of the classics; see *Shu tu hsiao chi* 數度小記 (Notes on arithmetic and mensuration; A, III, 654; L, 23).

12. Ch'ien Ta-hsin 錢大昕 (1728-1804). Wrote a textbook of mathematics and six technical treatises on early astronomy, including reconstitutions of Han computational methods. He was an editor of Benoist's account of Copernican cosmology, *Ti-ch'iu t'u shuo* 地球圖說 (see Chapter IV).

13. K'ung Kuang-sen 孔廣森 (1752-1786, disciple of Tai Chen). His *Shao-kuang cheng fu shu nei wai p'ien* 少廣正負術內外篇 (Problems in "diminishing the breadth" and positive and negative numbers, with appended writings) includes a dozen additional titles, three of them geometrical. Some were published separately (B, 628; L, 88).

14. Ling T'ing-k'an 凌廷堪 (1757-1809). Wrote on mathematics and astronomy; co-compiler of *Ch'ou jen chuan* (A, III, 647; L, 23).

15. Chiang Fan 江藩 (1761-1831). Wrote eight treatises on varied issues in mathematical astronomy (L, 52).

16. Chiao Hsun 焦循 (1763-1820). Wrote more than a dozen books on geometry and other areas of mathematics, including a monograph on ellipses, and a detailed investigation of the date of a solar eclipse in the Book of Songs, as well as a supplement to Benoist's *Ti-ch'iu t'u shuo* (IV, 679).[30]

17. Juan Yuan 阮元 (1764-1849). Principal editor of *Ch'ou jen chuan* and *Ti-ch'iu t'u shuo* (C, 749; L, 54).

18. Wang Yin-chih 王引之 (1766-1834). Author of *T'ai-sui k'ao* 太歲考, a monograph on the fictitious counter-rotating correlate of Jupiter used for dating in the late Chou period.

Bibliography

I. Wang's Writings. Wang Hsi-shan's extant writings are listed in detail in L. Carrington Goodrich (ed.), *Dictionary of Ming Biography* (New York: Columbia University Press, 1976), 1379–1382, and in Chinese in *Wu-an li suan shu-mu* 勿

30. See Wu Yü-bin 吴裕宾, "Chiao Hsun yü *Chia chien ch'eng ch'u shih* 焦循与加减乘除释" (Chiao Hsun and his Exposition of logistic operations), *Tzu-jan k'o-hsueh-shih yen-chiu* 自然科学史研究, 1986, *5*. 2: 120-128.

菴曆算書目 (Bibliography of Mei's writings on astronomy and mathematics, 1702; in *Pai-pu ts'ung-shu chi-ch'eng* 百部叢書集成).

Shortly after Wang died, P'an Lei 潘耒 (1646-1708), a younger adherent of the loyalist group, gathered a number of his essays, some of them already circulating collectively as *Hsiao-an hsien-sheng I shu* 曉庵先生遺書(Literary legacy of Wang Hsi-shan). P'an called his larger collection *Hsiao-an i shu* 曉庵遺書. The contents of greatest interest are *Yuan chieh* 圜解 (On trigonometry) and *Li fa* 厤法 (Astronomical methods, completed 1663), to which P'an appended 24 tables. *Li fa,*which embodied Wang's system for calendrical calculation, was included in the imperial *Ssu k'u chüan shu* 四庫全書 manuscript library under the title *Hsiao-an hsin fa* 曉庵新法, in 6 ch., without the tables. The *Shou shan ko ts'ung-shu* 守山閣叢書collection printed it in 1838, and Wang's equally important *Wu hsing hsing tu chieh* 五星行度解 (On the angular motions of the Five Planets) in 1839. They were reprinted in *Chung-hsi suan-hsueh ts'ung-shu* 中西算學叢書, 1st ser. (1896) and *Ts'ung-shu chi ch'eng* 叢書集成, 1st ser. (1926). The first eds. are cited here.

About 1890 Li Sheng-to 李盛鐸 combined the two treatises with *Ta-t'ung li fa ch'i-meng* 大統厤法啓蒙, an elementary introduction to the Great Concordance system used officially for most of the Ming period (1384-1644, along with the surviving short essays, to form the *Wang Hsiao-an hsien-sheng i shu* 王曉庵先生遺書, vols. XXXI-XXXV in *Mu hsi hsuan ts'ung-shu* 木犀軒叢書.

II. **Secondary Literature.** The most thorough study of Wang's life and astronomical work, based on unpublished as well as published sources, is Hsi Tse-tsung 席泽宗, "Shih lun Wang Hsi-shan te t'ien-wen kung-tso 试论王锡阐的天文工作" (An essay on the astronomical work of Wang Hsi-shan), *K'o-hsueh-shih chi-k'an* 科学史集刊, 1963, 6: 53–65. Its references provide an excellent starting point for further study. Earlier sources cited here are the MS "Wang Hsiao-an hsien-sheng i-shu pu-pien 王曉闇先生遺書補編" (Supplement to the posthumous works of Wang Hsi-shan) in the Beijing University Library, and a biography by Wang Chi 王濟 in *Sung-ling wen lu* 松陵文錄 (Literary records of Wu-chiang, 1874).

The first detailed account of Wang's work, based mainly on excerpts from his writings before they had been printed separately, was in *Ch'ou jen chuan* 疇人傳 (Biographies of mathematical astronomers, 1799; Shanghai: Commercial Press, 1935), II, 421–446. This programmatic compendium, which included European as well as Chinese figures, greatly influenced the style of nineteenth-century and later investigations in the exact sciences. I cite it and its sequels in the Appendix.

No more than isolated sentences from Wang have been published in translation. I have translated *Wu hsing hsing tu chieh* for eventual publication in a source book of Chinese science.

III. European Science in Seventeenth-Century China. Western Sinologists used to pay little attention to the early mathematical encounter of East and West. Several important books have crossed the watershed, studying the broad context of changes in intellectual life and demonstrating the role of European science: Jacques Gernet, *Chine et christianisme. Action et réaction* (Bibliothèque des histoires, 39; Paris: Éditions Gallimard, 1982), translated by Janet Lloyd as *China and the Christian Impact: A Conflict of Cultures* (Cambridge University Press, 1985); Benjamin Elman, *From Philosophy to Philology. Intellectual and Social Aspects of Change in Late Imperial China* (Cambridge, MA: Council on East Asian Studies, Harvard University, 1984), and John Henderson, *The Development and Decline of Chinese Cosmology* (New York: Columbia University Press, 1984). More technical is Keizo Hashimoto, *Hsü Kuang-ch'i and Astronomical Reform—The Process of the Chinese Acceptance of Western Astronomy 1629–1635* (Tokyo University Press, 1988). Hashimoto examines closely the process that led to the Ming court's not-quite-sponsorship of the Jesuits' writings, and provides fresh information on the European treatises that they adapted.

The only general history of the Chinese response to European exact sciences is in Wang P'ing 王萍, *Hsi-fang li-suan-hsueh chih shu-ju* 西方曆算學之輸入 (The introduction of Western astronomy and mathematics; Monographs, 17; Nankang, Taiwan: Institute of Modern History, Academia Sinica, 1966), summarized in *Journal of Asian Studies,* 1970, *29:* 914–917. This book draws heavily on the materials in *Ch'ou jen chuan* for the seventeenth and eighteenth centuries. There is a good deal of technically sophisticated information on China in Shigeru Nakayama, *A History of Japanese Astronomy. Chinese Background and Western Impact* (Harvard-Yenching Institute Monograph Series, 18; Cambridge, MA: Harvard University Press, 1969).

A useful tool for further study of both Jesuit and Chinese mathematical activities is Li Yen 李儼, "Ming-ch'ing chih chi Hsi suan shu-ju Chung-kuo nien-piao 明清之際西算輸入中國年表" (Chronology of the introduction of Western mathematics into China in the Ming-Ch'ing transition), in *Chung suan shih lun-ts'ung* 中算史論叢 ("Gesammelte Abhandlungen über die Geschichte der chinesischen Mathematik," vol. III; rev. ed., Beijing: K'o-hsueh ch'u-pan-she, 1955): 10–68. Yabuuchi Kiyoshi 藪内清 has ably surveyed Jesuit activity in *Chûgoku no temmon rekihô* 中國の天文曆法 (Chinese astronomy; Tokyo:

Heibonsha, 1969), 148–174, and "Kinsei Chûgoku ni tsutaerareta Seiyô temmongaku" 近世中國に傳えられた西洋天文學 (On the Western astronomy transmitted to China in modern times), *Kagakushi kenkyû,* 1954, *32:* 15–18. A rev. ed. of Yabuuchi's book appeared in 1990. See also the important group of studies in Yabuuchi and Yoshida Mitsukuni 吉田光邦, eds., *Min Shin jidai no kagaku gijutsu shi* 明清時代の科學技術史 (History of science and technology in the Ming and Ch'ing Periods; Research Report, Research Institute of Humanistic Studies; Kyoto: The Institute, 1970), 1-146. Joseph Needham, in *Science and Civilisation in China,* III (Cambridge University Press, 1959), 437–458, was the first to suggest that the limitations as well as the strengths of the Jesuit missionaries greatly affected the character of the Chinese response. His short and incidental discussion of Wang Hsi-shan (454) is, however, not very accurate.

The lives of the Jesuit missionaries and their publications in Chinese have been closely documented by historians of that order. See Henri Bernard, "Les adaptations chinoises d'ouvrages européens. Bibliographie chronologique depuis la venue des Portugais à Canton jusqu'à la Mission française de Pékin, 1514–1688," *Monumenta serica,* 1945, *10:* 1–57, 309–388; Joseph Dehergne, *Répertoire des Jésuites de Chine de 1552 à 1800* (Roma: Institutum Historicum S. J., 1973); and Louis Pfister, *Notices biographiques et bibliographiques sur les Jésuites de l'ancienne mission de Chine, 1552–1773* (2 vols., Shanghai: Imprimerie de la Mission Catholique, 1932–1934, completed before Pfister's death in 1891). Dehergne is a comprehensive guide to the extensive literature on missionaries, including archival sources; the last part includes aids to research. There are a number of useful essays in (Maur) Fang Hao, S. J., *Fang Hao wen lu* 方豪文錄: *Studies in the History of the Relations between China and the West* (Peiping: Institutum Sancti Thomae, 1948). Bernard may be supplemented by Erik Zürcher, Nicolas Standaert, S. J., & Adrianus Dudink, *Bibliography of the Jesuit Mission in China (ca. 1580-ca. 1680)* (Center of Non-Western Studies, Leiden University, 1991). The essays in Charles E. Ronan & Bonnie B. C. Oh (ed.), *East Meets West. The Jesuits in China, 1582–1773* (Chicago: Loyola University Press, 1988) are on the whole conventional and superficial.

On the background of the mission to China see Juan Casanovas, "Alle origini del Missionariato Scientifico nell'Asia orientale: Clavio e il Collegio Romano," in Isaia Iannaccone & Adolfo Tamburello (ed.), *Dall'Europa alla Cina: contributi per una storia dell'Astronomia* (Napoli: Università degli Studi "Federico II" & Istituto Universitario Orientale, 1990), 75–84. Only a couple of the papers in this collection are concerned with the astronomical encounter between Europeans and Chinese.

V

24

For writings in Chinese, see Hsu Tsung-tse 徐宗澤, *Ming Ch'ing chien Ye-su-hui-shih i chu t'i yao* 明清間耶穌會士譯著提要 (Annotated bibliography of Jesuit translations and writings in the Ming and Ch'ing periods; Taipei: Chung-hua Shu-chü, 1958), with indexes of authors, titles, and subjects. On missionaries as Chinese officials see Po Shu-jen 薄树人, "Ch'ing Ch'in-t'ien-chien jen-shih nien-piao 清钦天监人事年表" (Chronology of personnel in the Ch'ing Directorate of Astronomy), *K'o-chi-shih wen-chi* 科技史文集, 1978, 1: 86–101. For the history of the Directorate see Jonathan Porter, "Bureaucracy and Science in Early Modern China: The Imperial Astronomical Bureau in the Ch'ing Period," *Journal of Oriental Studies* (Hong Kong), 1980, *18:* 61–76. Huang I-nung 黃一農 narrates the remarkable ascent of Schall in Ch'ing officialdom in "T'ang Jo-wang yü Ch'ing ch'u hsi li chih cheng-t'ung-hua 湯若望與清初西歷之正統化" (Schall and the legitimation of the Western calendar at the beginning of the Ch'ing period), in *Hsin pien Chung-kuo k'o-chi-shih* 新編中國科技史 (New history of Chinese medicine; Taipei: Yin Ho Wen-hua Kung-ssu, 1990), II, 465–491.

European scientific works owned by the Jesuits in Beijing—one of the world's greatest collections of scientific writings of the sixteenth through eighteenth centuries—are listed in H. Verhaeren, *Catalogue of the Pei-t'ang Library* (3 vols., Beijing: Imprimerie des Lazaristes, 1944–1948). This collection is now in the Beijing Library, where I have examined its two copies of *De revolutionibus.* A list of 251 astronomical books has been excerpted from Verhaeren in Henri Bernard-Maitre, "La science européene au tribunal astronomique de Pékin (XVIIe-XIXe siècles)," in *Conférences du Palais de la Découverte,* ser. D, 9 (Paris: Le Palais, 1951). See also Boleslaw Szcześniak, "Note on Kepler's *Tabulae Rudolphinae* in the Library of Pei-t'ang in Pekin," *Isis,* 1949, *40:* 344–347.

Among studies of Chinese responses in the mathematical sciences, the book of Wang P'ing is additionally helpful because of its index, still unusual in Chinese scholarly books. There is a penetrating analysis in Jean-Claude Martzloff, "Space and Time in Chinese Texts of Astronomy and of Mathematical Astronomy in the Seventeenth and Eighteenth Centuries," with a valuable commentary by Jean Gernet, *Chinese Science,* 1993–1994, 11: 66–102.

For systematic annotated bibliographies, see Ting Fu-pao 丁福保 and Chou Yun-ch'ing 周雲青, *Ssu pu tsung lu suan-fa pien* 四部總錄算法編 (General register of the Quadripartite Library, section on mathematics; Shanghai: Commercial Press, 1957) and *Ssu pu tsung lu t'ien-wen pien* 四部總錄天文編 (General Register . . . Section on astronomy; *idem,* 1956), supplemented by Li Yen, "Chin-tai Chung suan chu-shu chi 近代中算箸述記" (Notes on books about

Chinese mathematics in modern times), in *Chung suan shih lun-ts'ung*, II (1954), 103–308; and "Ch'ing-tai wen-chi suan-hsueh lei lun-wen 清代文集算學類論文" (Articles that can be classified as mathematical in collected literary works of individuals in the Ch'ing period), *ibid.*, V (1955), 76–92.

The first resorts for biographies of the most prominent late imperial scientific figures are *Dictionary of Ming Biography* and Arthur W. Hummel, *Eminent Chinese of the Ch'ing Period* (2 vols., Washington: U. S. Government Printing Office, 1943–1944). *Ch'ou jen chuan* mainly provides long excerpts from technical writings, without much biographical data.

A few topical studies throw light on fundamental issues. Pasquale D'Elia describes the genesis, content, and distribution of major Jesuit scientific writings in "Presentazione della prima traduzione chinese di Euclide," *Monumenta serica*, 1956, *15:* 161–202, and Henri Bernard-Maître, "L'encyclopédie astronomique du Père Schall. *'Tch'ong-tcheng li-chou,'* 1629 et *'Si-yang sin-fa li-chou,'* 1645. La réforme du calendrier chinois sous l'influence de Clavius, de Galilée et de Kepler," *ibid.*, 1938, *3:* 35–77, 441–527. Willard J. Peterson summarizes early European writings in Chinese on the qualitative sciences in "Western Natural Philosophy Published in Late Ming China," *Proceedings of the American Philosophical Society*, 1973, *117:* 295–322.

The life, associations, and work of Wang Hsi-shan's contemporary Mei Wen-ting have been treated at length in Li Yen, "Mei Wen-ting nien-p'u 梅文鼎年譜" (Chronological biography of Mei Wen-ting), in *Chung suan shih lun-ts'ung*, III, 544–576; in Hashimoto Keizô 橋本敬造, "Bai Buntei no rekisangaku. Kôki nenkan no temmon rekisangaku 梅文鼎の暦算學。康熙年間の天文暦算學" (The mathematical astronomy of Mei Wen-ting. Mathematical astronomy in the K'ang-hsi Period), *Tôhô gakuhô* 東方學報 (Kyoto), 1970, 41: 491–518, and "Bai Buntei no sugaku kenkyû 梅文鼎の數學研究" (The Mathematical researches of Mei Wen-ting), *ibid.*, 1973, 44: 233–279; and in Liu Tun 刘钝, "Ch'ing ch'u li suan ta shih Mei Wen-ting 清初暦算大师梅文鼎" (Mei Wen-ting, master mathematician at the beginning of the Ch'ing), *Tzu-jan pien-cheng-fa t'ung-hsun* 自然辩证法通讯, 1986, 1: 52–64. On the traditional and European mathematical books that Mei read, see Martzloff, *Recherches sur l'oeuvre mathématique de Mei Wen-ting (1633–1721)* (Mémoires de l'Institut des Hautes Études Chinoises, 16; Paris: L'Institut, 1981), 25–31. On Mei Ku-ch'eng and his colleagues see Hashimoto's "Rekisho kôsei no seiritsu. Shindai shoki no temmon rekisangaku 暦象考成の成立。清代初期の天文暦算學 (The compilation of the *Li hsiang k'ao ch'eng*. Mathematical astronomy in the early Ch'ing period), in *Min Shin jidai no kagaku gijutsushi*, 49–92. Hashimoto completed his chronological account in "Daenhô no tenkai. Rekishô kôsei kôhen no naiyô ni tsuite 橢圓髮の展開。曆象考成後編の内容について" (The development of elliptical methods. On the content of the *Li hsiang k'ao ch'eng hou pien*), *Tôhô gakuhô* (Kyoto), 1971, *42:* 245–272.

The thought of Fang I-chih 方以智, probably the first Chinese to acquaint himself with the full spectrum of European sciences, has been examined by Sakade Yoshinobu 坂出祥伸, "Hô Ichi no shisô 方以智の思想" (The thought of Fang I-chih), in Yabuuchi and Yoshida, *Min Shin jidai no kagaku gijutsu shi,* 93–134; and by Peterson, "Fang I-chih: Western Learning and the 'Investigation of Things,'" in W. T. de Bary and the Conference on Seventeenth-Century Chinese Thought, *The Unfolding of Neo-Confucianism* (Studies in Oriental Culture, 10; New York: Columbia University Press, 1975), 369–411, an important volume for seventeenth-century thought. For a more general study see Peterson's "Fang I-chih's Response to Western Knowledge" (Unpublished Ph.D. dissertation, Harvard University, 1970). Sakade and Peterson are best read together. Sakade pays little attention to Fang's treatment of European ideas and techniques, and Peterson is not concerned with the Chinese scientific tradition or the evolution of European science. Peterson's biography *Bitter Gourd: Fang I-chih and the Impetus for Intellectual Change* (New Haven: Yale University Press, 1979), is mainly concerned with Fang as a humanistic philosopher. There is little overlap with his earlier work.

D'Elia narrates the introduction of cosmology into China in *Galileo in China, Relations Through the Roman College Between Galileo and the Jesuit Scientist-missionaries (1610–1640)* (trans. Rufus Suter and Matthews Sciascia; Cambridge, MA: Harvard University Press, 1960), but the emphasis on demonstrating Jesuit accomplishments obscures a number of basic issues. The same is true of Bernard, "Galilée et les Jésuites des missions d'Orient," *Revue des questions scientifiques,* 4e sér., 1935, *28:* 356–382. Somewhat more useful is Szcześniak, "Notes on the Penetration of the Copernican Theory into China (Seventeenth–Nineteenth Century)," *Journal of the Royal Asiatic Society,* 1945, 30–38. A rather black-and-white analysis with important additional information is Hsi Tse-tsung *et al.*, "Heliocentric Theory in China," *Scientia sinica,* 1973, *16:* 364–376. More limited in scope is Yen Tun-chieh 严敦杰, "Ga-li-lueh ti kung-tso tsao-ch'i tsai Chung-kuo ti ch'uan-pu 伽利略的工作早期在中国的传布" (The early dissemination of Galileo's work in China), in *K'o-hsueh-shih chi-k'an,* 1964, 7: 8–27. On the notion that Western mathematics originated in China, see Chiang Hsiao-yuan 江晓原, "Shih lun Ch'ing-tai 'Hsi-hsueh Chung yuan' shuo 试论清代西学中原说" (An essay on the Ch'ing-period claim that Western studies originated in China), *Tzu-jan k'o-hsueh-shih yen-chiu* 自然科学史研究, 1988, *7.*

2: 101–108, and Sivin, "On 'China's Opposition to Western Science During Late Ming and Early Ch'ing,'" *Isis,* 1965, *56:* 201–205. See also J. W. Witek, *Controversial Ideas in China and in Europe. A Biography of Jean-François Foucquet, S.J. (1665–1741)* (Rome: Institutum Historicum S. J., 1982). Foucquet wrote and circulated, but did not publish, a thorough critique of the first-generation Jesuit writings. A most interesting study is Harriet Zurndorfer, "Comment la science et la technologie se vendaient à la Chine au XVIIIe siècle: Essai d'analyse interne," *Études chinoises,* 1988, *7.* 2: 59–90.

Mikami Yoshio 三上義夫 was the first to note that seventeenth-century Chinese astronomers used European techniques mainly to revive their own tradition, in "*Chûjin den* ron 疇人傳論" (A study of the *Ch'ou jen chuan), Tôyô gakuhô,* 1927, *16:* 185–222, 287–333. Mikami's argument was repeated in "Chinese Mathematics," *Isis,* 1928, *11:* 125.

Retrospect

This biography is in large part a revision of the one published in *Dictionary of Scientific Biography*, XIV (New York: Charles Scribner's Sons, 1976), 159–168. Because so few Chinese astronomers of this crucial period in which European science received its first responses had been studied, that article was quite broad in scope. It emphasized Wang's technical work. Shortly afterward I accepted the invitation of L. Carrington Goodrich to prepare for a Sinological readership an essay for his *Dictionary of Ming Biography* that would be more attentive to Wang's life outside science. Fang Chao-ying, who wrote most of the articles on scientists in *Eminent Chinese*, reviewed the MS, and made several excellent suggestions for placing Wang in the intellectual milieu of his time. I requested his name be added to the second article as co-author.

I have integrated into this version the fruit of both explorations. It also incorporates the appendix and other portions of a version slightly expanded in 1976 for a book of essays that was never published. This new edition uses and cites studies published after 1976, and minimizes overlap with the Chapter IV. Readers will find it useful to consult the latter for evidence underlying my arguments here about the interaction of Chinese and European astronomy.

VI

Science and Medicine in Chinese History

The historical discoveries of the last generation have left no basis for the old myths that the ancestry of modern science is exclusively European and that before modern times no other civilization was able to do science except under European influence. We have gradually come to understand that scientific traditions differing from the European tradition in fundamental respects—from techniques, to institutional settings, to views of nature and man's relation to it—existed in the Islamic world, India, and China, and in smaller civilizations as well. It has become clear that these traditions and the tradition of the Occident, far from being separate streams, have interacted more or less continuously from their beginnings until they were replaced by local versions of the modern science that they have all helped to form.

Central to this clearing of the air has been the study of China. There the record of technical endeavor has been fullest, most continuous, most accessible, and most exactly dated. That is only what one would expect of a state that was administered by a full-fledged bureaucracy two thousand years ago and a country where books were being routinely printed four hundred years before the Gutenberg Bible.

Understanding the technical traditions of other cultures is not just a matter for exotic tastes. To the contrary, this wider awareness has transformed our sense of Europe's development. Consider, for example, Francis Bacon's influential attempt, shortly after 1600, to explain that great efflorescence of human knowledge and activity that we now call the Renaissance:

This chapter is a revision of "Science in China's Past," in *Science in Contemporary China,* ed. Leo Orleans (Stanford: Stanford University Press, 1980), 1–29.

> Again, it is well to observe the force and virtue and consequences of discoveries, and these are to be seen nowhere more conspicuously than in those three which were unknown to the ancients, and of which the origin, though recent, is obscure and inglorious; namely, printing, gunpowder, and the magnet. For these three have changed the whole face and state of things throughout the world; the first in literature, the second in warfare, the third in navigation; whence have followed innumerable changes, insomuch that no empire, no sect, no star seems to have exerted greater power and influence in human affairs than these mechanical discoveries.[1]

Bacon's understanding was still conventional in 1920, although by then the attention of historians had largely shifted back from inventions to politics and religion ("empires" and "sects") if not to astrology ("stars").

Today the origins of Bacon's three inventions are a great deal less obscure. None of the three was in fact a European discovery. Printing I have already mentioned. Movable clay type was used to mold inscriptions in Chinese cast bronze vessels of the mid second millennium B.C. The earliest extant texts printed on paper from carved wood blocks originated in Korea by 751 and in Japan about 770 (nearly one thousand years after the Chinese invention of paper and four hundred years before its introduction to Europe).

Because printers needed thousands of different characters, labor was cheap, and books were published in a series of small printings, printing with movable type developed slowly and did not replace the older process until the twentieth century. We have an account of Chinese ceramic movable-type printing in the 1040s. The oldest such book that still exists (set in wood type) dates from the 1350s. It is not in Chinese but in the Tangut script of the northwest frontier. Metal type was in use still earlier. Printing with it was brought to a high state of perfection in Korea through a succession of royally subsidized experiments that culminated in the early fifteenth century.

As printing evolved across East Asia, its potentialities for changing "the face and state" of literature were by no means unexplored. By 1100, for instance, the Chinese government had authorized the printing of standard collections of texts in an attempt—often repeated but always with limited success—to control education. This was the case not only in the humanistic classics but also in such fields as mathematics and medicine.

As for gunpowder, formulas for flare mixtures appear in Chinese alchemical books of the ninth century A.D. or a little earlier. The propor-

1. Francis Bacon, *The New Organon and Related Writings*, Library of Liberal Arts, no. 97 (New York: Liberal Arts Press, 1960), 118. Joseph Needham, *The Grand Titration: Science and Society in East and West* (Toronto: University of Toronto Press, 1969), 62–76, has several interesting observations on this passage.

tions that make gunpowder explosive were known by 1050, and up-to-date commanders over the next two centuries had at their disposal increasingly destructive flamethrowing devices, bombs hurled by trebuchets, and rocket weapons. Between 1270 and 1320 metal-barreled cannon appeared on the Mongols' main battlefronts in Europe, the Islamic west, and China. Although it is not beyond doubt that the inventions that constitute the cannon were finally assembled in Cathay, the prehistory of the crucial propellant and of weapons that used it to fire projectiles was clearly Chinese.

Magnetic attraction was not a matter of recent knowledge in Bacon's time. It is clear that he was speaking of its use in the navigational compass, for the polarity and directive property of lodestone were known and applied to steering by Europeans late in the twelfth century. In China it is likely that a pivoted lodestone spoon was used for divination in the first century A.D. Well before 1100 the true compass—apparently developed earlier for geomantic siting—was being used in navigation, and the declination of the needle was known to steersmen as an effect, if not as a concept.

Bacon's statement about the springs of change in the West reminds us, then, that for many centuries Europe was on balance a beneficiary of technology transfers. Through the long centuries in which Europe gradually relearned the arts of civilization after the Roman Empire collapsed, China continued to build on its high culture. Many important inventions that apply the power of water or wind to machines are first found in Europe more than ten centuries later than in China. A few Chinese priorities of other types are an efficient harness for draft animals, the drawloom, deep borehole drilling, the segmental arch bridge, rigging that allowed ships to sail to windward, the axial rudder, porcelain, and cast iron.

A Chinese visiting one of the great cities of Europe in 1400 would still have found it backward. In Bacon's lifetime some Occidental cities could no longer be considered backward. By 1840 what appeared glorious about China to its more sophisticated visitors were mainly vestiges of the past.

Some of the reasons for the reversal in technological preeminence are internal to China: centuries of disastrous fiscal and other administrative policies, the remorseless pressure of increasing population, and a large measure of social stability and cultural homogeneity that left traditional values and forms practically unchallenged as the creativity behind them was sapped by intellectual orthodoxies. Other reasons for the reversal arose in Europe, above all a universal quantitative and logical approach to empirical knowledge and practice that gradually redefined nature, reshaped society, and remade human consciousness. In the final reckoning the predominance of Western science and technology cannot be explained by contributing factors entirely within Europe or outside it, for it was built on the intercourse of civilizations.

CHINESE AND WESTERN, TRADITIONAL AND MODERN

It is almost impossible to think about traditional Chinese science and technology without comparing them, explicitly or implicitly, with their present-day analogues. Otherwise it would be difficult even to identify what past activities are of technical interest. Nevertheless, the transforming influence of the scientific and industrial revolutions was so great that the earlier sciences of China and Europe resemble each other more than either resembles the modern variety. It is important, if one is to think clearly about science and technology as worldwide phenomena, to avoid confusing differences between China and the West with differences between traditional societies and societies that have become essentially modern.

Certain aspects of today's technical activity that may appear to be universal are in fact peculiarly European or peculiarly modern. We consider normal, for instance, a sequence of sciences derived from classical Greek schemes of knowledge but formed by the gradual spread in modern times of system and quantification, beginning with physics and moving through chemistry, the life sciences, and ultimately (or so Auguste Comte believed) through psychology, the yet-to-be-achieved exact science of the human mind. In Europe this scheme and its precursors did not, when they arose, describe established structures of knowledge but justified projects to bring them about by creating new institutions and habits of thought.

In China there was, for example, no biology. Observations and theoretical perspectives on the manifestations of life were scattered through a very different grouping of knowledge that I will sketch out below. Alchemists, East and West, set down a great deal of information about chemical processes. How much they learned from physicians and illiterate craftsmen we do not know. In any case the designation of Chinese alchemy as early chemistry ignores the fact that alchemical goals were not cognitive but spiritual. Most topics of modern physics had not been imagined before the demise of traditional Chinese science, and the more old-fashioned subjects—for example, heat, sound, and magnetism—were in China a matter of dispersed experiences and reflections rather than coherent disciplines.

Another practically universal modern assumption is that technology is applied science, that technological progress and economic benefit are the natural end of new scientific knowledge. Before the mid eighteenth century, emerging modern science did not affect technology in Europe. Craft traditions were developed by people who, even as literacy began to diffuse through society, had little access to science and little reason to use it.

The old certainty that the industrial revolution in Western Europe and the United States grew out of the direct application of modern science has come so seriously into question since the 1960s that we are prepared to see a much subtler influence operating since the eighteenth century. Coming out of the new Western mentality shaped by theoretical science, this influence led the educated to take an active interest in manufacturing and led artisans (along with everyone else) to begin reasoning impersonally and abstractly about facts, processes, commodities, and labor to an extent unprecedented in human history. The design of economic systems to exploit new scientific knowledge is, at least outside the chemical industry, mainly a twentieth-century phenomenon. Familiar though the direct linkage of science to technology may be to readers of this book, it would have appeared wildly exotic to all except a visionary few three hundred years ago in Europe or Asia.

Since China's industrial revolution began less than two generations ago, what we find previously—as in early Europe—are sciences and manufacturing techniques that had little to do with each other. The sciences reflected the concerns of the tiny literate elite, their cosmologies, and the managerial problems they encountered in their careers and recorded in their writings. Technology was on the whole a matter of craft traditions, passed down privately from father to son or from master to apprentice. It was on these mainly oral and manual traditions rather than on cumulative science, recorded in writing, that the technological preeminence of China was built.

What we know about the early industrial arts comes not from those who did the work but from scholars writing for fellow dilettantes or from officials writing for their peers who happened to be curious about the work of the lower orders. As the learned compiler of the great technological encyclopedia of 1637 put it, "While the best rice is cooking fragrantly in the palace kitchen, perchance one of the princes would wish to know what farming implements look like; or, while the officers of the Imperial Wardrobe are cutting suits of brocade, another of the princes might wonder about the techniques of silk weaving." Those were the circumstances in which a civil servant might actually feel a need for the book. "Let the ambitious scholar toss this book onto his desk and give it no further thought; it is a work that is in no way concerned with the art of advancement in offialdom."[2]

2. Sung Ying-hsing, *T'ien-kung K'ai-wu: Chinese Technology in the Seventeenth Century*, trans. E-tu Zen Sun and Shiou-chuan Sun (University Park: Pennsylvania State University Press, 1966), xiii–xiv (translation modified). Another translation of the technological encyclopedia worth consulting is Li Ch'iao-p'ing et al., trans., *Tien-kung-kai-wu: Exploitation of the Work of Nature*, Chinese Culture Series, no. 3 (Taipei: China Academy, 1980).

Thus in both China and Europe before modern times science and technology went their own ways. Natural philosophers learned from the techniques being practiced around them, at least to an extent that their practical curiosity dictated, but what they read in books counted for a great deal more. The achievements of artisans did not depend on the enhanced scientific understanding of their betters.

SCIENCE AND SCIENCES IN CHINA

In Europe since classical times the various sciences were part of a single structure that included all systematic rational knowledge. They were part of *scientia,* the part that has to do with nature. When in the early seventeenth century Francis Bacon (1561–1626) in England began inventing a physics to which experiment was the key to learning what was fact and what was not, and Galileo Galilei (1564–1642) in Italy imagined a universe full of physical motions that could be measured and computed, these two styles of science—the experimental and the mathematical—developed separately for a while. But they were still seen as parts of one endeavor.

In China there was no single structure of rational knowledge that incorporated all the sciences. Knowing was an activity in which the rational operations of the intellect were not sharply disconnected from what we would call intuition, imagination, illumination, ecstasy, aesthetic perception, ethical commitment, or sensuous experience. The various sciences, unlike those of Europe, were neither circumscribed by the philosophies of their time nor subordinated to theology (which did not exist in East Asia). The sciences developed a great deal more independently of each other than in the West. The practitioners of each science extended and revised the concepts and assumptions about physical reality with which that science began (the particular sciences emerged between the second century B.C. and the first century A.D.). Over the centuries they seldom responded directly to contemporary philosophic innovations. For example, Chu Hsi (1130–1200), perhaps the most influential moral philosopher of the last two thousand years, was intensely interested in astronomy and cosmology, but what he knew of these fields was grossly outmoded; astronomers returned the compliment by ignoring his frequently odd opinions about the sky. They were free to exert that autonomy because no institution enforced the authority of Chu Hsi over them. There is an obvious contrast with the educational institutions of Europe, from Plato's Academy and Aristotle's Lyceum to the medieval and early modern universities, in which the natural sciences were kept subordinate to philosophy.

Although in China the sciences were relatively autonomous provinces of knowledge, there were definite limits to what was expected of them.

Scientists were well aware of the growth of understanding and of the increasing ability to predict, but we do not find the conviction that in the fullness of time all phenomena would yield their ultimate secrets. The typical belief was rather that natural processes wove a pattern of constant relations too subtle and too multivariant to be understood completely by what we would call empirical investigation or mathematical analysis. Scientific explanation merely expressed, for finite and practical human purposes, partial and indirect views of that fabric. This point was made with great clarity by Shen Kua (1031–1095), a polymath who served as astronomer-royal and who made lasting contributions to practically every science.

> Those in the world who speak of the regularities underlying the phenomena, it seems, manage to apprehend their crude traces. But these regularities have their very subtle aspect, which those who rely on mathematical astronomy cannot know of. Still even these are nothing more than traces. As for the spiritual processes described in the *Book of Changes* that "when they are stimulated, penetrate every situation in the realm," mere traces have nothing to do with them. This spiritual state by which foreknowledge is attained can hardly be sought through traces, of which in any case only the cruder sort are attainable. What I have called the subtlest aspect of these traces, those who discuss the celestial bodies attempt to know by depending on mathematical astronomy; but astronomy is nothing more than the outcome of conjecture.[3]

This understanding did not diminish astronomy in Shen's eyes; he devoted his best energies to improving the calendar. His perspective, far removed from that of the modern scientist, would have been quite comprehensible to his contemporaries in eleventh-century Christendom, although he was applying it to a level of scientific knowledge much more sophisticated than theirs.

To sum up, the Chinese sciences were able to attain a high standard—at times the highest in the world—without the overarching structure of natural philosophy that subsumed science in Europe and without the naive claims to universal knowledge that modern positivists have sometimes attempted to read back into the Western tradition.

THE CHINESE SCIENCES: QUANTITATIVE

It will be clear by this point that a catalogue of priorities torn out of context can only give a distorted impression of what is worth knowing about

3. Hu Daojing, ed., *Meng-ch'i pi-t'an chiao-cheng* [Brush talks from Dream Brook], modern variorum ed., 2 vols. (ca. 1090; Peking: Chung-hua Shu-chü, 1960), item 123. For a more extensive discussion of Shen's attitude, see Nathan Sivin, s.v. "Shen Kua," in *Dictionary of Scientific Biography*. An earlier astronomical passage of similar purport is translated in Nathan Sivin, *Cosmos and Computation in Early Chinese Mathematical Astronomy* (Leiden: Brill, 1969), 61–62.

science in China. Regardless of how clearly certain features of early science may appear to anticipate today's knowledge, their contemporary meanings had little in common with those of modern science. In describing the sciences below, therefore, I pay as much attention to context as to accomplishment. The Chinese sciences will be portrayed, in other words, not as a succession of triumphs of objective knowledge abstracted from a morass of superstition but as one important thread firmly woven into the fabric of culture.

The sciences described will be those the Chinese defined by applying their concepts of order to various areas of experience—the sky for astronomy and astrology, the bodies of humans and animals for medicine, and so on. As I have already pointed out, this is a very different demarcation from that of modern science.

Mathematics

It is an old Western habit (which those mathematical pioneers, the ancient Mesopotamians, did not share) to think of mathematics as a quintessentially pure science dedicated to the fundamental exploration of number and extension, with practical applications a matter of little interest to the most esteemed sort of mathematician. In China mathematics was not the queen of the sciences but their servant.

Nearly one thousand Chinese mathematical treatises, beginning in the second century A.D. or a little earlier, survive from previous centuries. The great majority have to do with practical matters of the kinds officials, their clerks, and landowners (and increasingly, in the last ten centuries, merchants) would encounter: surveying, determining areas and volumes, calculating exchange rates and taxes payable in money and commodities, and figuring costs of transportation, materials, and labor. Techniques remained numerical and algebraic, with trigonometric approximations that made it possible from the eleventh century on to do much of the work of geometry with no more need to visualize figures in space than one would need today when solving simple problems on a calculator.

A severely logical and axiomatic corpus of proofs like that of Euclid never emerged to provide a unifying standard of rigor in the quantitative sciences. One cannot argue that national character ruled such rigor out, for we find the beginnings of a universal set of definitions, spanning science, language, ethics, and other fields, in the surviving writings of the abortive Mohist school (ca. 300 B.C.). That some Chinese were capable of constructing geometric proofs is clear from surviving examples beginning in the third century A.D., but this remained as minor a concern as numerical procedures were in Europe before algebra was introduced from the Orient beginning in the thirteenth century.

Chinese mathematics was instrumental from the start. At first, problems were set up with computing rods on a surface ruled off in squares

Figure 7.1. Example of matrix algebra notation from a textbook of 1303. The leftmost of these three simultaneous equations corresponds to $-2x + 2y + z = 15$. The equations are written from top to bottom, and a negative coefficient is indicated by a diagonal slash. Figures 7.1 through 7.5 are reproduced from Joseph Needham, *Science and Civilisation in China* (Cambridge, 1954–), by permission of Cambridge University Press.

like a checkerboard. As the rods were manipulated in appropriate squares, the counting board could represent a two-dimensional array of numbers (figure 7.1). The potentialities of this device were gradually exploited until by ca. 1300 it was being used to manipulate the coefficients of equations in several powers of up to four unknowns. By that time the counting board was being replaced by the abacus, an instrument roughly as speedy as but no more flexible than an early twentieth-century adding machine, and extremely well suited to the routine needs of the growing urban merchant class. Because the abacus could only represent a dozen or so digits in a linear array, it was useless for the most advanced algebra until it was supplemented by pen-and-paper notation. There were few important innovations at the highest level of mathematics from the mid fourteenth century until the seventeenth century, when the Jesuit missionaries prompted an efflorescence of interest in European geometry, true trigonometry, logarithms, and so on. This hiatus may have been part of the price paid for the convenience of the abacus.

The predominantly practical orientation of Chinese mathematics made it neither inferior nor superior to the Western tradition. Its lack of development at the abstract geometric level was balanced by its strength in numerical problem solving. Algebra reached Europe ready-made after it had emerged from a long process of convergent discovery and interaction in China, India, and the Islamic world. Many comparatively late European "discoveries," such as Pascal's triangle of coefficients (published in 1665, but known in the sixteenth century) and Horner's method for solving numerical higher equations (1819), were, we now know, named in ignorance of their Chinese origins (the former ca. 1100 or earlier, the latter ca. 1245).

There was a theoretical and speculative side of Chinese mathematics that modern historians have generally ignored, at some cost to our understanding of what the art meant to its practitioners. The senses of both words used for mathematics before modern times, *shu* and *suan,* include numerology. They refer as well to a variety of divination techniques that identify regularities—not necessarily quantitative—underlying the flux of natural phenomena.

Often, especially in the early centuries of mathematics, prognosticating the future and divining the hidden were thought to be among the powers of master computators. There is an obvious parallel with the complementarity of mathematical astronomy and astrology, which we find not only in China, India, and Islam, but in the West as late as the time of Kepler. The form of the relationship of course varied with the intellectual and social character of the two activities in each culture.

Mathematical Astronomy

According to the Chinese theory of monarchy, the ruling dynasty remained fit to rule because of the accord the emperor maintained with the cosmic order. This accord depended on his personal virtue and his correct performance of certain rituals. His special status in the order of nature enabled him to maintain a corresponding order in the political realm, for the state was a microcosm. If the emperor lacked virtue or was careless in his duties, disorderly phenomena would appear in the sky or elsewhere in nature as a warning of potential disaster in the political sphere.

This theory divided celestial phenomena into those that were regular and could be computed and those that were irregular and unpredictable and thus omens. Astronomers had two tasks. First, mathematical astronomy (*li*) was to incorporate as many phenomena as possible in a correct calendar—actually an ephemeris that included, in addition to days and lunar months, predictions of planetary phenomena and eclipses. Second, astrology (*t'ien-wen,* which in modern Chinese has come to mean astronomy) was to observe unpredictable phenomena and to interpret their political meaning. Thus, the emperor could be warned that all was not well in his realm, so that he could mend his ways and take appropriate administrative measures—or be reassured if the omen was favorable.

The calendar, issued in the emperor's name, became part of the ritual paraphernalia that demonstrated his dynasty's right to rule (a function not entirely different from that of economic indicators in a modern nation). Astrological observations could easily be manipulated and thus could be dangerous in the hands of someone trying to undermine the current dynasty (the analogy with economic indicators is again obvious). It was therefore a principle of state policy that the proper place to do astronomy was the imperial court. In certain periods it was illegal to do it elsewhere.

The most sophisticated accomplishments of Chinese mathematics are largely concentrated in astronomical and chronological reckoning. Astronomers' ability to use what amounts to higher-order equations, to deal with apparent rather than mean celestial motions, and to determine astronomical constants with great precision grew steadily until the ephemerides reached their zenith shortly before 1300. Although lunar eclipses could be predicted with considerable accuracy by 100 B.C., the lack of the spherical geometry or trigonometry needed to calculate accurately the intersection of the moon's shadow cone with the earth's sphere made solar eclipses an abiding problem. The tendency of the imperial court to look abroad for technicians who could deal with it was fateful for the development of astronomy within China, as we will see below. Beneath the steady evolution of computational astronomy over two thousand years in China lay a foundation not only of observational instruments but of data recording systems for centuries on end. Joseph Needham has summarized what is characteristically Chinese in this foundation:

1. The elaboration of large and complex observational and demonstrational instruments, from at least the second century B.C. to the great bronze armillaries of 1421 still on display at the Purple Mountain Observatory, Nanjing (figure 7.2).
2. The invention of the clock drive, and perhaps of the clock escapement itself, in a long series of great astronomical clocks culminating in the eleventh century. A water-driven mechanism rotated an armillary sighting tube for observation, a celestial globe, and a variety of time indicators (figure 7.3).
3. The maintenance of accurate, dated records of such phenomena as eclipses, novae, comets, and sunspots over a longer continuous period than in any other civilization.
4. Early star catalogues embodying quantitative positional data (the earliest possibly from the fourth century B.C., with extant observations from ca. 70 B.C.). No less important, although not as old

Figure 7.2. "Abridged armillary sphere" of 1421, now at the Purple Mountain Observatory, Nanjing.

as Mesopotamian cuneiform records, are a recently excavated second-century ephemeris and a table of planetary motions from 244 B.C. to 177 B.C.

5. A coordinate system mainly oriented on the equator and the equatorial pole, unlike the ecliptic system prevalent in Europe and Islam until it was replaced in the late sixteenth century with a system like that used in China.
6. Among several early Chinese conceptions of the universe, one in which it was boundless and in which the stars floated in empty space. Perhaps because of its audacity, the details of this cosmic scheme were lost early, and no influence on astronomical practice has been documented.[4]

Despite the limitations of pretelescopic observation, ancient astrological records have proved useful in many ways to modern astronomers. The orbital periods of such celestial bodies as Halley's comet and the frequency of sunspot cycles have been determined with confidence;

4. Based on Joseph Needham, *Science and Civilisation in China,* 7 vols. projected (Cambridge: Cambridge University Press, 1954–), 3:458; for a fuller discussion and references see Leo Goldberg and Lois Edwards, eds., *Astronomy in China: A Trip Report of the American Astronomy Delegation* (Washington, D.C.: National Academy of Sciences, 1979), chap. 2.

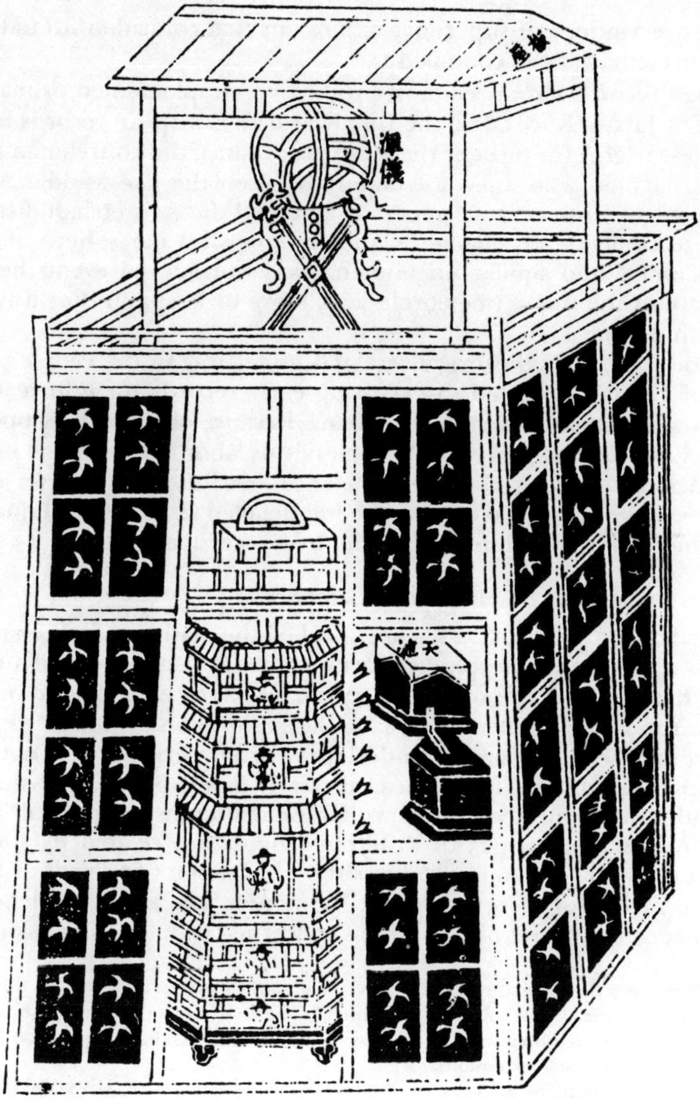

Figure 7.3. Water-driven astronomical clock from a book of ca. 1089.

detailed descriptions of supernova explosions that coincide with today's radio sources have been compiled.[5]

Chinese records have been most productive, supplemented primarily by those of Japan, Korea, and the Islamic world. European records have proved less useful for most of these purposes, since the churchmen and educated laymen who knew astronomy between the late Middle Ages and the seventeenth century generally accepted the Aristotelian dictum that there could be no change in the skies beyond the sphere of the moon. Comets and similar phenomena were considered events below the sphere of the moon (meteorological, more or less) and thus usually not worth noting.

A good part of the historical research under way in the People's Republic of China is devoted to exploring early records for information that bears on current scientific problems. Because of the wide scope of recorded astrological portents, the fruit of this labor has benefited other disciplines besides astronomy. For instance, over the past quarter century three major projects have resulted in detailed tables of earthquakes across the breadth of China and through its recorded history.[6]

Mathematical Harmonics

The third of the quantitative sciences studied the mathematical relations between sounds and the physical arrangements that produced them. The Pythagoreans in the sixth century B.C., it has often been said, motivated two and a half millennia of scientific exploration with their conviction that number and measure underlay the order perceptible in nature, even those aspects of it that appeared purely qualitative. The Pythagorean faith arose from the discovery that the main musical intervals produced by plucking strings of different lengths are related by simple ratios (a string half as long produces the octave, one two-thirds as long gives the fifth, and so on). Chinese harmonics was a similar blend of mathematics and numerology and was concerned with the dimensions of

5. For citations of Chinese research see Nathan Sivin, "Current Research on the History of Science in the People's Republic of China," *Chinese Science* 3 (1978): 39–58. David H. Clark and F. Richard Stephenson, *The Historical Supernovae* (Oxford: Pergamon Press, 1977), discuss ancient star explosions.

6. Chinese Academy of Sciences, Earthquake Working Committee, Historical Group, *Chung-kuo ti-chen tzu-liao nien-piao* [A chronology of materials on Chinese earthquakes], 2 vols. (Peking: Science Press, 1956); Chinese Academy of Sciences, Institute of Geophysics, *Chung-kuo ti-chen mu-lu* [A chronological list of Chinese earthquakes], 2 vols. (Peking: Institute of Geophysics, 1970; reprint, 1 vol., Washington, D.C.: Center for Chinese Research Materials, 1976); Hsieh Yü-shou and Ts'ai Mei-piao, eds., *Chung-kuo ti-chen li-shih tzu-liao hui-pien* [Collected materials on the history of Chinese earthquakes], 5 vols. (Peking: Science Press, 1983–). On the use of historical compilations in theoretical studies see J. Tuzo Wilson, "Mao's Almanac: 3,000 Years of Killer Earthquakes," *Saturday Review*, 19 Feb. 1972, 60–64.

resonant pipes. These pitch pipes provided the standards for ritual bells and stone chimes.

The approaches in East and West were by no means the same. Whereas the Pythagoreans proceeded by subdividing the octave, the Chinese multiplied the string lengths repeatedly by either 2/3 or 4/3 to generate a series (or "spiral") of fifths within the octave. The attempt to draw the most general significances of harmonics was equally ambitious at both ends of Eurasia. To give only one example, the Chinese adapted the pitch pipes as metrological standards of length, volume, and (indirectly) weight.

Harmonics in China is of interest not only for its use in connecting number and nature but for its application in musical performance, above all in the ceremonial music of the court. Music, like astronomy, was cultivated as an aspect of the imperial charisma.

In 1978 a set of sixty-five elaborate bronze chime bells with gold-inlaid inscriptions was unearthed in a tomb dated from the fifth century B.C. The bells produce tones over a range of five octaves; each bell produces two related tones when struck in two different places, a feat that involved remarkably sophisticated trial-and-error design by almost certainly illiterate craftsmen.[7]

THE CHINESE SCIENCES: QUALITATIVE

Most of the aspects of natural order studied in early science were not governed by number and measure. The patterns of function and dysfunction in the human body, for instance, could be accounted for only by qualitative theories. In the early West such theories incorporated a variety of explanatory entities, among them Empedocles' four elements and schemes of pneumata, exhalations, and humors. Such concepts labeled the aspects of a given thing or activity and thus divulged what it had in common with, and how it differed from, other things or activities that could be so analyzed.

By A.D. 200 at the latest, Chinese thinkers had elaborated two main ways to distinguish the phases of a process in time or a configuration in space. Temporal processes in nature were generally considered cyclical, and configurations were thought of as finite. One important theoretical entity was the complementary pair yin and yang, which when applied to processes stood for their taking and giving, abiding and transforming,

7. See Sinyan Shen, "Acoustics of Ancient Chinese Bells," *Scientific American* 256, no. 4 (April 1987): 104–10. For the theory of tuning to which these bells were adapted, see Cheng-Yih Chen (Ch'eng Chen-i), "The Generation of Chromatic Scales in the Chinese Bronze Set-Bells of the −5th Century," *Science and Technology in Chinese Civilization,* ed. Cheng-Yih Chen (Singapore: World Scientific, 1987), 155–97.

retracting and expanding, relaxing and stimulating aspects, and when applied to configurations stood for the ventral and dorsal, lower and upper, inner and outer aspects and analogous functions that could be thought of as feminine and masculine pairs. Yin and yang were without exception relational conceptions. As a modern textbook of traditional medicine puts it, "Considering the relation of chest and back, the chest is yin and the back yang; but when associating chest and abdomen, the chest, being above, is assigned to yang, and the abdomen, below, to yin." Similarly, yin and yang did not necessarily correspond to female and male; an old man may be yin when compared with a young man, and a young woman yang when compared with an old woman. Something is yin or yang only in reference to an ensemble of which it is a part.[8]

In addition to resolving any whole into its yin and yang aspects, one could understand it in terms of a similar system of five phases (*wu-hsing*, often mistranslated as "five elements" by false analogy with the Greek four elements). This understanding was not different from yin and yang but merely finer textured. In cycles of change the phases labeled "water" and "fire" stood for the most intensive aspects of yin and yang; "metal" and "wood" stood for the less intensive aspects; and "earth" stood for the aspect in which the opposed tendencies were balanced and in effect neutralized each other. The use of the five phases in the study of spatial relations was analogous. Most commonly four of the five stood for the cardinal points of the compass and the quarters of the sun's annual path corresponding to the four seasons, and earth stood for the central point on which others pivoted. Thus, for instance, in scientific discourse earth always implied balance and the neutral center; it did not refer to particles of earth as ultimate constituents.

Yin-yang and the five phases were not primarily technical concepts. Like the European notion of cause and effect, they belonged to everyday language and were likely to be used whenever anyone tried to explain structure and change. At the same time, like cause and effect they had more specialized meanings in learned discourse. As each of the qualitative sciences assumed its classical form, yin-yang and the five phases were given special definitions related to the subject matter of that field and were supplemented with other technical conceptions to provide a language adequate for theory. Because the sciences evolved independently, the exact meanings of common concepts tended to differ considerably from one discipline to another.

8. Kwangchou Armed Forces, Rear Support Units, Medical Administrative Organization, *Hsin-pien Chung i-hsueh kai-yao* [New essentials of Chinese medicine] (Peking: People's Hygiene Press, 1972), 2. This work is translated in Nathan Sivin, *Traditional Medicine in Contemporary China*, Science, Medicine, and Technology in East Asia, no. 2 (Ann Arbor: Center for Chinese Studies, 1988), 203.

Astrology

Mathematical astronomy has been described above as the science of celestial phenomena that can be predicted and thus need not be observed. Astrology, its complement, depended on data available only through contemplating the sky. It sought to discern the significance of unpredictable phenomena for current politics, interpreting observed data in part by precedent and in part by theoretical analysis.

In the standard histories published by successive dynasties we find records of a great variety of phenomena involving solar eclipses, sunspots, the stars and planets, and comets, as well as what would now be considered atmospheric phenomena and terrestrial prodigies. The character of the record and its contemporary use are perhaps best revealed by example. Here is the account of the great supernova of 1006 from the treatise on astrology in the official history of the Sung dynasty. It is one of several accounts in Chinese sources that, when pieced together, provide rich data on the supernova and allow its identification with a present-day radio source.

> On the fifteenth sexagenary day, fourth month, third year of the Luminous Virtue reign period [May 6, 1006], a Chou-po star appeared. It emerged in the south of the lunar lodge Base, 1° west of the constellation Mounted Guard [twenty-seven stars, mostly in Lupus]. Its form resembled that of the half-moon, and it had pointed rays. It glowed so brightly that objects could be distinguished by its light. It passed to the east of the constellation Treasury-and-Tower, and in the eighth month, following the rotation of the sky, "entered the turbid zone" [i.e., set with the sun and thus became invisible]. In the eleventh month it appeared once again in Base. Thereafter it was seen regularly heliacally rising in the east in the eleventh month and setting in the southwest in the eighth month.

Earlier in the same treatise the astrological characteristic of a Chou-po star, one of five canonical classes of auspicious stars, is noted in long-established language: "A Chou-po star is yellow in color and brilliant; the state in correspondence to which it appears will greatly prosper." The last phrase refers to a system of interpretation in which each part of the sky corresponds to an area of China. Where the manifestation appears indicates the part of the country affected. This instance corresponds roughly to modern Honan province.

Astrological interpretations are neither mumbo jumbo nor unsuccessful science. They are best understood, like modern economic indicators, as a technical framework for policy debates, resolved, as often as not, on other grounds. Faith in the validity of astrological categories, like confidence in extensively manipulated statistics today, persists despite their repeated failure to deliver accurate predictions.

In the biography of Chang Chih-po, an imperial commissioner in

1006, there is a fragment of the discussion that followed the appearance of a new star. We find Chang attempting to turn the attention of the court away from the auspiciousness of the omens and the details of interpretation toward the moral vigilance of the monarch, which Chang urged should not be relaxed at moments of good tiding:[9]

> When the Chou-po star appeared, the astronomer-royal reported it as an auspicious portent. The court officials prostrated themselves at the palace gates to offer congratulations. Chang expressed the view that the Ruler of Men ought to cultivate his virtue in response to the celestial phenomena, but that the appearances and disappearances of such stars had no particular significance [literally, "were not tied to anything"]. He proceeded to outline the essentials for mastering the Way [of correct government].

These essentials, we can be sure, supported the politics that Chang favored.

The inseparability of astrology from politics should not be taken to imply that the former was less pure a science than mathematical astronomy. Their institutional setting was the same, at least until classicists outside government circles in the seventeenth century took up the computation of ancient eclipses and other phenomena to understand early technical writings, correct the chronology of the Confucian scriptures, and test their authenticity. In early astronomy the great concentration of effort on solar eclipses, the relative neglect of such interesting topics as latitude theory and apparent planetary longitudes, and the use of gnomon-shadow observations for many centuries after armillary instruments had surpassed the gnomon in accuracy all represent the direct imprint on mathematical astronomy of court ritual patterns and the social norms behind them. The difference between astronomy and astrology was a contrast of emphasis on the quantitative as opposed to the qualitative and on objective motions as opposed to the correlation between celestial and political events.

Medicine

The data collected over the centuries about the body, health, and disorders were structured by the concepts of nature described earlier, form-

9. The three translations are from *Sung shih* (Standard history of the Sung period) (Peking: Chung-hua shu-chü, 1977), 56:1226, 52:1076, and 310:10,187. For additional sources and excellent discussions based on freer translations see Bernard R. Goldstein and Ho Peng Yoke, "The 1006 Supernova in Far Eastern Sources," *Astronomical Journal* 70 (1965): 748–53. For a worldwide study drawing on Goldstein and Ho, see Clark and Stephenson, *The Historical Supernovae,* 114–39. The lunar lodges were twenty-eight divisions of the sky along the equator that were used for recording locations of astronomical events. Base, one of these lodges, was centered about R.A. 14h 20m, and the brightest stars in Mounted Guard (mostly in the Western constellation Lupus) were between −35 and −40 degrees in declination. Treasury-and-Tower was a group of about ten stars in Centaurus.

ing a coherent body of theory used to diagnose and treat illness.[10] Classical medicine deserves the adjective "scientific" no less (but no more) than its counterparts in Western culture until recent times. It provided health care for a small portion of the Chinese populace. The majority of its patients and its more eminent practitioners belonged to the upper crust of society. Most of the afflicted among the Chinese population over the course of history had no access to the few fully qualified physicians. They depended on a great variety of less educated healers, ranging from herbalists to priests—a situation that would have been perfectly familiar in eighteenth-century France.[11]

What we call medicine incorporates and imposes order on experience related to every aspect of health, disease, and injury. One Chinese scheme of the major divisions of medicine included theoretical studies of health and disorder; therapeutics; the theory and practice of longevity techniques, including sexual hygiene; pharmacognosy; and veterinary medicine. Pharmacognosy, the study of vegetable, animal, and mineral substances used in therapy, brings together so much information on the sources and characteristics of thousands of drug ingredients that its literature was studied not only for therapeutic purposes, but also as compendia of natural history.[12] Prescriptions made up of both crude drugs and extracts were commonly used in combination with a great variety of other therapeutic means, including acupuncture and moxibustion, dietary regulation, calisthenics, breathing exercises, and massage.

In acupuncture needles were inserted into the flesh at certain points; in moxibustion cones of punk were burnt on the skin at those points. These stimuli affected what traditional physicians considered a circulation of vital fluid throughout the body (figure 7.4) and what clinical researchers today in China and abroad are more inclined to see as peripheral nerve endings and receptors. On the whole, acupuncture was classically considered a minor component of therapy, usually effective

10. The general level of Western publications on Chinese medicine is extremely low. The only penetrating analysis of classical medical concepts in any European language is Manfred Porkert, *The Theoretical Foundations of Chinese Medicine,* MIT East Asian Science Series, no. 3 (Cambridge, Mass.: MIT Press, 1974). For a Chinese outline of medical doctrine see Sivin, *Traditional Medicine in Contemporary China,* 201–427. The sections of *Science and Civilisation in China* on medicine are not yet in press. Several excellent essays by Needham in collaboration with Lu Gwei-djen have been gathered in *Clerks and Craftsmen in China and the West: Lectures and Addresses on the History of Science and Technology* (Cambridge: Cambridge University Press, 1970), in particular "Medicine and Chinese Culture," 263–93.

11. The only substantial study that looks beyond classical medicine to the diversity of health care in early times is Paul U. Unschuld, *Medicine in China: A History of Ideas* (Berkeley: University of California Press, 1985).

12. Paul U. Unschuld, *Medicine in China: A History of Pharmaceutics* (Berkeley: University of California Press, 1986), gives information about a large part of this literature.

only for certain disorders and when continued over a long period. In modern Chinese clinical practice it is of course impossible to evaluate acupuncture separately from the other Chinese and Western remedies with which the patient is treated.

Acupuncture has been promoted to the status of wonder cure by the American media in their restless quest for novelty. It has proved attractive out of proportion to the rest of traditional therapy. It is exotic to Americans, who were very late in learning about its long-standing popularity in East Asia. Few know about its use in Europe for roughly three hundred years or its vogue among physicians in early nineteenth-century America.[13] It is usually confused with acupuncture anesthesia (or analgesia), a new technique used in modern surgery since the late 1950s. In traditional surgery (limited to external medicine, amputation, and bonesetting), the only anesthetics we know to have been used were drugs—mainly alcohol, datura, and cannabis.

Classical Chinese medicine has often been represented as an empirical science based on the clinically sound use of effective natural drugs and other remedies. In this view, theories served primarily as mnemonic devices or as mystifications to confuse the untrained. Some modern physicians who have not troubled themselves to study classical medical doctrines dismiss them as futilities of the feudal past. Other authors have portrayed classical medicine as a remarkable corpus of theory—based on adaptations of the yin-yang and five-phases concepts—that succeeded in understanding the body as a many-leveled system and treated its ills holistically. They accordingly recommend it as a corrective to the impersonal, excessively lesion-centered and nihilistic tendencies of modern biomedical therapy. The mild and usually nonspecific remedies of traditional medicine are sometimes, but by no means always, expected to contribute to this reform. A closer acquaintance with the literature of classical medicine and with its practice in today's China suggests that these are partial pictures of a more complicated reality.

Before 1920 the strengths of medicine everywhere lay predominantly in the care of mild and chronic disorders. There was little physicians could do for most acute emergencies beyond strengthening the patient's defenses and preparing his family for the worst. With that emphasis on mild and chronic cases in mind, we are better able to judge the importance of gentle, gradual remedies, practices aimed at evaluating and improving the physical, mental, spiritual, and social circumstances of the patient, and theories used to relate symptoms to a multitude of therapeutic measures, both to design a flexible program of therapy and to

13. James H. Cassedy, "Early Uses of Acupuncture in the United States, with an Addendum (1826) by Franklin Bache, M.D.," *Bulletin of the New York Academy of Medicine*, 2d series, 50 (1974): 892–906.

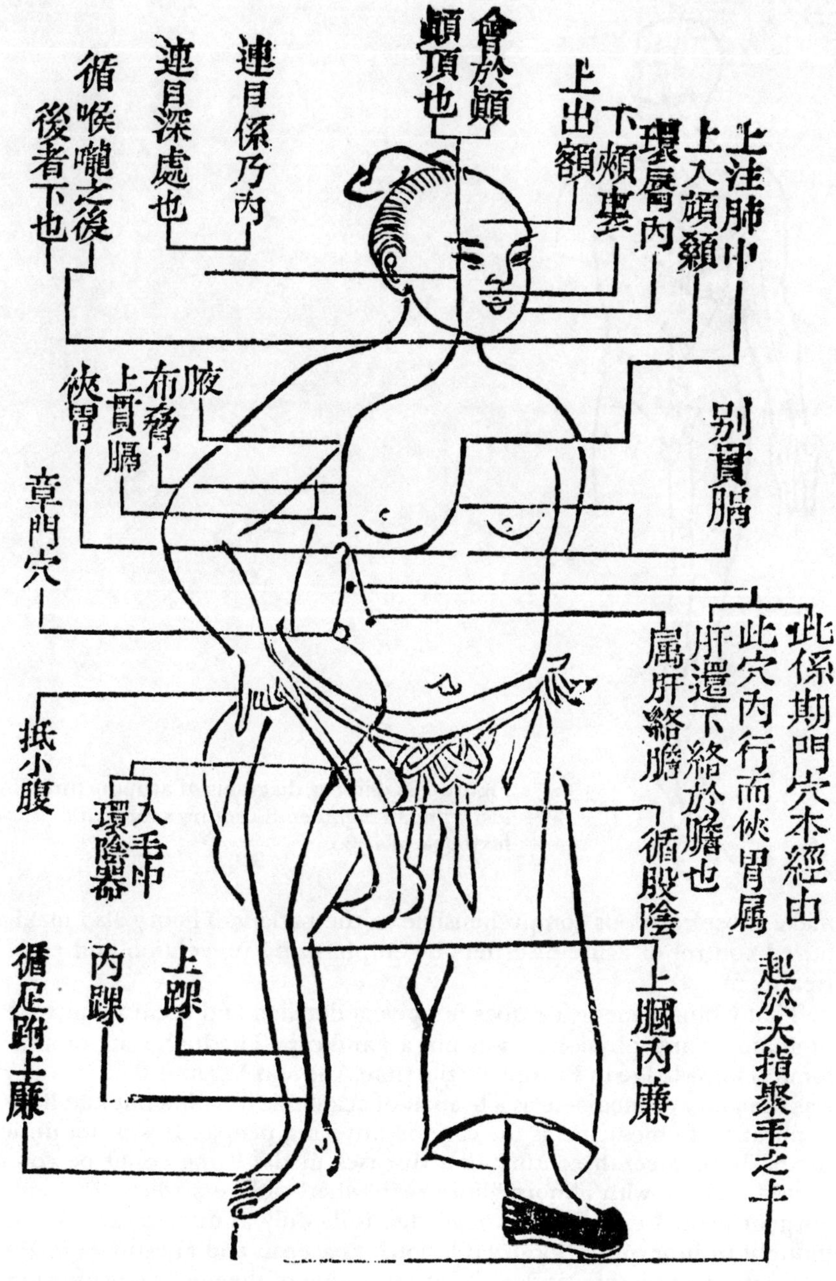
會於顛
顛項也
連目係乃內
連目深處也
循喉嚨之後
後者下也
上出額
下頰裏
環唇內
上入頏顙
上注肺中
別貫膈
腋
布脅
上貫膈
挾胃
章門穴
此係期門穴本經由
此穴內行而挾胃屬
肝還下絡於膽也
屬肝絡膽
循股陰
上膕內廉
起於大指聚毛之上
抵小腹
入毛中
環陰器
上踝
內踝
循足跗上廉

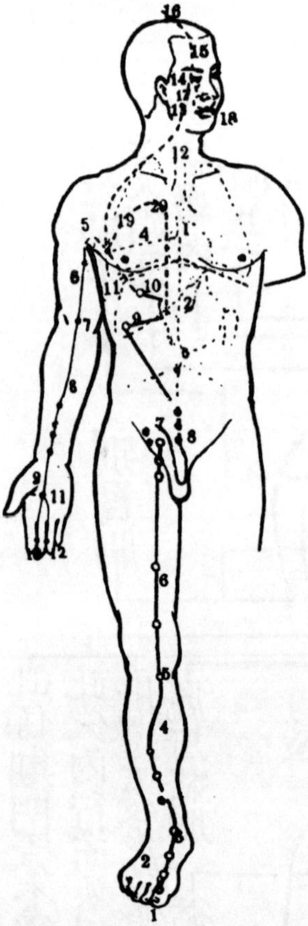

Figure 7.4. Similar diagrams of acupuncture loci from an eighteenth-century source and a textbook of 1960.

make those relations comprehensible to the patient. Theory also maximized control of acute disorders by emphasizing prevention and early treatment.

That Chinese medicine does not give a detailed and accurate picture of anatomy and physiology was not a handicap. The high state of anatomical knowledge in Europe by the time Andreas Vesalius died in 1564 was the glory of medicine as a branch of academic learning but had little application to medicine as the care of suffering people. It was not until the early nineteenth century that diseases in the living could be connected reliably with abnormalities seen when cadavers were dissected. Surgeons could exploit this knowledge fully only as disease came to be thought of in terms of localized lesions, as asepsis and anesthesia made local intervention safe, and as the organization of the medical profession imposed a single high standard of qualification. In the United States, for instance, the changes that made surgery safe and routine were not well under way until about 1920.

Unlike modern biomedicine, the traditional Chinese art could not draw on anything comparable to modern biology, chemistry, or physics. For that reason, its concepts look more like those prevalent in Europe four hundred years ago. They are not concerned with microorganisms or details of the body's organs and tissues. The strength of classical Chinese medical discourse lay rather in its sophisticated analysis of how functions were related on many levels, from the vital processes of the body to the emotions to the natural and social environment of the patient, always with therapy in mind. Chinese medicine is best evaluated in the light of this strength rather than according to criteria that could not have been applied anywhere until half a century ago.

Alchemy

Chinese alchemy used chemical techniques to prepare elixirs, which were perfected substances that brought about personal transcendence and eternal life. Elixirs could also be used for medical purposes and for transforming ordinary metals into gold. That is how an alchemist might have defined "external alchemy" (*wai-tan*); its analogue, "internal alchemy" (*nei-tan*), used the language of the laboratory to teach meditational (or sometimes sexual) disciplines in which the adept's body was visualized as the reaction vessel and furnace. In the first millennium the two alchemies were regularly practiced together, but after 1200 little activity in the external art was recorded.

The materials and apparatus of alchemy were on the whole the same as those of pharmacology, with some contributions from metallurgy and other practical chemical arts (figure 7.5). Certain developments, such as elaborate distilling vessels, appear so exclusively in alchemical literature that they may have originated there. The same may be said of gunpowder. What may be the earliest mention of flare mixture composition appears, oddly enough, in a list of external alchemists' misguided activities in a treatise on internal alchemy written not later than the end of the ninth century: "There was a case in which sulphur and realgar were mixed with saltpeter and honey, and burnt. Flames leapt up, burning the alchemist's hands and face and incinerating the building."[14]

The roughly one hundred remaining *wai-tan* books are probably the world's richest source for what was known about reactions and their products up to 1200. They reveal, in fact, that alchemy was not entirely

14. *Chen yuan miao tao yao-lueh* [Essentials of the mysterious Tao of the true origin], in *Cheng-t'ung tao tsang* [Collected Taoist works], 596:3a. On this book see Needham, *Science and Civilisation in China*, vol. 5, part 3, 78–79.

VI

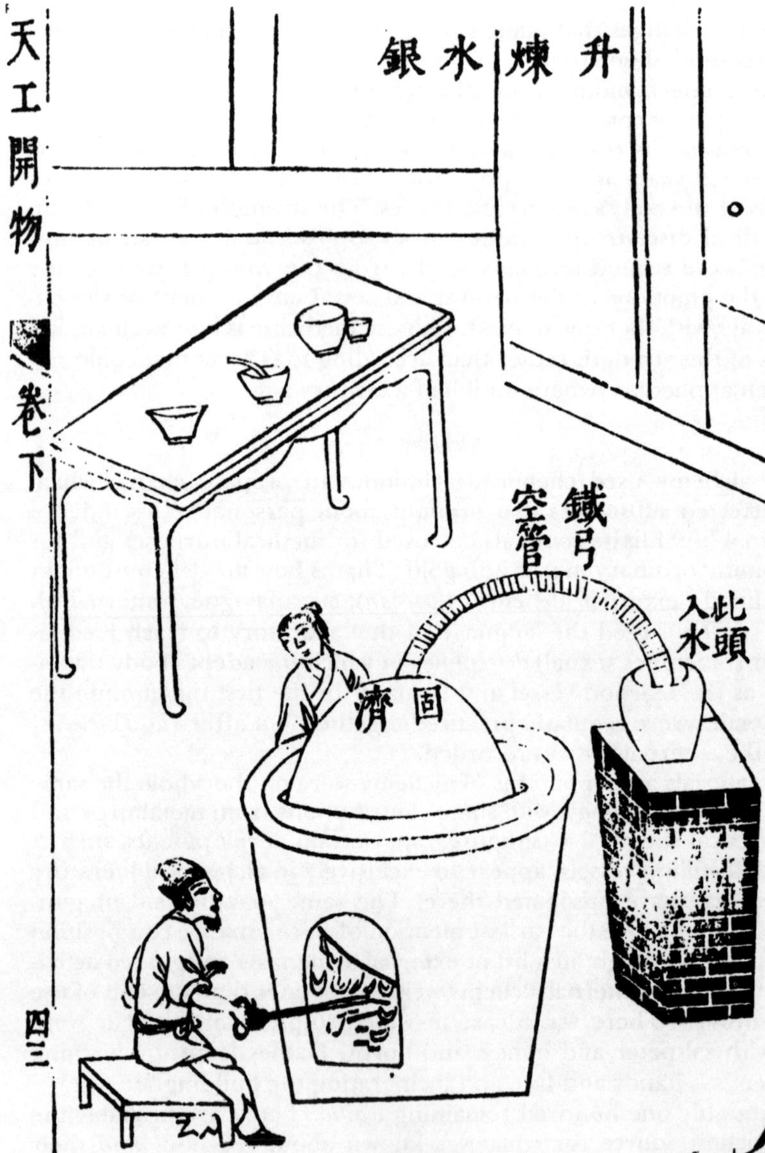

Figure 7.5. Commercial distillation of mercury, from the *Technological Encyclopedia of 1637*.

qualitative; some adepts took a lively interest in what weights of reagents would combine to form a new substance.[15]

Knowledge of chemical change was a means and a by-product, not the aim, of external alchemy. For some practitioners the goal was hardly distinguishable from that of medicine. Others were less interested in a product that would bring health or immortality than in the alchemical process, which they designed to serve as a model of the great cycles of nature, the rhythms of the tao.

No mortal could experience the cosmic cycles in their millennial sweep. These alchemists accelerated the scale of time, using theories based on yin-yang, the five phases, and numerology, to create, in a laboratory procedure that might require a few weeks to a year, an object of mystic contemplation. They believed, like the Pythagoreans before them, that to grasp the constant patterns that underlie the phenomenal chaos of experience is to be freed from the bonds of mortal finitude. As in the other Chinese sciences, the motivation that led to chemical discovery was connected to the deepest values of the seekers.

Siting

Masters of geomancy, or siting (to use the more informative term proposed by Steven J. Bennett), analyzed topographic features to find balanced land configurations in which to place houses and tombs.[16] Alchemy, as we have seen, used the yin-yang and five-phases concepts to understand and manipulate time as it shaped certain laboratory processes; siting adapted the same concepts to the study of space and form in the landscape. Some practitioners depended on their experience to judge the visual balance of topography, and some used readings taken with the famous geomantic compass. This instrument, which among its eighteen or more concentric dials (figure 7.6) incorporated indications of magnetic declination, was apparently a forerunner of the sailor's navigational instrument.

Once its contributions to the evolution of the compass are acknowledged, siting is sometimes shrugged off as a mass superstition. There is indeed not a great deal to be said for it if the sole criteria of evaluation are those of science today. But those are not the criteria that lead to understanding, since siting was not as narrowly focused as modern sci-

15. Nathan Sivin, in Needham, *Science and Civilisation in China*, vol. 5, part 4, 300–305. This essay is summarized in "Chinese Alchemy and the Manipulation of Time," *Isis* 67 (1976): 513–26, especially 521; reprinted in Nathan Sivin, ed., *Science and Technology in East Asia*, History of Science, Selections from Isis (New York: Science History Publications, 1977), 117.

16. Steven J. Bennett, "Patterns of the Sky and Earth: A Chinese Science of Applied Cosmology," *Chinese Science* 3 (1978): 1–26.

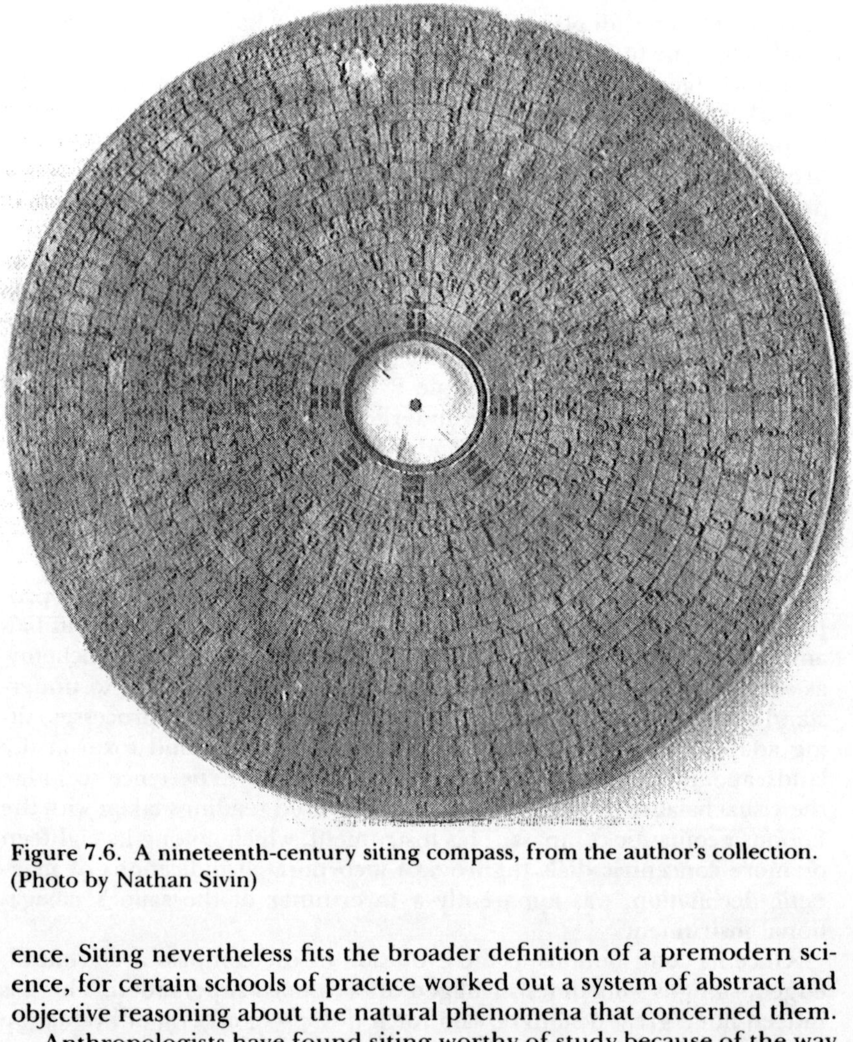

Figure 7.6. A nineteenth-century siting compass, from the author's collection. (Photo by Nathan Sivin)

ence. Siting nevertheless fits the broader definition of a premodern science, for certain schools of practice worked out a system of abstract and objective reasoning about the natural phenomena that concerned them.

Anthropologists have found siting worthy of study because of the way it was used (and is still being used) to resolve conflicts of status between practitioners' clients. Geographers and landscape architects have expressed interest in siting theory because it accomplished for centuries what modern theorists have been trying to do: provide a systematic framework for reasoning that will reliably yield beautiful sites. Because the balance that siting strove for was dynamic and complex, the benefits of the science are in large part aesthetic—not a matter of purely scientific interest, but of no small utility nevertheless. That may explain why siting has proved so durable. It and medicine are the only traditional sciences still widely practiced in Chinese communities today.[17]

Physical Studies

"Physical studies" is a grab bag of traditions that considered a great range of natural phenomena in the light of fundamental concepts. So far these traditions have been little studied, but a great deal in their literatures throws light on early knowledge of change and interaction, whether chemical, physical, or biological.

A few items from the section on trees in one of the most popular treatises suggest the wealth of information to be found there: "If peach is grafted onto persimmon, the fruit will be golden peaches. If peach is grafted onto Japanese apricot, the fruit will be meaty. If male and female gingko trees are planted together, the females will bear fruit. If pomegranate is grafted onto 'flowering cassia' [*Osmanthus fragrans Lour.*], the blossoms will always be red." Because we do not have records left by those who mastered grafting, it is from sources such as this one, as well as writings on materia medica, that the history of Chinese arboriculture must be reconstructed. An analogously broad range of documents must be studied by those who wish to understand many other ancient activities that required technical knowledge.

It would be misleading to leave the impression that the physical studies literature is a trove of information all of which is pertinent to modern science. The mixed character of the tenth-century book just cited, for instance, is obvious from two consecutive items in the section on "human affairs": "When a mother cries and her tears fall into the eye of her son, they will cause damage to its pupil and give rise to a cataract. In epidemic warm-factor disorders [a class of diseases involving fevers], if the clothing of the first person to fall ill is steamed in an earthenware steamer, the rest of the family will not catch the sickness."[18]

The abstract ideas in these extensive compilations also deserve to be investigated. In a study of how yin-yang and the five phases bear on alchemy, P. Y. Ho and Joseph Needham called the ancient application of these ideas "a hitherto unrecorded chapter in the prehistory of the conception of chemical affinity," by which they meant that it "seems to take its place in the linear ancestry of the idea that things can be arranged in

17. Siting is officially considered superstitious and is discouraged in the People's Republic, but it is still practiced. On siting in the New Territories of Hong Kong see Baruch Boxer, "Space, Change and Feng-shui in Tsuen Wan's Urbanization," *Journal of Asian and African Studies* 3 (1968): 226–40.

18. Tsan Ning, *Ke-wu ts'u-t'an* [Simple discourses on the investigation of phenomena] ed. Pai-pu ts'ung-shu chi-ch'eng (ca. 980), 1:4b and 2:16b. Compare the translation in Needham, *Science and Civilisation in China*, vol. 5, part 2, 149.

chemical classes the members of which are susceptible of chemically similar processes."[19]

The Transmission of Science between Civilizations

Things and ideas have flowed between China, Europe, and the other great civilizations regularly since the New Stone Age. Whether such characteristics of the earliest Chinese civilization as writing, the manufacture and use of the wheel, and bronze and iron metallurgy diffused into China from abroad or whether they were reinvented from scratch is still a matter for heated debate. These two hypotheses need not be considered mutually exclusive, for a mere rumor that something exists vastly simplifies making it. In any case the wheel and metalworking did not appear in China fully formed, as we would expect had they been imported.

Wrought iron, for instance, was being made in the mid seventh century B.C., perhaps six centuries later than the wrought iron of Anatolia. Cast-iron objects from the sixth century B.C. have been excavated. Cast iron could not be produced at will in the West for another seventeen centuries. Steel was produced at first from wrought iron, but the industrial processes for making it from cast iron were perfected in the second century B.C. This feat was possible because all the necessary elements were available: efficient furnaces derived from ceramic kilns, reciprocating bellows to provide an air blast, good refractory clays, and phosphorus-rich ores or fluxes to lower the melting point of iron.[20]

Science was part of the flow between East and West from early times. The Greek medical doctrine of the four elements reached China via India by about A.D. 500. Physicians from Byzantium or Syria, possibly Christians, are said to have cured a Chinese emperor in the mid seventh century. Jesuit missionaries in China published a treatise on Western physiology and an anatomical atlas in the mid seventeenth century, but the two books were very little circulated and received practically no attention. The first European book on Chinese medicine was published in 1682, the first on acupuncture in 1683. Alchemical elixirs of Western origin are noted in seventh-century China, and there is some evidence that Chinese alchemical gold was traded or sold abroad. But the greatest foreign scientific influence on China in early centuries, as in the seven-

19. Ping-Yü Ho and Joseph Needham, "Theories of Categories in Early Mediaeval Chinese Alchemy," *Journal of the Warburg and Courtauld Institutes* 22 (1959): 173–210, especially 201; a revised version appears in Needham, *Science and Civilisation in China,* vol. 5, part 4, 322.

20. For a summary of recent research see Tsun Ko [K'o Chün], "The Development of Metal Technology in Ancient China," in *Science and Technology in Chinese Civilization,* ed. Cheng-Yih Chen (Singapore: World Scientific, 1987), 225–43.

teenth-century heyday of Jesuit missionary activity, was exerted in astronomy. A fuller discussion of that field is in order.

Before the middle of the seventh century A.D., after Buddhism had become rooted in China, Indian astronomers worked in the Chinese capital. Their techniques were partly derived from Greek astronomy. They were more reliable for predicting solar eclipses than those current in China. The political significance of solar eclipses led the Chinese court from the turn of the eighth century to depend on resident foreign astronomers. When the Mongols brought China under their rule in the second half of the thirteenth century, their astronomical officials were Islamic, from Persia and Central Asia. Their computational procedures were more accurate than those of their Indian predecessors. Muslims were still performing the same services for the court when Jesuits began competing with them three hundred and fifty years later.

The Jesuit missionaries in China in the early seventeenth century were there not to propagate astronomical science but to convert the empire from the top down to Roman Catholicism. The only established access to the top was in the Astronomical Bureau, which had provided court positions to foreigners for nearly one thousand years. By 1645 the Europeans had gained operational control of the bureau after submitting to the throne a series of treatises that set out in Chinese the mathematical and cosmological foundations of European astronomy and winning several dramatic eclipse prediction contests. They maintained their status in the astronomical civil service past the middle of the eighteenth century.

The church's injunction against teaching the Copernican doctrines of the central sun and the planetary earth in 1616 and the condemnation of Galileo in 1633 made it impossible for the missionaries to disseminate the state of the art as it developed over the next century or so or to explain fully what the accomplishments of Copernicus, Kepler, and Newton had been. As the new astronomy evolved, discussions of it in Chinese by Jesuit writers before 1760 were thus full of gaps and contradictions, which were never explained. What is often considered the great watershed in European scientific consciousness was not revealed in China until late in the nineteenth century, long after it had become commonplace in the West. Nor were the Jesuit writings sufficiently technical and detailed to permit the Chinese to advance world knowledge. Nevertheless, they stimulated a high pitch of astronomical activity.

The geometric and trigonometric approaches and the cosmological framework of the Jesuit writings in Chinese, obsolete though they became as the seventeenth century wore on, precipitated what can only be called a scientific revolution. The best Chinese astronomers of the time (who mainly worked outside the imperial court) adopted new concepts, tools, and methods. They changed their convictions about what constituted an astronomical problem and what significance astronomy could have for the ultimate understanding of nature and even of human society.[21]

This metamorphosis of astronomy did not lead to the fundamental changes in thought and society that are naively supposed to be the inevitable outcomes of a scientific revolution. Conceptual revolutions, like political revolutions, occur at the margins of societies. The astronomers who responded to the Jesuit writings were members of the educated elite who above all felt the responsibility for strengthening and perpetuating traditional ideas. They were, in other words, at the center of their society. It is scarcely surprising that they used what they learned from the West to rediscover and carry to new heights the astronomical techniques of their greatest Chinese predecessors.

Only after the Opium Wars of the early 1840s could the Chinese receive a systematic education in the exact sciences as then taught in Europe. This time the educators were not individual priests dependent on the toleration of their hosts. They were mostly Protestant missionaries exempt from Chinese laws, their right to operate missions and schools guaranteed by imposed treaties and enforced by gunboats. They were no longer appealing, as the Jesuits had done, to an elite intent on adapting new techniques to traditional ends. The Protestant missionaries educated mostly the poor and people of modest means. Even their richer converts came to them because changes in the civil service gave their children little chance of conventional success in the old society.

Late-nineteenth-century Chinese astronomers trained in Western institutions had no reason even to be curious about what their compatriots in earlier times had done. By 1880 Protestant missionaries, generally working with Chinese, had translated from European languages a number of basic textbooks in astronomy, mathematics, and physics and made them generally available at low prices. Their schools, libraries, bookshops, and other institutions were founded to instigate change, not to preserve Chinese civilization.

As the threat of dismemberment by the colonial powers became more imminent, the Chinese government was belatedly persuaded to begin educating modern scientists. In 1866 a department of mathematics and astronomy was added to the T'ung-wen-kuan in Peking, which had previously been a college for interpreters. In 1867 a translation department was added to the Shanghai Arsenal, which had been established two

21. John B. Henderson, *The Development and Decline of Chinese Cosmology* (New York: Columbia University Press, 1984); Benjamin Elman, *From Philosophy to Philology: Intellectual and Social Aspects of Change in Late Imperial China,* Harvard East Asian Monographs, no. 110 (Cambridge, Mass.: Council on East Asian Studies, Harvard University, 1984); Nathan Sivin, "Copernicus in China," *Studia Copernicana* (Warsaw) 6 (1973): 63–122.

years earlier. There Chinese foreign employees of the imperial government systematically undertook the translation and publication of modern works in science, engineering, medicine, law, and so forth. These and the less systematic technical publications of the missionaries were widely distributed and eagerly studied by amateur groups that maintained the tradition begun by the Jesuit writings. The new translations often played a part in the education of statesmen and reformers, for whom they provided a window on the world. The future lay, however, not with those who saw modern technology as a tool to breathe new life into an empire and a traditional culture, but with those at the margin of the old society, educated in modern schools and given employment in imported institutions.

From about 1900 onward, Chinese astronomers began to emerge who were fully prepared to benefit from advanced training abroad. They were educated in missionary institutions as well as in government universities as these appeared. When ten foreign powers extracted heavy indemnities from China after the Boxer Uprising in 1900, the United States used income from its share to support students during their technical training abroad. Scientists trained in the United States, Europe, and Japan, as well as those educated in China, created the first large-scale system of research institutions and an educational system to train scientists. A few of these institutions were founded by foreigners in the image of their own, but—to redirect the words of a leading historian of China—"on all sides of this gleam of Western light, China was being torn apart by forces so powerful that they made the Westerners' efforts poignantly irrelevant."[22]

Since 1949 China has by fits and starts invented policies toward education and science that reflect its own priorities rather than the expectations of other nations.

THE HISTORY OF SCIENCE IN CONTEMPORARY CHINA

In China today it is normal for scientific journals to publish historical studies, for research centers, observatories, and other organizations to house research groups for ancient science and technology, and for modern scientists to be knowledgeable about their country's scientific heritage.[23] This situation contrasts so greatly with that of most other countries that it calls for an explanation.

I have already noted that early astronomical and earthquake records

22. Jonathan Spence, *To Change China: Western Advisors in China, 1620–1960* (Boston: Little, Brown, 1969), 172.

23. Recent astronomical publications and research are summarized in Sivin, "Current Research."

are a valuable resource for current scientific research. But this is only one motivation for the awareness of history. Other reasons are related to China's place in the world. For millennia the Chinese considered their land the one true center of civilization. Over the past century China has had to make an entirely new place for itself as only one member of a large family of nations. For most of the last one hundred years it has been dependent on and has been looked down on by foreigners for that reason. The present government is resolved to make China independent within limits set by the imperatives of survival and is trying to enlist the energies of every citizen.

China's recent policies for technological development have been unique in many respects. They have made unique demands for adaptation on the part of the whole scientific sector of society, which until a couple of decades ago was considered quintessentially Western. Science and engineering were what one learned from foreigners in order to safeguard oneself against them. This view has gradually been changed over the past twenty years by popularizing the history of Chinese science. Children's books, postage stamps, museum exhibits, and school lessons have all carried the message that science is not European but a world enterprise and that over most of history China was one of the great contributors to that enterprise.

It is not difficult to find this point explicitly stated. Consider the afterword to a book for teenagers on every aspect of technical history, published just after the Cultural Revolution: "The achievements of China's ancient science and technology prove that the Chinese people have the ability needed to occupy their rightful place among the world's peoples. These achievements will also encourage our faith and strengthen our resolve, so that in the shortest possible time we may catch up to and surpass the world's most advanced levels of development. China has yet greater contributions to make to humanity."[24] Scientists and medical researchers experience this consciousness as a close linkage between scientific work and political activity.

Enhanced consciousness of Chinese scientific history is not entirely an internal matter. The importance and fascination of the Chinese scientific tradition have long been known in Europe, Japan, and the United States. Scholars in many countries have contributed to understanding it as well

24. Chinese Academy of Sciences, Institute for the History of Natural Sciences, ed., *Chung-kuo ku-tai k'e-hsueh ch'eng-chiu* [Achievements of ancient Chinese science and technology] (Peking: Chung-kuo Ch'ing-nien Ch'u-pan-she, 1978), 706. Written for teenage readers, this comprehensive work contains contributions by China's best-known historians of science and technology. The sentences translated in the present essay do not appear in the English version, *Ancient China's Technology and Science* (Peking: Foreign Languages Press, 1983).

as to making the work of many great Chinese historians of science accessible in other languages.[25]

Educated people all over the world are now prepared to respond to new revelations about Chinese scientific traditions—whether they be new applications for the ancient art of acupuncture or the unique archaeological finds that have been appearing without interruption since the 1950s. The heightened interest has meant a small but perceptible rise in the world's esteem for China. More to the point, it has meant that scientists all over the world are increasingly involved in the give and take that help Chinese scientists to be fully involved in the international scientific community.

25. Historical scholarship, mainly since 1980, is summarized in Nathan Sivin, "Science and Medicine in Imperial China—The State of the Field," *Journal of Asian Studies* 47, no. 1 (1988): 41–90.

VII

WHY THE SCIENTIFIC REVOLUTION DID NOT TAKE PLACE IN CHINA —OR DIDN'T IT?

When people learn that there were many scientific traditions in the ancient world, they usually begin wondering why the fateful transition to modern science first happened where it did.[1] Joseph Needham has given the "Scientific Revolution problem" its classic formulation: "Why did modern science, the mathematization of hypotheses about Nature, with all its implications for advanced technology, take its meteoric rise *only* in the West at the time of Galileo?" This

[1] This essay incorporates my current views on a historical issue to which I have returned regularly for some years. No doubt my views on this topic will be different in another decade; all I mean to accomplish with these ephemeral reflections is to transmit the idea that the issue is worth thinking about, to suggest how one might think about it, and to point out that certain ways of thinking about it are so burdened by suspect assumptions that they do not encourage clear explanation. I have addressed one aspect or another in previous writings, to which the reader is referred for documentation: "Copernicus in China," *Studia Copernicana*, 1973, *6*: 63-122; "Shen Kua" and "Wang Hsi-shan," in *Dictionary of Scientific Biography, s.v.;* "Next Steps in Learning about Science from the Chinese Experience," *Proceedings*, XIVth International Congress of the History of Science (Tokyo and Kyoto, 19-27August 1974), I, 10-18; Sivin (ed.), *Science and Technology in East Asia* (New York: Science History Publications, 1977), pp. xi-xxi; and "Chinesische Naturwissenschaft: Weber und Needham," in Wolfgang Schluchter (ed.), *Max Webers Studie über Konfuzianismus und Taoismus. Interpretation und Kritik* (forthcoming, München, 1982). Joseph Needham has also provided a summary of our conversations and correspondence on the "Scientific Revolution problem," an interesting attempt to specify differences and similarities in our views, in *Science and Civilisation in China*, vol. 5, part 2, *Spagyrical Discovery and Invention: Magisteries of Gold and Immortality* (Cambridge, England: At the University Press, 1974), pp. xxii-xxvii. Here I set out my own somewhat different view of the divergences that accompany our very broad areas of agreement. In several points regarding the Scientific Revolution problem I have been anticipated by Wing-tsit Chan, "Neo-Confucianism and Chinese Scientific Thought," *Philosophy East and West*, 1957, *6*: 309-332.

I use "Scientific Revolution" to refer primarily to the transition in the exact sciences between Galileo and Laplace and its wider repercussions by 1800. This is one of several definitions in current use. I adopt it for the purpose of this essay not because it is the best possible definition, but because it is the one most commonly presupposed by Sinologists and laymen who set out to compare developments in China and the West. Needham's usage of the term "Scientific Revolution" is often, but not consistently, broader. No definition can be considered better than a historiographic expedient. Lack of a consensus about the significance of the term has led some historians of science recently to reject its use altogether.

affirmation implies that one must investigate the absence of such a revolution elsewhere, and indeed page after page of *Science and Civilisation in China* is given over to "why modern science had not developed in Chinese civilization . . .?" He adds a second question that bears on this absence, enhancing the interest of the larger inquiry: "why, between the first century B.C. and the fifteenth century A.D., Chinese civilization was much *more* efficient than occidental in applying human natural knowledge to practical human needs."[2]

In two decades of study, teaching, and public lecturing on Chinese science and medicine, I have encountered no question more often than why modern science did not develop independently in China, and none on which more firmly based opinions have been formed on the basis of less critical attention to available evidence. Since those who put forth these opinions are on the whole intelligent and thoughtful, I have gradually been led to suspect that there is more to the Scientific Revolution problem than meets the eye. In this essay I will turn it inside out in order to ask what assumptions about the European tradition of science—assumptions by no means confined to Europeans and Americans—encourage us to take this problem more seriously than its intrinsic merits justify.

Issues

In that millennium and a half European civilization was first experiencing a general collapse and then recovering from it. It is obvious that we ought to be looking at the Western end of Eurasia to account for European inferiority in technology over a span of fourteen hundred years. But there are still other doubts to be expressed in connection with this second question. The natural knowledge that was being applied to human needs was not what we usually call Chinese science.

Early technology did not succeed or fail according to how well it applied the insights of early science. Science was done on the whole by members of the minority of educated people in China, and passed down in books. Technology was a matter of craft and manufacturing skills privately transmitted by artisans to their children and apprentices. Most such artisans could not read the scientists' books. They had to depend on their own practical and esthetic knowledge. What that knowledge was like we can only reconstruct from the artifacts they left and from the scattered written testimony of literate people. Literacy spread considerably outside the elite over the last several centuries, but this did not lead to the substantial use of books to teach craft skills. The classical tradition of medicine was an applied science, but not a mere technology, in the sense that therapeutic decisions were based on a cumulative structure of systematic theory applied to experience. Less exalted kinds of curing were sometimes a matter of techniques, sometimes of rituals, neither of which depended on the abstractions of the classical tradition.

It also seems to me that comparing all of the scientific and engineering activity of one civilization with all that of another conceals more than it reveals, since it is only in modern times that these various kinds of work became closely connected.

[2]Needham, *The Grand Titration: Science and Society in East and West* (Toronto: University of Toronto Press, 1969), pp. 16 and 190.

It is true that between the end of the Roman period and 1400 or so, a Chinese visiting Europe would have found it in many respects technologically backward. At the same time there was probably not a great deal to choose between Chinese and European medical practice before about 1850 (knowledge of anatomy and physiology had little therapeutic application earlier). Mathematical astronomy in China by its last high point about 1300 did not quite reach the general level of predictive accuracy that Ptolemy had mastered eleven hundred years earlier.

I need not dwell on comparisons of this kind. They tell us nothing at all about what we can expect to learn from one culture or the other. After all, no one is proposing that we give up the study of Hellenistic alchemy just because it has become clear that the Chinese alchemical literature is richer in chemical knowledge.[3] What matters is that we are now able to begin comparing several strong traditions of science and technology based on the ideas and social arrangements of different civilizations. All of them must be attentively studied if we want to understand the general relations through history and across the globe between science and culture, science and society, science and individual consciousness. Without that understanding we will remain trapped in our own parochial viewpoints. Historians have more urgent work to do than trying to prove the inferiority of every other culture to the one in which they specialize.

Science and Sciences

As an example of how studying the Chinese experience can suggest clues about the character of early science in general, I offer the case of Shen Kua 沈括 (1031-1095), one of the most versatile figures in the history of Chinese science and engineering. Just to give a few examples, he is famous for the first discussion of magnetic declination and of printing with movable type, the only application of permutations in traditional Chinese mathematics, a proposal for nightly measurements of the lunar and planetary positions, the first suggestion in East Asia of a purely solar calendar, an explanation of the process of land formation by both deposition of silt and erosion, and an important book on the theory and practice of medicine. In addition to his technical activities, his writing has to be consulted by every student of early Chinese archeology, music, art and literary criticism, economic theory, and diplomacy. He made his early reputation as a land reclamation expert, and was deeply involved as a high official in the 1060's in the most important political reform movement for some centuries.

Shen's combination of unlimited curiosity and involvement in the affairs of his time had a special interest for my own education. For some time, through a series of studies roaming through different historic periods and technical disciplines, I have been trying to piece together bits of answers to a large question that I find boundlessly interesting. How did Chinese scientists in traditional times explain to themselves what they were doing? In other words, what was their understanding of nature and of their relation to it as conscious individuals living in a society? How did the insights of the various sciences hang together to form this understanding? I had gradually formed a general idea of the sciences as defined in early China, but I couldn't see how their insights were combined to form that general

[3] See in particular Needham, *Science and Civilisation in China*, vol. 5, parts 2-4 (1974-1980).

understanding. It occurred to me that I might do well to study how the sciences fit together in the mind of a person who was involved in all of them. The obvious person to study was Shen Kua.

The pattern that emerged was familiar in its details but unexpected in its overall shape. One aspect was that there does not seem to have been a systematic connection between all the sciences in the minds of the people who did them. The sciences were not integrated under the dominion of philosophy, as schools and universities integrated them in Europe and Islam. Chinese had sciences but no science, no single conception or word for the overarching sum of all of them. Words for the level of generalization above that of the individual science were much too broad. They referred to everything that people could learn through study, whether of Nature or human affairs (*hsueh* 學), or even more broadly to any pattern that could be apprehended through any form of cognition (*li* 理 and *tao* 道). All of these terms could, for instance, include ethical or religious principles discovered through reflection on authoritative texts. *Li* and *tao* could be grasped through mystical illumination.

Let us consider for a moment the connection between items of knowledge about Nature in Shen Kua's writings. We can trace through a thousand years of comprehensive encyclopedias in China the old division between heaven, earth, and man—the cosmos and its phenomena; the earth and its features, territorial divisions, creatures, and products; man and his institutions, usages, and accomplishments. In this conventional classification astronomy and astrology fell obviously enough under the sky; alchemy and medicine usually fell next to divination under technical skills (a rubric that includes painters and sometimes hired assassins). On the other hand, drugs fell sometimes among products of the earth and sometimes among commodities.

In Shen Kua's memoirs, *Meng ch'i pi t'an* 夢溪筆談 (Brush talks from Dream Brook), there is a rubric called "regularities underlying the phenomena 象數." Under this heading he like many others grouped together physical and numerological aspects of astronomy, astrology, cosmology, and divination, which refract the pattern of physical reality in their various ways. A section called "technical skills 技藝" puts medicine, engineering, and mathematics (including astronomical mathematics), which share purely instrumental value, alongside architecture and games. His chapter on "strange occurrences 異事" sets out his thoughts on the origin of plant fossils, the first recorded description of a tornado in East Asia, an account of his experiment on the formation of rainbows, and similar gems, arrayed cheek by jowl with unlikely hearsay and ephemeral curiosities.

What makes us think of Shen Kua as a scientist was widely scattered through his own scheme of human knowledge (see Appendix). That scheme cohered not on the level of science, but on a much more general level. In his writing, there are no clear boundaries between material that fits the modern conception of science and material that does not. That conception cannot carry us far in the effort to understand what Shen Kua was getting at.

Shen Kua, in the second half of the eleventh century, made his turn on the stage of history at a time when a great upsurge in social mobility was broadening the group that ruled China. Many of these new men were interested in all sorts of practical affairs that well-born people in earlier times would have considered beneath them. It was to some extent a matter of greater versatility in public service. This was after all a time when merit ratings of officials were being based on quantitative measures of efficiency in collecting taxes, reclaiming land, and so on, instead of entirely on virtue, breeding, and orthodoxy, as had been the case earlier. At leisure too, this large group that Shen Kua belonged to was free to indulge curiosity—in an amateur way, of course—about anything in the universe, including technical matters that earlier were fit only for clerks or artisans. Only after Shen's lifetime did this evolving amateur ideal settle on philosophy, the arts, and literature as the appropriate realms to be universal within, once again leaving the study of the earth and sky largely to the mere technicians. In the eleventh century Shen was only one of a number of polymaths whose scientific and technological interests, however amateur, all emerged in connection with their varied official responsibilities. Su Sung 蘇頌 (1020 - 1101) and Yen Su 燕肅 (d. 1040) are also well-known examples.[4] The intellectual consistency of Shen's style in scientific thought seems to reflect the consistency of his public career, in which that style was formed.

The astronomer in the court computing calendars to be issued in the emperor's name, the doctor curing sick people in whatever part of society he was born into, the alchemist pursuing archaic secrets in mountain haunts of legendary teachers, had no reason to relate their arts to each other. Philosophers were in no position to define a common discipline for all of them, as Aristotle and his successors had done in Europe, and so philosophers had practically no influence on the development of these special pursuits. I do not mean to deny that people who worked in the sciences embedded their special knowledge in world-views and social philosophies. To the contrary, until very recent times the prospect of being a mere functional unit of scientific manpower, unconcerned with the larger significance of one's work, would have been unimaginable. But these wider perspectives to which each field related were passed down within the sciences themselves. Mathematics, geomancy, and so on, each came equipped with its own cosmic meanings. These were seldom quickly or deeply influenced by contemporary philosophy, but continued to develop in their divergent ways out of the theoretical perspectives with which the sciences began in the Han period (206 B.C. - A.D. 220).

If anyone were going to seek out the common ground of the sciences in China, it was people like Shen Kua, who were mastering them all. But Shen put his own understanding together in ways that did not closely link the various traditions that studied physical nature, and in ways that did intimately associate what today would be considered scientific with what would be called grossly superstitious. The distinction between science, proto-science, pseudo-science, and non-science simply gets in the way of understanding the articulation of Shen Kua's thought. Surely it is necessary to understand thought before one begins to evaluate it.

[4] In addition to frequent references in *Science and Civilisation in China*, see S. Miyasita, "Su Sung," III, 969-970 in Herbert Franke (ed.), *Sung Biographies* (3 vols., Wiesbaden, 1976); Teng Kuang-ming 鄧廣銘 & Wang Chen-to 王振鐸, "Su Sung," pp. 123-134 in Institute for the History of Science, Chinese Academy of Sciences (ed.), *Chung-kuo ku-tai k'o-hsueh-chia* 中国古代科学家 (Ancient Chinese scientists; Beijing, 1959); and Wang Chin-kuang 王錦光, "宋代科学家燕肅," *Hang-chou ta-hsueh hsueh-pao* 杭州大学学报, 1979, *3*: 34-38.

I would have to say that I failed to find the internal unity of Chinese science that I was looking for in the mind of Shen Kua. By way of compensation, I did learn the importance of an issue that I hadn't paid enough attention to before, that is, the relations of the sciences to other kinds of knowledge.

Let me give one example. Shen devoted about as much space in his memoirs to stories that involve strange happenings, predestination, prognostication, and divination, as to accounts that we would relate to science and technology. He himself does not seem to have viewed these enthusiasms as in conflict with his studies of Nature.

His rational explanation of divination is not very different from the view of its most sensible modern students. It is more sensible, I would say, than Carl Jung's famous attempt to explain the notions behind the Book of Changes by his hazy and un-Chinese notion of synchronicity.[5] Shen seems to have believed that the divination techniques used in this time did not really describe the future or what was happening far away, so much as provide counsel about the unknown and inaccessible. For instance, he explained why the same technique gives different outcomes when used by different people:

> **In his discussions of divination to determine locations of houses and tombs and to foretell length of life, Lü Ts'ai 呂才 [d. 665] believed that techniques cannot yield reproducible results. It is quite true that technique is not reliable in this sense. Still he did not realize that every kind [of divination] is a matter of 'substitutes 寓.' "The ability to respond spiritually and make the truth manifest depends on the person"; thus if two people use a single technique the result of their divinations will be different. The human mind is by nature spiritually responsive, but since it is unavoidably burdened, one must, in order to gain access to it, use as a substitute some thing that does not have a mind. The result of divination can only be explained by what makes one's own spiritual response possible. . . . In fact anything that can be seen, heard, thought about, or speculated upon can be used as a substitute for this purpose. If this seems irrational, is not all good and bad fortune, are not all the mutations of life and death, irrational? Only with someone able to understand the pattern common to all this can one discuss the spiritual response that makes foreknowledge possible.**

Let me now rather freely paraphrase this statement in language more familiar to modern readers: Understanding is a matter of introspection. The questions we ask in divination are the ones we have tried to deal with reflectively but can't, because at the time our minds are too burdened for introspection to work. Divination techniques ritually manipulate passive material objects, yarrow stalks and so on, providing an external process or set of images on which to concentrate as a

[5] Jung, "Foreword" to *The I Ching or Book of Changes* (tr. Richard Wilhelm & Cary F. Baynes; Bollingen Series, XIX; New York: Pantheon Books, 1950), I, iii-vii. The quotation below is from *Meng ch'i pi t'an*, item 145. Lü Ts'ai was a famous diviner and theoretician of divination. The word I have translated "substitute" means more literally "to make something serve as a temporary abode." The sentence in quotation marks is quoted by Shen from the Great Commentary (*Hsi tz'u* 繫辭) to the Book of Changes, A. 12 (compare Wilhelm tr., I, 349).

way round that blockage. In other words, prognostication draws indirectly, through ritual, on the power of self-examination.

Access to one's personal future, whether by visionary foresight or by divination, is a perfectly natural phenomenon. It merges into the moral faculties, whose choices condition the future. It merges at the other end into the rational comprehension of the natural order as it is reflected in any authentic experience. Shen did not confuse introspection and observation, nor did he draw a clear line between them. Nor did he need to compare the importance of these two ways of knowing. What finally united the sciences, in other words, was the universal system of knowledge—uniting intellection, imagination, and intuition—of which they constituted only a part. I am arguing that sometimes to look at the sciences alone is to look too closely.

Assumptions in the Scientific Revolution Problem

Now back to the Scientific Revolution problem. It is striking that this question —Why didn't Chinese beat Europeans to the Scientific Revolution?—happens to be one of the few questions that people often ask in public places about why something didn't happen in history. It is analogous to the question of why your name did not appear on page 3 of today's newspaper. It belongs to an infinite set of questions that historians do not make part of their research programs because they have no direct answers. They translate into questions about the rest of the world. The one that concerns us, for instance, translates into "in what circumstances did the Scientific Revolution take place in the seventeenth and eighteenth centuries in Western Europe?"

Why do people keep asking why the Scientific Revolution did not take place in China when they know enough not to explain why their names did not appear on page 3 of today's newspaper? Because the question encourages exploration of a fascinating topic and provides some order for thinking about it. It is, in other words, heuristic. Heuristic questions are useful at the beginning of an inquiry. As we comprehend enough to deal with complicated patterns, heuristic questions tend to grow murky, and finally to lose their interest compared with the emerging clarity of what did happen.

So much for heuristic questions in general. Why do we tend to take this one more seriously than the general run? Somehow the Scientific Revolution problem holds a special urgency.

That urgency is there, I suggest, because this way of putting the problem contains and supports certain Western assumptions, assumptions that ordinarily we do not question. Above all we usually assume that the Scientific Revolution is what everybody ought to have had. But it is not at all clear that scientific theory and practice of a characteristically modern kind were what other societies yearned for before they became, in recent times, an urgent matter of survival amidst violent change. In fact we have made very little progress so far in understanding how Europeans originally came to want modern science and its concomitants, since the attention of historians has been concentrated on how these innovations came about. Nor has much attention been given to what accounts for the gradual and uneven diffusion of modern science within Europe—an issue that does not differ essentially, I submit, from that of its spread to other civilized parts of the world.

There is usually the further assumption that civilizations which had the potential for a scientific revolution ought to have had the kind that transpired in the West and encompassed the sorts of institutional and social changes that are identified with modernization.

These assumptions are usually linked to a belief—or a faith, if you prefer—that European civilization all along was somehow in touch with reality in a way no other civilization could be, and that its great share of the world's wealth and power comes from some intrinsic fitness to inherit the earth that was there all along. Many of those like myself who reject this assumption argue that the privileged position of the West comes instead from a head start in the technological exploitation of nature and the political exploitation of societies not technologically equipped to defend themselves.

Finally there is the assumption that, since modern science has so quickly and thoroughly become international, it transcends European historical and philosophic biases, and is as universal, objective, and value-free as the Nature that it seeks to understand and manipulate.

What seems to be common sense in that last assumption (or in the self-conception that all the assumptions I have mentioned are part of) does not stand up to thoughtful examination. Modern science is still too marked by the special circumstances of its development in Europe to be considered universal.

Chinese science got along without dichotomies between mind and body, objective and subjective, even wave and particle. It is not that, for instance, physical and mental acts were confused, but they were not considered mutually exclusive, to be accounted for by sharply different varieties of discourse. In the West the first two dichotomies were entrenched in scientific thought by the time of Plato. Galileo, Descartes, and others carried them into modern times to mark off the realm of physical science from the province of the soul, which was decidedly off limits to secular innovators like themselves. These distinctions let scientists claim authority over the physical world on the ground that purely natural knowledge could not conflict with the authority of established religion.

Science and religion have long since learned to coexist, but we are still living with these distinctions. If they are European peculiarities, and perpetual sources of trouble at that, why hasn't modern science managed to rid itself of them? It is evidently not a simple matter to root them out. Until we do, there is something to be said for frankly admitting a certain parochialism in the foundations of science. The mathematical equations may be universal, but the allocation of human effort among the possibilities of natural knowledge is not.

Science and technology have spread throughout the world, but that has not made them universal, in the sense of transcending European patterns of thought. In one society after another the encounter between old and new ideas has been abortive, resolved by social change and political fiat. Traditional ideas are simply excluded (on the grounds that they are primitive, superstitious, regressive, fit only for the lower classes, etc.) from the educational systems created to teach a new technical and managerial elite the values of technology alongside its theory and practice.

Modern technology is clearly more powerful than that of traditional societies;

but to a larger extent than we generally realize, its strength emerges in application to needs and expectations that do not exist until it generates them.[6] True universality would require modern technology to coexist with and serve cultural diversity rather than consistently serving as a tool to standardize it out of existence.

I am arguing that the notion of a universal and value-free modern science, which has somehow become independent of its social and historical origins, is wishful thinking. It is easy even for an intelligent reader to be led astray on this point. The kernel of certainty from which it arises is seldom defined carefully by those who set out to explain science to non-scientists. It would be foolish to deny that modern science has attained a verifiability, an internal consistency, a taxonomic grasp, a precision in accounting for physical phenomena, and an accuracy in prediction that no other kind of activity shares, and that lay far outside the grasp of early sciences. The rigor that makes these remarkable characteristics possible quickly disappears, however, once the formulation of a law or theory in mathematical equations, matrices of categories, or exactly defined technical concepts and models has been translated into the ordinary language and general discourse of a given culture. That translation into analogies and metaphors steeped in values must precede all public discussion of science, and most philosophic discussion. It even precedes most reflection by scientists on fields outside their own disciplines.

Beyond the narrow, abstract realm in which exactitude is possible, we are no longer insulated from the values and subjective judgments that shape every activity situated within a society. There are, for instance, profound differences between the character of modern scientific activity in the contemporary People's Republic of China and United States which reflect different predominant convictions about the relations between basic and applied science, the relation of both to general culture, the roles of scientists in defining research programs, procedures for planning and supporting individuals' research projects, expectations about the social aims to which scientific work will contribute, the organization and status of professional scientists, the connections of political ideas and scientific knowledge, and the division of national resources between science and other priorities, and between various scientific activities. That certain equations and models are invariant between the two societies is a factor in all these consensuses, but so is the ubiquity of opposable thumbs. Despite the invariance, a given constellation of values will determine that certain laws and hypotheses can be developed further, and that others will be abandoned unless they are among the very few that individuals can explore at their private discretion and their private expense. That great disparity in Chinese and American definitions of psychology is only one particularly obvious example that affects the life and death of particular theories in one society or the other.

So long as there is variation of such magnitude in the balance between the cognitive, practical, normative, and social dimensions of science, such words as

[6] This point has been most persuasively argued in Langdon Winner, *Autonomous Technology: Technics-out-of-control as a Theme in Political Thought* (Cambridge, MA, 1977).

"international" and "universal" are out of place. When applied to the narrow, rigorous technical realm of scientific cognition alone, they constitute a modest claim indeed–so modest that the two adjectives can hardly be understood in their customary senses.

Nor can one accept uncritically the idea that modern science is in every essential respect European in its social and historical origins. To those familiar with the science of other cultures, any account of the early history of science is lopsided, and misleading on the most fundamental issues, if it restricts itself substantially to discoveries made and understandings worked out at the Western end of Eurasia; if it loses sight of the constant movement of ideas back and forth between civilizations from the New Stone Age to the present; if it does not adequately consider what Europeans had learned by the seventeenth century about Islamic, Indian, and Chinese science; or if it ignores the impact of exotic technologies and materials on the experiences of Europeans.

Fallacies of Historical Reasoning

Growing awareness of the high level of science and technology in ancient China has led to cascades and avalanches of hypotheses from one scholar or another about factors that inhibited the evolution of modern science in China, or characteristics unique to the West that made possible or furthered a major scientific revolution.[7] These often incorporate elementary fallacies of historical reasoning that deserve notice.

[7] The most significant early contributions to this literature, in order of their appearance, are Jen Hung-chün 任鴻雋, "說中國無科學之原因," (The reason for China's lack of science), *K'o-hsueh* 科學, 1915, *1*: 8-13; Yu-lan Fung, "Why China Has No Science–An Interpretation of the History and Consequences of Chinese Philosophy," *The International Journal of Ethics*, 1922, *32*: 237-263; Homer H. Dubs, "The Failure of the Chinese to Produce Philosophical Systems," *T'oung Pao*, 1929, *26*: 96-109; Derk Bodde, "The Attitude toward Science and Scientific Method in Ancient China," *T'ien Hsia Monthly*, 1936, *2*: 139-160; and Rhoads Murphey, "The Nondevelopment of Science in Traditional China," *Papers on China*, 1947, *1*: 1-30 (for others see the bibliographies of *Science and Civilisation in China*, esp. vol. II). Jen claims that science failed to develop in China after the Han period because of inattention to "the inductive method." Fung claims "it is because of the fact that the Chinese ideal prefers enjoyment to power that China has no need of science . . ." (p. 261). Dubs refutes the silly prejudice that the character of the Chinese language made systematic thought impossible, but argues that "the result of the absence of mathematical systems was that the Chinese philosophers attacked the world piecemeal . . . by empirical rather than by rational methods" (p. 108). He has nothing whatever to say about Chinese scientists. Bodde considers a number of aspects of attitudes toward science, and is aware of a few isolated scientific accomplishments despite his disregard for the technical literature, but suggests that the most important "retarding effect upon scientific innovation . . . has been the ideographic nature of the Chinese written language" (p. 158). Murphey, dependent upon Western-language sources and influenced by the stereotypes of F. S. C. Northrup, concludes "a naturalistic philosophy which might be called a reliance on the aesthetic continuum . . . clearly had no place for the inductive hypotheses necessary for science" (p. 15). Writings of this sort are full of acute observations, particularly about philosophic attitudes expressed in the early classics, but demonstrate a failure on the parts of their authors to acquaint themselves with the literature of the Chinese scientific traditions. They may well strengthen our conviction that Lao-tzu or Hsun-tzu would have been mediocre biologists or mathematicians, but they do not help us account for the theoretical analyses, mathematical proofs, and programs of empirical discovery so profusely documented in the writings of those actually engaged in studies of nature.

For roughly two-thirds of a century, it has been argued that although Ch'ing dynasty thinkers took the world as observable, nominalistic fact, just as Sir Francis Bacon (1561-1626) did, unlike him they did not develop a scientific methodology. Despite the positivist assumptions of such arguments, whether Bacon's scientific method has survived in the practice of contemporary science was not even considered.[8] It was, in fact, largely Scholastic in its origins, concerned with taxonomies rather than theories of natural phenomena, and resolutely unconcerned with mathematical measurement. Of the major early modern attempts to define how physical science might fruitfully proceed it was probably the most sterile, in contrast to Bacon's very influential convictions about the organization and ideology of scientific activity.

This pattern of thought that faults the Chinese for not developing a scientific method that later proved abortive in the West crops up in many other forms. Another example is a well-known sociological study of Han astronomy when it attempts to explain the failure to develop a "unified scientific system." One reason given is that Chinese astronomers "were not interested in applied technical sciences, e.g., in developing theoretical tools which could be used to control the flight of a cannon shell or to direct ships safely across the sea."[9] So much for the first civilization to note the declination of the compass needle. So much for the astronomy of an era more than a millennium before the invention of the cannon. The same lack of interest is prominent in the impetus theoreticians from John Philoponus (fl. ca. 530) to Jean Buridan (ca. 1295 - ca. 1358) and others of the School of Paris whose investigations furnished much of the basis for Galilean mechanics. How then did what is presented as a disastrous shortcoming in China fail to prevent in Italy—in fact, according to the conventional wisdom, help directly to bring about—the mathematical study of bodies in motion?

Because of his knowledge of the Chinese sciences and the breadth of his hypotheses, Needham's is the earliest discussion of the Scientific Revolution problem that still commands attention, and is still the best. The most useful critiques of Needham's writings on this subject are, from Sinologists, Bodde, "Evidence for 'Laws of Nature' in Chinese Thought," *Harvard Journal of Asiatic Studies*, 1957 (publ. 1959), *20*: 709-727, and "Chinese 'Laws of Nature': A Reconsideration," ibid., 1979, *39*: 139-155, Chan, and A. C. Graham, "China, Europe, and the Origins of Modern Science: Needham's The Grand Titration," pp. 45-69 in Shigeru Nakayama & Sivin (ed.), *Chinese Science. Explorations of an Ancient Tradition* (Cambridge, MA, 1973); from a historian of science, Nakayama, "Joseph Needham, Organic Philosopher," ibid., pp. 23-43; from a philosopher, Robert S. Cohen, "The Problem of 19 (k)," *Journal of Chinese Philosophy*, 1973, *1*: 103-117; and from sociologists, Benjamin Nelson, "*Sciences* and Civilizations, 'East' and 'West.' Joseph Needham and Max Weber," *Boston Studies in the Philosophy of Science*, 1974, *11*: 445-493, and Sal Restivo, "Joseph Needham and the Comparative Sociology of Chinese and Modern Science," *Research in Sociology of Knowledge, Sciences and Art*, 1979, *2*: 25-51.

[8]Jen, loc. cit.; Joseph R. Levenson, *Confucian China and its Modern Fate. The Problem of Intellectual Continuity* (London, 1958), pp. 3-14; and David E. Mungello, "On the Significance of the Question 'Did China Have Science,'" *Philosophy East and West*, 1972, *22*: 467-478, and my comments on his article in the same journal, 1973, *23*: 413-416. Fung also refers to Bacon in connection with the Scientific Revolution problem, but there the issue is not method but, more pertinently, the relation of science and power (Fung does not specify what kind of power).

[9]Wolfram Eberhard, "The Political Function of Astronomy and Astronomers in Han China," pp. 33-70, 345-352 in *Chinese Thought and Institutions* (ed. John K. Fairbank; Chicago, 1957), p. 66.

Considered generally, this fallacy amounts to claiming that if an important aspect of the European Scientific Revolution cannot be found in another civilization, the whole ensemble of fundamental changes could not have happened there. The flaw of reasoning that underlies it is the arbitrary assumption, never explicit, never discussed, that a given circumstance amounts to a necessary condition. It is almost invariably arbitrary because if we trace the prehistory of the actual Scientific Revolution backward far enough, in most cases we can find a point when the circumstance is absent in Europe. In that case, on what grounds can it be considered a necessary condition? In most cases one need not go back very far. That is why, despite their currency among Sinologists, in the past generation necessary conditions have practically disappeared from the armamentarium of discriminating historians of science.

The mirror image of this fallacy may be seen in an influential estimate of the *Chou i* 周易, the Book of Changes, as a deterrent to science. Here is the way Joseph Needham put it in 1956:

> while the five-element 五行 and two-force 陰陽 theories were favourable rather than inimical to the development of scientific thought in China, the elaborated symbolic system of the Book of Changes was almost from the start a mischievous handicap. It tempted those who were interested in Nature to rest in explanations that were no explanations at all. The Book of Changes was a system for *pigeon-holing* novelty and then doing nothing more about it.

Nearly two decades later Ho Peng Yoke assured us that if Chinese scientists "were fully satisfied with an explanation they could find from the system of the *Book of Changes* they would go no further to look for mathematical formulations and experimental verifications of their scientific studies. Looking at the system of the *Book of Changes* in this light, one may regard it as one of the inhibiting factors in the development of scientific ideas in China."[10]

In these instances one is tempted to counter the arguments with matters of fact. Although Needham's extended discussion treats the Book of Changes predominantly as a static classificatory system of concepts, if we examine its applications to natural philosophy we find that it was most often used to construct dynamic explanations of change. One also looks in vain for a habit among Chinese scientists of constructing mathematical formulations and experimental verifications; if one cannot prove that this tendency was evolving steadily to a certain point, if there is no tangible evidence that without the Book of Changes they would have "gone further," there seems to be no warrant for introducing from modern biology the metaphor of inhibition.[11]

[10] Needham, *Science and Civilisation in China, II*, 336 and 340; Ho, "The System of the Book of Changes and Chinese Science," *Japanese Studies in the History of Science*, 1972, *11*: 23-39. Attempts to explain scientific revolutions by lists of positive and negative factors abstracted from context have been criticized by Robert K. Merton in *Science, Technology and Society in Seventeenth-Century England* (New York, 1970), p. x.

[11] Although Needham has given considerable weight to the notion of inhibition, as one would expect of a first-rate biologist he is cautious about using it in relation to processes that he cannot prove were under way. Writers who draw on his work do not tend to be so discrim-

Exactly what does "inhibiting factor" mean in such contexts? Consider one of these often adduced to explain the failure of China to beat Europe to the Scientific Revolution despite an early head start, namely the predominance of a scholar-bureaucrat class immersed in books, faced toward the past, and oriented toward human institutions rather than toward Nature as the matrix of the well-lived life. But in Europe at the onset of the Scientific Revolution we are faced with the predominance of the Schoolmen and dons, immersed in books, faced toward the past, and oriented toward human institutions rather than toward Nature. They did not prevent the great changes that swept over Europe. It would take a more imaginative historian than myself to say whether those changes would have taken place sooner had Scholasticism never existed.

The confusion about "inhibiting factors" is no less a confusion when it has to do with ideas or techniques. One might just as well call Euclidean geometry an inhibiting factor for the development of non-Euclidean geometry, since so long as people were satisfied with it they didn't move on to a new step. But can one argue that non-Euclidean geometry would have developed sooner without it? It is unfortunate to see the remarkably interesting technical language of the Book of Changes, so powerful in systematically relating broader ranges of human experience than modern science attempts to encompass, written off as an obstacle before anyone has taken the trouble to comprehend it thoroughly.

The first fallacy confuses for a cause or necessary condition what is merely a description of an earlier state of a culture, or of a culture's way of doing something. In its complement, as can be seen by the examples just given, the absence of the subsequent state is confused with an inhibitor. One who commits this second fallacy is metaphorically stopping growth that may not have been taking place. Both of the confusions I have described–blaming the earlier state for delaying the later state, and using the absence of something modern at one point to explain the unattainability of modernity later–confound continuity with stasis. They are bad history because they are bad philosophy.

I recur to the assumptions about ourselves that I have discussed earlier, for they are at the root of both these fallacies. They turn the history of world science into a saga of Europe's success and everyone else's failure, or at best inherently flawed and transitory success, until the advent of redemption through modernization.

---

inating. This point is easily demonstrated by examining the perceptive list of 29 "factors inhibiting the emergence of modern science in China and Western Europe" compiled from his writing in Restivo (see note 7), pp. 46-47. In only four of these 29 does Needham actually invoke the concept of inhibition, and all are tautologous or too vague to challenge (e.g., "it is a matter for reflection how far Chinese algebra was inhibited from developments of post-Renaissance type by its failure to produce a sign which would permit the setting up of equations in modern form," *Science and Civilisation in China*, III, 115). In a half-dozen other places Needham uses wording which suggests inhibition, generally in a similarly vague way (e.g., Confucian rationalism and humanism as "fundamental tendencies which paradoxically helped the germs of science on the one hand and injured them on the other," II, 12). In the remaining score of instances inhibition has been read by the sociologist into statements about failure, lack, and inadequacy.

Joint use of the pair of fallacies makes it easy to prove that the European breakthrough is not simply a fact of history, but was inevitable since history began. Was the horse and buggy a necessary preliminary to the invention of the automobile, or did it delay that invention? Would the automobile have emerged sooner if the buggy had never been invented, so that people would have been dissatisfied with less adequate vehicles? If we find some analogue of the horse and buggy in Europe, by fallacy 1 its absence in China made the invention of some analogue of the automobile impossible. If we find some analogue of the horse and buggy in China, we apply fallacy 2 and make it an inhibiting factor. Thus medieval European impetus theory, abstract and unconcerned with application, was a stage in the evolution in inertial guidance; the presumed unconcern of Chinese thinkers for application proves that inertial guidance could never have originated in East Asia.

This is an infallible formula for reading the strength and power of modern science into the historic past—but only the past of Europe. For the past of other civilizations the test is always anticipation of or approximation to some aspect of early European science, or modern science. Why does the science of early Europe not need to be tested? Because of the *assumption* that its parentage of the Scientific Revolution was unique. Other civilizations shine only as they reflect the light of the European tradition. Or so the prophets of modernization suppose.

I claim, therefore, that the fallacies that so often accompany discussions of the Scientific Revolution problem reflect a set of disastrous assumptions that lie beneath the obvious interest and charm of its surface. They are disastrous because they encourage us to devaluate, without troubling ourselves first to comprehend on their own terms, scientific quests other than the one from which modern science most directly sprang.[12] We now find these assumptions accepted not only in Europe but to some extent in every country in which the history of China is studied.

Why should intellectuals in a non-European country, that owes little of its culture before modern times to Western influence, accept this bias? That is perhaps inevitable, considering that modern education establishes itself (as it did originally in Europe) by teaching the rejection of the traditional past or its demotion to a cultural exhibit that may be of use for nurturing nationalism (and, in the era of cheap package tours, for enticing tourists). Since Japan has had a century's experience with a modern educational system and the self-consciousness it produces, Nishijima Sadao's acute analysis does not come as a surprise:

[12] Although Needham consciously assumes "that there is only one unitary science of Nature, approached more or less closely, built up more or less successfully and continuously, by various groups of mankind from time to time," he sees this as a reason to study, rather than to ignore, non-European traditions. See his discussion cited in note 1 above.

The sorts of scholars who affirm, without troubling themselves to peruse the Chinese scientific literature, that it could not possibly have any value (see notes 7 and 8) have recently provoked a reaction, equally uninformed, that claims European science was markedly inferior to that of China as recently as three hundred years ago. See John Gribbin, "Did Chinese Cosmology Anticipate Relativity?" *Nature*, 1975, *256*: 619-620, and for a critical discussion, Sivin, "Chinese Cosmology," *ibid.*, 1976, *259*: 249.

> The 'static character' hypothesis holds that Chinese society lacked the capacity to form progressively a new era through its own efforts. This hypothesis was afforded particular emphasis by the viewpoint that the modernization of Chinese society was retarded. . . . Originally the 'static character' hypothesis, in company with that of 'Oriental despotism,' was advocated, in contrasts with Western European society, as a notion in polar opposition, for the sake of validating the self-consciousness that came into being with the formation of modern Western European society. That is, it was a postulate to serve as an element in the recognition of the value of modern Western European society. . . . In our country, when we deal with Chinese society from the point of view that makes the formation of the modern ego identical with the equal valuation of individuals in Western European civilization, we are led uncritically to use the 'static character' hypothesis. This has brought about our tendency to be controlled by the inverted logic that makes the goal of understanding Chinese society equivalent to grasping the origins or even the mechanism responsible for the persistence of its 'static character.'[13]

In other words, if one begins with the assumption that the paramount issue in the study of China is accounting for the inevitability of backwardness, one is unlikely to question whether backwardness was inevitable, to ask whether there were not in her history prominent patterns of success from which we might learn, or to reexamine the assumptions about the modernized West that organize European history as a crescendo of success (with setbacks, to be sure, adding to the complexity and thus the charm of the crescendo), and that of other civilizations as a tableau of failure. Thus Nishijima states the intellectual convictions that justified and supported not only esteem for European civilization but also Japan's political aspirations in East Asia before and during the Sino-Japanese War.

One more fallacy often appears in connection with the Scientific Revolution problem, when historians select the aspects of the European experience that are appropriate for comparison with other civilizations. I mean the fallacious assumption that one can make sense of the evolution of science by looking at intellectual factors alone, or socio-economic factors alone, according to preference. Some people think of science predominantly as an intellectual quest after truths hidden in nature. They tend to think of China's failure to beat England to modern science as an intellectual failure. Other people, who think of science as primarily a social or economic phenomenon, tend to see the defeat as a matter of Chinese social or economic backwardness. But neither of these exclusive approaches to explanation is adequate. The distinction between intellectual and social factors or between internal and external factors is not out there in the events we study, but in the mental habits and professional associations, in the division of labor, of historians.

[13] Nishijima Sadao 西嶋定生, *Chūgoku keizaishi kenkyū* 中國經濟史研究 (Studies in Chinese economic history; Tokyo, 1966), pp. 3-4. Nishijima's remarks are part of an effort to explain the slow development of studies in Chinese agriculture.

Dimensions of the Scientific Revolution

The Scientific Revolution and its consequences cut across the boundaries of historical specializations. Let me make this clear by defining its important dimensions.

To begin at the intellectual end, the Scientific Revolution was a transformation of our knowledge of the external world. It changed the questions we asked, the means we used to explore them, and the character of the answers. It established for the first time the dominion of number and measure over every physical phenomenon. It not only "combined, with effective logical precision, a theoretical search for common forms of explanation with a practical demand for accurately reproducible results," but envisioned precisely controlled action as the outcome of understanding.[14]

Ernst Gellner pointed out not long ago a particular way in which the European Scientific Revolution is more than a leap to a new form of knowing. It is natural to assume that in science the crucial test has always been "is it true?" But earlier that was only one of several equally important questions: Is it beautiful? Is it conventional? Is it morally improving? Does it lead to perception of the Good? Does it conform to certain esthetic patterns that all truth must, as astronomers up to Kepler believed that celestial orbits must be compounded of perfect circular motions? In science the test of truth has displaced most of these and redefined the others. This demand for truth above all was an appeal to fact—fact that was in principle public, verifiable, morally neutral, invariant with the social circumstances of the observer, immune from interference by magician or god. But the new science did more than appeal to facts. It created facts of that kind for the first time. That is an awesome creation. It took place in Europe between the time of Copernicus and Laplace and has spread across the world since.[15]

The same leap was not taken in seventeenth-century China. Could it in principle have been taken anywhere but in Europe, or any time but then? It would be unbecoming to hide our practically total ignorance.

The Scientific Revolution also meant a continuing redefinition of the connections of natural philosophy (i.e., science) to other kinds of knowledge. It meant a redefinition of man's orientation toward the past and the future. It meant a redefinition of what authority should determine what uses may be made of knowledge. It meant a redefinition of what knowledge of nature is socially desirable, and what socially undesirable. It meant a redefinition of how knowledge ought to be related to human individuality and to the active relations of man and nature.

Galileo and his friends and successors could not have got round the authority of the Church on the strength of ideas alone. That message was conveyed to Galileo by the Congregation of the Index in 1616, and then with drastic finality

[14] Alistair C. Crombie, "Science and the Arts in the Renaissance: The Search for Truth and Certainty, Old and New," *History of Science*, 1980, *18*: 233-246, esp. p. 235.

[15] Gellner, *Legitimation of Belief* (Cambridge, England: At the University Press, 1974), *passim.* The physicist-philosopher Robert Cohen makes the same point incidentally but a little more broadly when he speaks of "the Galilean turn" as a "rush toward dynamic, functionalized mathematics and abstract quality-stripped epistemology" (see note 7), p. 114.

when he was condemned in 1633. But he and his fellow spirits had begun constructing a new intellectual community outside the old establishment. A hundred years earlier there had been no organized alternative to the Church and its scholastic educational system; then even Galileo himself might have died an archbishop.[16] But in the Counter-Reformation the Church was beginning to be less attractive to the most talented and ambitious (and of course there was less room for those who were attracted). A variety of new careers was emerging. Among them the profession of scientist was being invented. This profession could not provide structures that paid for careers in research for more than an occasional genius until about 1800.[17] Nevertheless its creators assumed from the start, for its amateurs, devotees, and enthusiasts, independent authority to formulate the laws of nature—took it away, in fact, from the Scholastics, for whom science could never be more than a collaborator of faith. Secular learning remade the universities and displaced other ancient institutions as over several centuries of evolution and revolution it formed a technical establishment. As Francis Bacon forecast in his *New Atlantis* at the beginning of the process, the institutions of science ultimately inherited the established church's charisma and its public claims to authority over all valid knowledge. The early modern concession to the Church of authority over matters of cognition, rational decision, emotion, etc., to which I have already referred, was essential but only temporary, as these came no longer to be considered activities of the soul. The soul's province has been narrowed to that of faith, when it appears in scientific textbooks at all.

Freidson and others have shown the formation of a profession depends on a grant of autonomy from society—autonomy not only to set its own standards for admission, competence, and compensation, but also to evolve bodies of knowledge and universes of meaning that only specialists can understand.[18] In Europe natural science could contend for this status because other learned groups already had it. Physics could follow medicine in creating new institutions with the right to regulate their internal affairs. I might remark in passing that no such institutions existed in China. The guilds had no such status. Despite the secure bureaucratic status of imperial physicians and astronomers, by 1600 there was no occupational group sufficiently autonomous or coherent to be called a profession. Scientific knowledge, although highly technical, was widely studied by elite amateurs, and the social consensus forbade corporate bodies governed by norms distinct from those to which all subscribed (I will refer below to later changes in this situation).

All of the mental and social changes in Europe that I have summarized above were one process, and none of its parts seem to me separable. I remember how puzzled a classicist colleague once was at why Archimedes didn't set off a scien-

[16] On the earlier position of the Church as a locus of careers open to talent see Alexander Murray, *Reason and Society in the Middle Ages* (Oxford, 1978), pp. 282-314.

[17] See, for instance, Arnold Thackray, "Natural Knowledge in Cultural Context: The Manchester Model," *The American Historical Review*, 1974, *79*: 672-709, esp. p. 692.

[18] On professional autonomy, Eliot Freidson, *Profession of Medicine. A Study of the Sociology of Applied Knowledge* (New York, 1970), esp. pp. 23-46; on "conceptual machineries of universe-maintenance" and the organizations that support them, Peter L. Berger & Thomas Luckmann, *The Social Construction of Reality. A Treatise in the Sociology of Knowledge* (Harmondsworth, England, 1971, first publ. 1966), pp. 110-146.

tific revolution. My colleague was convinced that Archimedes had the mathematical tools to invent dynamics; why did he stop short, eighteen hundred years before Galileo? Why didn't someone form a committee and fund an enormous research project so that Archimedes' dynamics and his talent for invention could have saved the Hellenistic world from its enemies? Unfortunately, neither new ideas without human organization to carry them out, nor new associations rearranging clichés, can change the world.

This outline of the Scientific Revolution's many dimensions is meant to suggest how much we are likely to miss if we care only about social factors, or only about intellectual factors, as we survey the situation in China. Until recently, for instance, people concerned with that topic, including myself, have overlooked a significant piece of the Chinese picture, which I will now consider.

Scientific Revolution in Seventeenth-Century China

By conventional intellectual criteria, China had its own scientific revolution in the seventeenth century. This is a point of no small interest if we are meditating about why China could not have had one.

Western mathematics and mathematical astronomy were introduced to China beginning around 1630—in a form that before long would be obsolete in those parts of Europe where readers were permitted access to current knowledge (in post-Galilean Italy they were not). Several Chinese scholars, among them Mei Wen-ting 梅文鼎 (1633-1721), Hsueh Feng-tso 薛鳳祚 (ca. 1620-1680), and Wang Hsi-shan 王錫闡 (1628-1682), quickly responded and began reshaping the way astronomy was done in China. They radically and permanently reoriented the sense of how one goes about comprehending the celestial motions. They changed the sense of which concepts, tools, and methods are centrally important, so that geometry and trigonometry largely replaced traditional numerical or algebraic procedures. Such issues as the absolute sense of rotation of a planet and its relative distance from the earth became important for the first time. Chinese astronomers came to believe for the first time that mathematical models can explain the phenomena as well as predict them. These changes amount to a conceptual revolution in astronomy.

That revolution did not generate the same pitch of tension as the one going on in Europe at the same time. It did not burst forth in as fundamental a reorientation of thought about Nature. It did not cast doubt on *all* the traditional ideas of what constitutes an astronomical problem, and what significance astronomical prediction can have for the ultimate understanding of Nature and of man's relation to it.

Most important, it did not extend the domain of number and measure in astronomy until it embraced every terrestrial phenomenon (the Jesuits were obliged to conceal from the Chinese that development in Europe). What happened in China bears comparison with the conservative revolution of Copernicus rather than with the radical mathematization of hypotheses Galileo precipitated. In a sense the Galilean breakthrough had nothing to break through in China, where, since no Aristotle had convinced students of nature that mathematical precision could not apply to quotidian terrestrial events, it was freely so applied. Still we do not find anything corresponding to Galileo's conviction that physical realities must be isolated from the flux of sensation by measurement (as in experiments) so that physical truth can be made to manifest itself in straightforward quantitative relations. Shen Kua, in the eleventh century, once argued that measure is an artifact, and possible only when phenomena are isolated from the continuum of nature in observational instruments; but his concern was the design of armillary spheres, and he saw no repercussions for the study of those aspects of nature that man can manipulate.[19]

The most striking long-range outcome of the mid-seventeenth-century Chinese encounter with European science, in fact, was a revival of traditional astronomy, a rediscovery of forgotten methods, that were studied once again in combination with the new ideas and that supported what might be called a new classicism.[20] Rather than replacing traditional values, the new values implicit in the foreign astronomical writings were used to perpetuate traditional values.

Why didn't this conceptual revolution have the social consequences that historians of Western science have insisted we should expect? By the mid-seventeenth century European civilization had had no appreciable political or social impact in China, and astronomy had to make its way on its own merits. The old and new astronomy ceased to be in antagonistic competition, once the predictive superiority of the European techniques was acknowledged.

One is tempted to see the process by which Western astronomy put down its first roots in China as the last major face-to-face encounter of non-Western and European science in world history. By the eighteenth century modern science was crossing national boundaries on the coattails of Empire, and competition between sciences, literatures, religions, etc., on the basis of their abstract merits had become a thing of the past. Even in the seventeenth century, despite the high drama of eclipse prediction contests in the court, the fact remains that the triumph of European computational techniques came about not through a consensus of great minds but by an imperial decision to hand over operational control of the Astronomical Bureau to Jesuit missionaries.

Revolutions in science as well as in politics take place at the margins of society, but the people who made the one in seventeenth-century China were firmly attached to the dominant values of their culture.[21] At the time there could be no students of astronomy motivated to cast off traditional values, and willing to follow ideas where they led even if the society around them fell apart. The foreign techniques, powerful though they were, offered Chinese students no alternative route to security and fame, and the civil service examination system hardly left leeway for one. The only astronomers who could respond to the Jesuits' writings were members of the old intellectual elite. They were bound to evaluate innovations in the light of established ideals that they felt an individual responsibility to strengthen and pass on to the next generation.

The most influential early champions of Western astronomy, such as Mei and Wang, were men of the lower Yangtze region who lived through the Manchu

[19] See "Shen Kua" (note 1), p. 385.

[20] Yoshio Mikami, "The Ch'ou-Jen Chuan of Yüan Yüan," *Isis*, 1928, *11*: 125.

[21] See "Wang Hsi-shan," pp. 159 and 164, and for further details, Sivin, "Wang Hsi-shan," *Dictionary of Ming Biography*, *s.v.*

invasion and adopted the traditional role of the loyalist who would not serve a new dynasty不貳臣. Having refused to strive for conventional careers in a society that in their view *had* fallen apart, they were motivated to spend their lives studying and teaching the new mathematics and astronomy while they used them to master the neglected techniques of their own tradition. They rejected the Ch'ing present not for a modernist future but to keep alive the lost cause of the Ming for one more generation. Wang Hsi-shan even avoided using the Ch'ing dating system. The ancient model he adopted provided a role for only one generation, the one that was maturing during the transition between two dynasties; he could have no permanently marginal posterity. Despite his superb critical acumen he was the opposite of Descartes, for whom every ancient institution had to justify itself by the new criteria of clear and distinct ideas or be considered a mere vestige.

The principal Chinese proponents of European astronomy in the late seventeenth and early eighteenth centuries–the successors of Wang and Mei–argued that its archaic foundations had originated in China and made their way westward, so that studying it could not be considered a rejection of tradition. They made it one of the foundations of the Evidential Research (*k'ao-cheng*考證) movement. That movement substituted for the intense moral inquiry and concern with statecraft of the Ming-Ch'ing transition a sustained cumulative application of philology and quantitative research to recover the pristine forms of the classics. An important analysis of this movement by Benjamin Elman demonstrates that, especially in the lower Yangtze region, it did evolve career patterns independent of the civil service, patterns in which success and financial security depended on the regard of peers for one's technical publications. But the great philologists, despite their concern for rigor, were not Cartesians either: the goal of their striving was authenticated, definitive editions of their civilization's most ancient documents.[22]

In summary, over most of the past thousand years Chinese society was sufficiently unified and permeable in its upper reaches that few who had the learning and ambition to formulate alternatives to conventional ideas were relegated to the margins. The classical mode of education ensured that to be an intellectual at all was to take up the burdens of elite culture. The contrast with Japan, beginning in the late eighteenth century, is instructive. There astronomy for some time had been done only by functionaries, who could not even perceive the basic ideas that animated Western science. On the other hand, medicine provided a practically unique avenue of social and economic mobility *within* the rather broad margins of a society that as a matter of principle excluded non-hereditary talent from the top. One finds advocates of Western ideas among medical men from the *rōnin* and townsman classes, whose stake in the Chu Hsi synthesis and other aspects of shogunal orthodoxy was minimal. It was the doctors who championed foreign ideas

[22]Elman, "The Unravelling of Neo-Confucianism: The Lower Yangtze Academic Community in Late Imperial China," unpublished Ph. D. dissertation, Oriental Studies, University of Pennsylvania, 1980 (UM8107740). Another signal contribution on the influence of astronomy is John Henderson,"The Ordering of the Heavens and Earth in Early Ch'ing Thought," unpublished Ph. D. dissertation, History, University of California at Berkeley, 1977 (DDK-77-31393).

in other fields of science as well.[23] In the very different social circumstances of China before the twentieth century, physicians were as unimportant in the propagation of European science as astronomers were important.

If then we seek in China those for whom science was not a means to conservative ends, for whom a proven fact outweighed the whole body of millennial values, we do not find them until the late nineteenth century, when they became the first modern scientists in that country. By that time foreigners exempt from Chinese law and backed by gunboats had constructed new institutions and new career lines. These were most attractive to those they had educated, who had no other prospects. We can no longer talk about the encounter of the old and new astronomy. Social and political change had left nothing for the old to do. It became rare as time passed for modern scientists to be aware that their country had had its own scientific tradition. Only in the last generation has that awareness been revived.

Conclusion

My frustrations in trying to make sense of science in China arise partly because of the many levels of human activity that have to be encompassed over such a great sweep of time and human experience. They arise partly because the European Scientific Revolution seems to call for an understanding in greater breadth and depth than its historians have insisted upon. Once we keep in mind the many dimensions of scientific change and their complex relations it becomes less surprising that the Scientific Revolution took place only when and where it did. The process increasingly comes to resemble historic evolution—the sum of human decisions and acts, always to some extent arbitrary and wrongheaded—rather than fate, inexorable determinism, teleology, manifest destiny, an inexorable logic unfolding from ancient Greek rigor, or the hidden operation of some World Spirit.

Looking at these three scientific revolutions—the one we think we know so well in Europe, the one that didn't take place in Archimedes' lifetime, and the one that wasn't what we expect it to be in seventeenth-century China—suggests that we have a great deal to learn about the specific circumstances of each, seen in all its dimensions, before we are ready to tell the world why the first of the three couldn't have happened in other times or places.

I believe that the breakthroughs coming up in the study of Chinese science will be of another kind altogether. They will have to do with understanding in depth and in an integral way the circumstances of people who did science and technology: how their technical ideas related to the rest of their thought; what the scientific communities were—that is, who formed a consensus that certain phenomena were problematic, and that certain kinds of answers were legitimate; how those communities were related to the rest of society; how the responsibility of men of knowledge to their colleagues in science was reconciled with their responsibility to society; what larger ends the sciences served, that kept their

[23] Shigeru Nakayama, *Characteristics of Scientific Development in Japan* (New Delhi, 1977), pp. 20-23, which duplicates to some extent his "Japanese Scientific Thought," *Dictionary of Scientific Biography XV*, 728-758, esp. pp. 742-744; Masayoshi Sugimoto & David L. Swain, *Science and Culture in Traditional Japan. A.D. 600-1854* (Cambridge, MA, 1978), ch. 5.

laws conformable to the laws of Chinese painting and to the basic principles of moral conduct.[24]

These are issues about which we understand very little with respect to China or to Europe. It will take much further study and reflection on both sides before the comparative history of science is ready to take off. My prognostication is that by that time we will no longer be asking why the transition to modern science did not first take place in China.

APPENDIX

CHAPTER HEADINGS IN SHEN KUA, BRUSH TALKS FROM DREAM BROOK

Number	Title	Begins With Jotting No.
1	Ancient Usages	1
3	Philological Criticism	42
5	Music and Mathematical Harmonics	82
7	Regularities Underlying the Phenomena	116
9	Human Affairs	151
11	Civil Service	189
13	Wisdom in Emergencies	224
14	Literature	245
17	Calligraphy and Painting	277
18	Technical Skills	298
20	The Supernormal	338
21	Strange Occurrences	357
22	Errors	388
23	Wit and Satire	401
24	Miscellaneous	420
26	Materia Medica	480

There are a total of 507 jottings in Brush Talks from Dream Brook, or 609 including its two sequels. For a classification of the book's contents according to field of knowledge treated, see Needham, *Science and Civilisation in China,* I, 136.

[24]The language in which I pose these questions is more or less that of the sociology of knowledge. It is interesting that Crombie (see note 14) has phrased a very similar set of topics, also intended to provide an integrated view of the Scientific Revolution problem in terms familiar to intellectual historians and thus to the majority of historians of science: "conceptions of nature and of science, of scientific inquiry and scientific explanation, of the identity of natural science within an intellectual culture, and the intellectual commitments and expectations that affect attitudes to innovation and change" (p. 234).

VIII

OVER THE BORDERS:

TECHNICAL HISTORY, PHILOSOPHY, AND THE SOCIAL SCIENCES

It might be worth while to ponder the fact that our work is part of a larger enterprise, namely the history of science in all places at all times. I would like to dwell on a simple idea, namely that there is a research frontier, a state of the art, in the history of science. I want to remind you that if we are not aware of it, it will not be aware of us.

Let us, then, scout the three borders of the history of science as it is now practiced: first the border with scientific practice, then the border with history and philosophy, and then the border with the social sciences, mainly anthropology and sociology. As we survey each border region, let us see how well it is populated by scholars of Chinese science, technology, and medicine, and think about what the outcome suggests.

The Three Borders

The history of science began to emerge as a coherent, self-conscious discipline about a century ago, as scientists searched for their predecessors. That was not a new quest. In more general terms the historical study of science began with Book I of Aristotle's Metaphysics. There Aristotle reviewed the ideas of the Hellenic students of Nature before him. Setting the pattern for countless generations after him, he argued that his predecessors had been asking the same questions he asked, but not answering them as well as he was about to do. He did not point out that thinking critically about their writings had helped him to redefine the questions, or that what he considered better answers than theirs were largely answers to questions they had had no reason to pose.

The first border, then, is that between history and science (in which for the moment I include medicine and technology). Technical researchers began looking into their heritage long before technical history became a discipline. Scientists still do much of the historical research today. Much of it is a matter of translating ancient sources into modern language (or modern notation). This makes valuable documentation accessible, but ignoring the context makes much of its content look more up to date than it really is.

Such studies tend to be a chronicle of progress and a celebration of triumphs. Failures are usually condemned without being studied. People whose thought was typical of their time and place are often described as though they were modern scientists. Why?

Some scientists have no use for the past, and are interested only in the future. They see their work as replacing, not continuing, that of their predecessors. They are amused that historians should want the books that scientists discard.

But other scientists recognize that what they do builds on millennial quests. They write history, as Aristotle did, to construct, out of the piled-up accomplishments of the dead, a line of predecessors, a genealogy for themselves. They do it in a way that embodies their ideals, and, of course, makes it clear that, like Aristotle, they have found better answers than their forerunners. They seldom admit how greatly the questions have changed. Those explorations of the past are not objective scientific research, but disciplined, value-laden creative acts of the imagination.

The second border, the one with the history and philosophy of science, became a hotbed of activity in Europe and the United States in the 1950's, as degree courses in the history of science began to appear in the universities. In the academic world prosperity had come to imply specialization. The new-model historian of physics was no longer ready to grant that any experienced physicist with a good knowledge of languages could write state-of-the-art history. Newly adapted tools were mandatory, especially philosophic and philological tools that could be used to dissect scientific ideas in an unprecedentedly precise way.

Historians in several countries showed that many key concepts of the Scientific Revolution were not at all modern. They saw no *historic* value in the scientist's quest for his own precursors. They argued persuasively that it made better sense to show how Galileo's dynamics evolved out of Archimedes' statics and medieval impetus theory than to think of seventeenth-century physics as striving but failing to be that of Einstein. In this view, people who claim they have found precursors are simply using oversimplified definitions or ignoring how utterly the questions had changed. Ancient scientific theories were too complicated, too imbued with the world view of their own time, to be precursors of modern ones in any sense that is not trivial. Arguing that Democritus' theory of matter is atomic in any modern sense, or translating *ch'i* as "matter-energy," with its implications for relativity, asserts relationships that do more to muddle than to clarify. A weakness common in this border region was a failure to appreciate the large documentary contributions that the scientists were continuing to make.

Unlike the scientists on the first border, whose obsession was reconstructing what their predecessors had achieved and evaluating it from the modern point of view, the historians wanted above all to explain change. Like Galileo in physics, they transformed the history of science from kinematics—a sequence of individual precursors connected only by the influence of their ideas—into dynamics—a process driven by human experiences and decisions.

This new approach was not based on the scientist's faith in an inexorable march of progress. It accepted instead the philosopher's more abstract faith that the ability of man's rational mind to comprehend Nature was leading science, generation by generation, to a more and more adequate account of the physical world. That was in fact an odd faith for the 1960's. Thinkers at least since Immanuel Kant (1724-1794) had been doubting that external reality is directly knowable, but those who influenced historical studies were not about to dismiss science as an illusion or a mere social artifact. In a famous confrontation, a squad of august philosophers savaged Thomas Kuhn for suggesting in *The Structure of Scientific Revolutions* that knowledge of reality could be qualitatively affected in any significant way by the organization of the human beings that produced it. Joseph Needham, who took social history seriously long before it was fashionable, warned in 1964 about thinking "solely in terms of the internal or autonomous filiation of ideas, theories, mental or mathematical techniques and practical discoveries, handed on like torches from one great man to another."[2]

These historians of science who came and went across the border with philosophy created most of the academic posts and departments that now exist, and moved the center of gravity of the profession into them. This activity had little effect on the historians of medicine, half of whom remained safely within the fold of the medical schools, where the modern physician and his ancestors are not to be treated irreverently. Nor did it impress the historians of technology in the engineering departments, who were becoming irritated at the philosophers' assumption that technologists applied scientific ideas because they had none of their own.

---

2. For a critical discussion of the reaction to Kuhn, with references, see Barry Barnes, *T. S. Kuhn and Social Science* (New York, 1982). Needham is cited from "Science and Society in East and West," in Maurice Goldsmith & Alan Mackay (ed.), *The Science of Science* (London, 1964), reprinted in Needham, *The Grand Titration. Science and Society in East and West* (London, 1969), p. 215.

The third border, shared with anthropology and sociology, did not emerge from the mists until the late 1960's and early 1970's. Its emergence was partly due to historians inspired by the French *Annales* school, who analyzed large bodies of social and statistical data with astonishing results. It was gradually delineated further by structural and symbolic anthropologists, who explained patterns of human motivation and behavior in very new ways.

The new anthropology was so powerful, in fact, that within a decade or so it had thoroughly undermined the intellectual barriers between anthropology and sociology. It used to be that anthropologists studied what they called savages, and sociologists studied "us." "Us" almost always meant Western Europeans or Americans. Neither profession paid much heed to those who were neither savage nor "us," for instance the immense population of China. As anthropology and sociology drew together, the same expanding body of methods, insights, and theories could be applied to everyone. Now we decode the meanings of Oxford academic dress by the same methods that reveal why the Yanomamo of the Amazonian jungle decorate their own skins.[3]

The social sciences, as they influenced history, inspired the main innovative thrust of the history of science from about 1970 to the mid-1980's. What finally succeeded this commerce across the third border was not a new frontier, but a new approach across all three established borders, especially the second and third, at once.

Perhaps the most influential insight that historians took from the social sciences has to do with what is called "the social construction of reality." This idea was applied to science as early as 1935.[4] Anthropologists and sociologists, observing an increasing variety of peoples over many decades, found that their perceptions of Nature differed in every conceivable way. The cosmos might be run by gods or impersonal forces. It might be easy to understand, or open only to mystical insight, or simply illusory. It might be controllable by knowledge, or manageable by ritual, or chaotic and treacherous. Its structure might be holistic, binary or multiplex.

Most people make sense of their own experience, all of their experience at once because they are not specialists, using what resources they inherit from those around them. What we see as their world view or cosmology or

---

3. For the Oxford dress code see Mary Douglas (ed.), *Rules and Meanings* (Penguin Modern Sociology Readings; Harmondsworth, 1973), pp. 209-211.

4. The classic exposition is Peter L. Berger and Thomas Luckmann, *The Social Construction of Reality* (Harmondsworth, 1967). See Ludwik Fleck, *Entstehung und Entwicklung einer wissenschaftlichen Tatsache. Einführung in die Lehre vom Denkstil und Denkkollectiv* (Basel, 1935), tr. Fred Bradley & Thaddeus J. Trenn as *Genesis and Development of a Scientific Fact* (Chicago, 1979). On cultural variations in cosmology see the influential hypotheses in Douglas, *Natural Symbols* (1970; Harmondsworth, 1973).

science is simply a portion of the single reality that people construct as they grow up. As part of that larger construct, cosmology is not adventitious. The notions of order that people observe in their relations with others shape it. They adapt the social order they know to make sense of a welter of phenomena outside society that otherwise would be meaningless.

For most historians thirty years ago the differences between one cosmology and another were nothing more than randomly distributed errors, of no interest to the study of science. If Nature is an objective reality, there is only one rational way to see it, and the rest are fantasies. But the fallacy is obvious to the anthropologist. The distinction between the social, spiritual, and physical worlds may be obvious to the modern scholar, but it does not necessarily subsist in the minds of those he studies. Such a distinction was not at all clear in the minds of Europeans in the Middle Ages. Its unthinking use by modern medieval historians led to a great deal of confusion that we are only beginning to resolve. Even for Newton, Boyle, Hooke, and many of their contemporaries—all of them devout Christians—space and time were the sensorium of God and, as orthodox theology insisted, spirits pervaded and activated the cosmos.[5] Fifty years ago historians reading their works found this content embarrassing and ignored it. Now we see it as central to understanding the Scientific Revolution.

Since anthropology no longer restricts itself to the study of what we no longer call savages, comparable insights must apply even to the history of modern chemistry, mining, and surgery. If they do not work there, how can we argue that they are valid anywhere?

We have learned that they do work there. There is by now a long roster of examples to draw from. The sociologist Joseph Ben-David's comparative study of how scientific disciplines evolved in Europe has led to a small library of historic studies. Michel Foucault's *The Birth of the Clinic* is indispensable for the radical changes in "what doctors looked for and what they sought to cure" in the early nineteenth century. Peter Galison's study of accelerator physics from ca. 1900 on emphasizes the social process that led teams of researchers to resolve endless disagreements about experimental conclusions. All we have to do is observe scientists closely at work

5. The most thorough discussion is in Simon Schaffer, "Godly Men and Mechanical Philosophers: Souls and Spirits in Restoration Natural Philosophy," *Science in Context*, 1987, *1*. 1: 55-85. A significant comparative study of seventeenth-century France and England is Peter Dear, "Miracles, Experiments, and the Ordinary Course of Nature," *Isis*, 1990, *81*. 309: 663-683.

to learn that the final consensus is not dictated by some inevitable logic, but is negotiated a step at a time.[6]

Historians have set aside the idealized pictures of science and scientific method that have been used since the time of Galileo to impress laymen. We now look at what scientists actually do—not only theoreticians but those who describe, those who classify, and those who build equipment. It is no longer necessary to ask whether social norms affect the actual content of their work. It is now old hat to speak of theory as value-laden, recognizing that some aspects are affected only by quantitative data, but others are formed by a profession's conventional wisdom about what makes sense. The intensity and focus of research are greatly affected by what values motivate individuals, what topics are fashionable and, if I may restate the obvious, by what work will be paid for. Some scientists still piously deny that all this is true, but their personal papers become splendid evidence for social influence. Studies of medicine have yielded even richer fruit. We are beginning to harvest analogous studies of engineering.

These, then, are the three borders of the history of science. Most of its practitioners spend their careers in one border region. Many forget that the other borders exist, or think of them as the wrong place for anyone to be. Until a decade or so ago, even in the best universities, research students were encouraged by their mentors to think that technical content and social context are mutually exclusive, and that sensible people choose one or the other and become staunch internalists or externalists. It took until the mid 1980's for a significant number of scholars to remember that both intellectual and social history are ways of looking at the same phenomena, and that both are essential for a rounded understanding. Recently we have had many stunning examples of such rounded understandings, based on close attention to what scientists do as well as to the circumstances in which they do it. They have become more frequent as a result of free passage across all three borders.

Comparing Claims

But that raises an interesting theoretical problem. The realms across the three borders encompass more than research strategies. Each contains a distinct vision of reality and convictions about how one can come to know it. In its own realm, physical science is an increasingly comprehensive, deep,

---

6. Ben-David, *The Scientist's Role in Society: A Comparative Study* (Englewood Cliffs, 1971); Michel Foucault, *Naissance de la clinique* (Paris, 1973), tr. by A. M. Sheridan Smith as *The Birth of the Clinic. An Archeology of Medical Perception* (New York, 1973); Galison, *How Experiments End* (Chicago, 1987). Shigehisa Kuriyama has made pioneering use of Foucault's approach in "Visual Knowledge in Classical Chinese Medicine," an unpublished paper to be discussed at the "Epistemology and the Scholarly Medical Traditions" conference, McGill University, 13-16 May 1992.

and precise description of a world that exists independently of what human beings think about it. In the realm of philosophy and intellectual history, science is how we reason about our experience of certain phenomena. In the realm of sociology and anthropology, science is cosmology, created by applying notions of order learned in society (which is after all where one learns them) to experience of the extra-social world.[7] If these views are incompatible, as they look to be, why should anyone take all three seriously? What is the point of frequenting all three realms if only one of them is realistic about its object of study?

Few first-rate scientists are prepared to consider the possibility that scientific knowledge is in any meaningful sense a social construct, or that any amount of human interaction would in the long run change the laws of Nature. A basic value of science affirms that it differs from other spheres of human activity, which are shaped by value and intention. Historians and philosophers often raise the obvious objection, namely that the notion of a science untouched by social norms and values is patently absurd when applied to bygone concepts, theories and laws. Scientists are quite ready to grant that social norms or situations are useful for explaining *erroneous* knowledge—as found in other cultures or in earlier stages of European science. An inherent characteristic of scientific change, however, is that much current knowledge turns out sooner or later to be wrong. The belief that it becomes eligible for social explanation at that point, and not before, is flagrantly irrational. The scientific hard-liner may insist that his creed is justified for those aspects of current science that will always be correct and up to date. Since it is impossible to know which aspects are timelessly true, this turns out to be a personal faith in something unspecifiable, a species of mysticism.

On the other hand, it is fashionable in other circles to deny that science differs in any significant way from theology, or to affirm that anyone who accepts a given theory of heredity or ballistics is as dependent on institutional and specialist authority as, say, a Catholic who affirms triune Godhead. This skepticism has been useful in canvassing possible ways of understanding science that scientists themselves have found too threatening to explore. But even the most incisive attempts to dismiss the cumulative depth, span, and precision of technical knowledge require an ignorance that severely restricts their influence. It is, in fact, a counterpart of the ignorant faith of others that the thought world of "pure" science is completely unconnected to the society in which scientists live and work.

7. This view of cosmology has received its strongest statement—which sees cosmology as a projection of social notions on the natural world—from Mary Douglas in *Natural Symbols* (note 3 above); for work by others see Douglas (ed.), *Essays in the Sociology of Perception* (London, 1982).

I claim that, despite the tendency of specialists to talk past each other, the epistemologies of all three realms are complementary.

Leaving aside the claims of some spokesmen that modern science is objective and value-free, there is no doubt that it is able to explain and predict phenomena and to allow for verification to an extent that no other cognitive activity even approaches. Its explanations, generalizations, and predictions are publicly demonstrable, to the extent that the situations it describes are publicly accessible. In this sense (although the extent is problematic in some areas of modern physical science) it differs from private fantasies, no matter how detailed, and from the concrete public fantasies of literature and music. How, then, can a technological enterprise so powerful be related to the messiness of social norms and values?

One can put this question in perspective, as an anthropologist does, by looking at science, not alone, but in context. Its explanatory power is precisely a function of its narrow scope. As a philosopher of computation puts it, "The narrowing of vision that characterizes modern scientific investigation, just like the narrowing of the field of view accomplished by a microscope, can only be justified, and sustained, because it permits us to see things we could otherwise not see."[8]

If we reflect back over the evolution of science, we see immediately that this narrowness is a relatively recent characteristic. For the Pythagoreans at the beginning, number (as qualitative as it was quantitative) was the key not only to knowledge of Nature, but to all certain knowledge. For Plato and Aristotle, as for Tung Chung-shu 董仲舒 and the authors of the Inner Canon of the Yellow Lord (*Huang ti nei ching* 黄帝内經), *all* knowledge of physical reality was esthetic and moral. As every undergraduate learns, in Aristotle's notion of causation, means, ends, and intentions were equally important. The geometry and statics of a building were no more focal than the architect's wish to create or the patron's desire for admiration. What could be explained quantitatively was no more important than the human desire to create. To explain something it was necessary to say not only how it came to be but what it might become as it grew in perfection. As a Chinese thinker would put it, to understand something was to know what its Way was.

From early in the seventeenth century to the end of the eighteenth, science bought depth—rigor and power—at the cost of breadth. The cost was paid by drastically retracting the scope of physical explanation. Descartes led the way by dividing all human experience into two parts. One part could be fully described by specifying its size, shape, position, and motion, and could be explained mechanically. This realm of extension was that of

8. Joseph Weizenbaum, *Computer Power and Human Reason. From Judgment to Calculation* (San Francisco, 1976), p. 158.

the body. The other, that of the soul, was left to the Church to explain. Science has more recently claimed the realm of the soul by redubbing it psychology, but the few "objective, value-free" generalizations it has offered are only about the individual's empirical encounter with the outside world. Ecstatic, religious, sensuous, esthetic, imaginative experience still elude it.

But the authority that science eventually won by its conquest of Nature has expanded its arc of the spectrum of experience to fill more and more of the part formerly occupied by those other aspects. That was an explicit aim of the Enlightenment. As Condorcet hopefully put it, "In meditating on the nature of the moral sciences, one cannot indeed help seeing that, based like the physical sciences upon the observation of fact, they must follow the same methods, acquire an equally exact and precise language, attain the same degree of certainty."[9]

Alas, modern ethics as an exact science has been a dismal failure. Despite the tidal wave of rationality that has swept over the world since Condorcet's day, the power of the emotions to shape public life, which turned the French Revolution into a bloodbath and ended his life, have hardly diminished.

Just as in a theocratic society what is no concern of the gods ceases to concern humans, in the modern world what is not objective becomes (if no less compelling) increasingly unreal. That the explanatory power of science seems to be endlessly expanding beyond the physical is in part because what cannot be numbered and measured is being dropped from consciousness (or maintained unofficially in dissent from this ideal of the bottom line, as we see in the spread of religious sects and the perennial vogue of astrology). But despite its expansiveness within its methodological boundaries, science has no power to move them or to blur them without forfeiting its rigor. It can tell us more only *because* it tells us about less.

That is why the anthropologists are bound to complement the physicists. They do not claim that in some parts of the world cosmology is a cultural artifact. This is true of all cultures, or none. Modern science does not admit miracles, denies that the physical universe is operated by a clan of capricious gods, and assumes that it is unaffected by incantation. But that does not release science from the grip of culture.

It simply turns out to be another kind of artifact. Surveying everything that human beings have been known to think, feel, and do, we can hardly take seriously the notion that the realm of quantitative measure and empirical generalization automatically earns a status above all others.

---

9. Marie-Jean-Antoine-Nicolas Caritat, marquis de Condorcet (1743-1794), *Selected Writings*, ed. Keith Michael Baker (The Library of Liberal Arts, vol. 159; Indianapolis, 1976), p. 6.

The generally accepted status of science in certain modern societies was not attained automatically, but evolved through great travail and intellectual turmoil in the seventeenth century. Galileo argued that authority for determining the truth about the empirical world could no longer be entrusted to the scholastic philosophers of the Counter-Reformation, who had nothing new to say, but should be turned over to those who go directly to Nature, experiment and calculate. By 1800 a handful of Europeans could even earn a living as experts on physical reality. By 1900 "natural philosophy" had become natural science, and was a potent institutional force. That force was put to use generating social benefits in return for prestige, institutional autonomy, and a growing role in the ongoing discussions that redefine culture.

The result may look to the naive like a culture that has needed no definition. But the insights of anthropology converge at this point with those of history and philosophy. However ingeniously it hides its craft, an artifact is an artifact.

The Three Borders and the History of Chinese Science

Now let us consider what these new approaches have yielded for the history of East Asian science.

Looking at our field as a whole, most scholars in it were educated as scientists, and had no training in historical analysis and the understanding of large-scale social and cultural phenomena. They are understandably grouped near the first of the three borders, where they document the march of progress and translate ancient data and methods into modern technical terminology. This is true, but not true only, in East Asia, where intellectual history has not been much cultivated, and especially in China, where anthropology and sociology of the kind being done elsewhere have no significant presence.

Some important areas of East Asian research, on the other hand, lead the world in the area of the first border. The study of ancient metal artifacts as practiced by K'o Chün (Tsun Ko), Hua Chueh-ming, and their colleagues, is a central area of archeological metallurgy. Critical study outside China of the enormous ancient medical literature is insignificant in scope compared with the work of such scholars as Ma Chi-hsing 马继兴.

By now we know a great deal about what praiseworthy things Chinese scientists, engineers, and physicians did earlier than anyone else. When Chu K'o-chen 竺可桢, Li Yen 李俨, Ch'en Pang-hsien 陳邦賢, Liu Hsien-chou 劉仙洲, Yabuuti Kiyosi 藪内清, and Joseph Needham began work, few people in China, and fewer elsewhere, had any idea that there had been native scientific traditions too important in world history to ignore. Now schoolchildren, not only in Asia, but all over the world, know it. Even people who learn all they know from television know it. There is no point in striving to prove it now. The game has long since been won.

Quite aside from who did what first, we can learn a great deal from the study of China that every historian of science needs to know. As the worldwide effort continues to understand how human capacities for technical innovation are productively organized, the experience of China, Japan, and Korea have as much to contribute as that of Europe and of that latecomer, the United States. In our preoccupation with stale issues—priorities, the Scientific Revolution problem, and so on—we have been missing this opportunity to be heard. It is time that, as our colleagues in the general history of science have done long since, we move on to productive but neglected research issues.

We still have only a handful of concrete studies of constraints on technical change. We are largely ignorant about what ancient scientists themselves saw as the goals of their work, what they believed was its significance, and how they explained it to each other. We have no idea of how practitioners influenced each other outside of the formal relations of teachers and disciples. We don't know how their relations with colleagues varied from mathematics to medicine to building. We have hardly any idea of how social and political change influenced theory or practice, or how science influenced philosophy, literature, or for that matter the economy. We cannot even guess how technical specialists who were not officials earned a livelihood, or how much medical care cost through the ages.

Most scholars stress that in Chinese medicine theory and practice were inseparable, but others deny it outright. Some historians assume that traditional therapy can cure every disorder, some that it is effective only for certain disorders, some that it depended almost entirely on the patient's ability to recover regardless of (or in spite of) a physician's ministrations. Such opinions tend to be strongly held but based on guesswork, not on systematic study. We are losing the opportunity for deep and meaningful comparisons that the better understanding of these issues in Europe now makes possible.[10]

If we are intent on catching up, and I believe we are, the next moves are obvious. The first step is for us as individuals to become familiar with all three of the frontiers I have outlined. We can read the most innovative journals, such as *Isis, Science in Context, Technology and Culture, The British Journal for the History of Science*, the *Bulletin of the Society for the Social*

---

10. For more on these questions see Sivin, "Science and Medicine in Imperial China—The State of the Field," *Journal of Asian Studies*, 1988, *47*. 1: 41-90. On the matter of theory and practice in medicine see the important forthcoming anthropological study by Judith Farquhar, *Knowing Practice: The Clinical Encounter of Chinese Medicine*.

Studies of Science, and *Social History of Medicine.* We can begin using the tools, already well broken in, that will help us incorporate the state of the art in our own contributions. We can learn from some of the remarkable work being done on science in other East Asian countries, for instance that on the social context of Tokugawa mathematics and on the anthropology of Japanese medicine.[11]

It may also be useful to keep in mind that our field has become firmly international and multidisciplinary. It has been obvious in recent years that Western scholars who do not follow current Chinese and Japanese publications, Chinese scholars who do not keep abreast of recent research in Japan and the Occident, Sinologists who cannot bring themselves to read scientific books, and technical historians who do not keep up with the quickly evolving forefront of Chinese history, are no longer making the contributions that redefine the field.

The problem is not, of course, merely individual. We all face formidable difficulties because most scholars in China have no way to read foreign books, and often find those published in other Chinese cities inaccessible. Scholars elsewhere cannot buy or find in libraries many important Chinese books. It is very difficult even to find out what books have been published before they go out of print. Access to Japanese books and journals is another obstacle course. One of the most elementary strategies for developing innovation in any discipline is to increase opportunities for young scholars. That is why at conferences in most history of science specialties a large proportion of the participants are under thirty-five years old. Conferences on the history of Chinese science have not done well in that respect.

I would call it auspicious that this conference will include the inaugural meeting of an international society. It has emerged as the result of hard work by a large number of scholars who feel that the time has come to find collective remedies for some of the difficulties that have been holding us back. The problems of wider intellectual exchange, of access to publications worldwide, of encouraging innovation, and of recognizing promise and talent: there is no end of work to do. This will, I hope, be remembered as the meeting at which we began coming to grips with it.

---

11. On mathematics see, for instance, Hirayama Akira 平山諦, Shimodaira Kazuo 下平和夫, and Hirose Hideo 広瀬秀雄, *Seki Takakazu zenshu* 関孝和全集 (Osaka, 1974), and Philip C. Brown, "The Mismeasure of Land. Land Surveying in the Tokugawa Period," *Monumenta Nipponica*, 1987, *42*. 2: 115-155, and "Never the Twain Shall Meet: European Land Survey Techniques in Tokugawa Japan," *Chinese Science*, 1989, 9: 53-79. On medicine see the writings of Margaret Lock, e.g., *East Asian Medicine in Urban Japan. Varieties of Medical Experience* (Comparative Studies of Health Systems and Medical Care, 4; Berkeley, 1980) and those of Emiko Ohnuki-Tierney, e.g., *Illness and Culture in Contemporary Japan. An Anthropological View* (Cambridge, England, 1984).

IX

SELECTED, ANNOTATED BIBLIOGRAPHY OF THE HISTORY OF CHINESE SCIENCE

SOURCES IN WESTERN LANGUAGES

This annotated bibliography covers science in traditional and modern China. A similar list for medicine appears in *Medicine, Philosophy and Religion in Ancient China.* All except the last category are concerned with imperial China:

This list emphasizes recent publications, for two reasons. With respect to China before ca. 1800, the bibliographies in Joseph Needham's *Science and Civilisation in China* (see below) are extremely rich, and merely need to be supplemented. Although they are unannotated, the reader can easily use the indexes to find evaluations of sources. There is no correspondingly thorough survey for the last two centuries.

The obvious differences in the subdivisions of this bibliography reflect the varying character and extent of the literature in each category. Historians have conspicuously neglected recent technology and science. Most publications are concerned with policy rather than the work and the people who did it. Much of that literature published more than a decade ago is already obsolete.

Publications are included mostly because of their quality and usefulness, a few in order to warn readers that the promise of their titles is specious.

Reference Works

Beryozkina [=Berezkina], Elvira. 1981. Studies in the History of Ancient Chinese Mathematics. In *Science and Technology: Humanism and Progress. Soviet Studies on the History of Science, II*, 162–178. Problems of the Contemporary World, 96. Moscow: USSR Academy of Sciences. Full coverage of publications in Russian. Not well informed on Western European work, but most of the items in this essay have not been cited outside the Soviet Union either.

Cullen, Christopher. 1983. Science and Medicine in China. In *Information Sources in the History of Science and Medicine,* ed. Pietro Corsi & Paul Weindling, 476–499. Butterworths Guides to Information Sources. London: Butterworth Scientific, 1983. Introductory historic survey, oriented toward bibliography of secondary sources.

Ho Peng Yoke. 1977. *Modern Scholarship on the History of Chinese Astronomy.* Occasional Papers of the Faculty of Asian Studies, 16. Canberra: The Australian National University. A bibliographical essay on the state of the field, with a bibliography of about 400 items. See review in *Chinese Science,* 1982, 5: 42–44.

Loewe, Michael, editor. 1993 (publ. 1994). *Early Chinese Texts. A Bibliographical Guide.* Early China Special Monograph Series, 2. Berkeley: The Society for the Study of Early China and the Institute of East Asian Studies, University of California. Includes essays on early scientific texts.

Mungello, David E. 1979. Some Recent Studies on the Confluence of Chinese and Western Intellectual History. *Journal of the History of Ideas, 40*: 649–662. Discusses a number of contributions published in the 1960's and 1970's bearing on the flow of philosophic and scientific ideas in both directions. Mungello sensibly stresses the value of broadly conceived studies, but in other respects his approach is not very critically acute.

Needham, Joseph, et al. 1954- . *Science and Civilisation in China.* 17 vols. to date. Cambridge University Press. The massive bibliographies in each volume provide the fullest documentation of most topics.

Pregadio, Fabrizio. 1994. Chinese Alchemy: An Annotated Bibliography of Western-Language Studies. IBM-compatible computer text file, 75KB. September 1994. Available from the author, Kyoto. Covers from mid-nineteenth century to 1994.

Revue bibliographique de sinologie, 1955- . 1957- . Paris: Mouton. Valuable analytical annual bibliography of current sinological articles and books in humanities and social sciences published in Chinese, Japanese, and European languages. Wide coverage, but far from exhaustive. Abstracts, 40–250 words in length, in English or French, are signed by experts. Section VII, "Histoire des sciences," includes astronomy, medicine, and technology.

Selin, Helaine. 1992. *Science across Cultures: An Annotated Bibliography of Books on Non-Western Science, Technology, and Medicine.* Garland Reference Library of the Humanities, 1597. London: Garland Publishing. Useful for cross-cultural studies, but not a first choice for China. Entries are neither comprehensive nor carefully chosen. Notes generally give an accurate idea of topics, but do not evaluate. It is impossible to tell whether an item is a classic or incompetent, esp. a serious problem with respect to medicine.

Sivin, Nathan. 1981. Some Important Publications on Early Chinese Astronomy from China and Japan, 1978–1980. *Archaeoastronomy, 4.* 1: 26–31. Annotated, with analytic discussions. Includes collections of essays.

Sivin, Nathan. 1988. Science and Medicine in Imperial China. The State of the Field. *Journal of Asian Studies, 47.* 1: 41–90. Mainly covers publications on science and medicine since 1980, with discussion of trends.

Sivin, Nathan. 1989. Chinese Archeoastronomy: Between Two Worlds. In *World Archeoastronomy. Selected Papers from the 2nd Oxford International Conference on Archeoastronomy Held at Merida, Yucatan, Mexico 13–17 January 1986,* 53–64. Cambridge University Press. Summary of recent work and research issues, with bibliography.

Tsien Tsuen-hsuin; James K. M. Cheng. 1978. *China: An Annotated Bibliography of Bibliographies.* Boston: G. K. Hall. Lists over 2500 bibliographies, mainly in English, Chinese, and Japanese, with coverage of science and medicine. Author, title, and subject indexes.

Vogel, Kurt. 1977. Bericht über neuere, in westlichen Sprachen erscheinene Arbeiten zur Mathematik der Chinesen. In ΠΡΙΣΜΑΤΑ. *Naturwissenschaftsgeschichtliche Studien. Festschrift für Willy Hartner,* ed. Y. Maeyama & W. G. Saltzer, 423–430. Wiesbaden: Franz Steiner Verlag GmbH. Broad coverage of publications on Chinese mathematics in European languages, including Russian.

Xi Zezong. 1981. Chinese Studies in the History of Astronomy, 1949–1979. *Isis, 72:* 456–470.

Xi Zezong. 1983. Chinese Researches in the History of Science and Technology, 1982. *Chinese Science,* 6: 59–83.

Xia Nai. 1980. Bibliography of Recent Archeological Discoveries Bearing on the History of Science and Technology. *Chinese Science,* 4: 19–52. Supplement by Donald Blackmore Wagner, 53–60. Good evaluative coverage of Chinese publications, with some references to related studies published elsewhere.

General

Bodde, Derk. 1953. Harmony and Conflict in Chinese Philosophy. In *Studies in Chinese Thought,* ed. Arthur F. Wright, 19–80. University of Chicago Press. A penetrating attempt at a synthesis of Chinese cosmology and values.

Bodde, Derk. 1957. Evidence for 'Laws of Nature' in Chinese Thought. *Harvard Journal of Asiatic Studies, 20:* 709–727. Issue published 1959. Takes issue with Needham, who doubts that the word "law" is applicable to Chinese natural conceptions. Concludes "at least a few early Chinese thinkers viewed the universe in terms strikingly similar to those underlying the Western concept of 'laws of nature.'"

Chan Wing-tsit [=Ch'en Jung-chieh]. 1957. Neo-Confucianism and Chinese Scientific Thought. *Philosophy East and West, 6:* 309–322. Challenges Needham's overview, but ignores non-philosophical literature.

Chen, Cheng-Yih [=Ch'eng Chen-i], editor. 1987. *Science and Technology in Chinese Civilization.* Singapore: World Scientific. Specialized papers, mostly by prominent Chinese scholars, from a 1985 workshop.

Chinese Science. Irregular, roughly annual. Actual scope is history of science, technology, and medicine in East Asia. Published by the International Society for the History of East Asian Science, Technology, and Medicine. Occasional reviews.

Elvin, Mark. 1973. *The Pattern of the Chinese Past.* Stanford University Press. A stimulating but problematic attempt to interpret Chinese social history in the light of the economics of technology, with some attention to science. The book's use of evidence pertaining to science and technology is evaluated in Sivin, "Imperial China: Has Its Present Past a Future?" *Harvard Journal of Asiatic Studies,* 1978, *38:* 449–480.

Elvin, Mark, editor. 1980. Symposium: The Work of Joseph Needham. *Past and Present,* 87: 17–53. Essay reviews of Needham's writing on philosophy (W. Peterson), mathematics (U. Libbrecht), and astronomy (C. Cullen).

Elzinga, Aant; Andrew Jamison. 1981. *Cultural Components in the Scientific Attitude to Nature: Eastern and Western Modes?* Technology & Culture, Occasional Report Series, 2. Lund: Research Policy Institute. Reviews Western discussions of Chinese natural philosophy and reevaluations of the seventeenth-century European Scientific Revolution, with only four inadequate pages on views of Chinese. Argues for Chinese dialectics vs. Western determinism, with due qualifications.

Graham, A. C. 1973. China, Europe, and the Origins of Modern Science: Needham's *The Grand Titration.* In Nakayama & Sivin 1973: 45–69. A thorough and original critique of assumptions in Needham 1979.

Graham, A. C. 1978. *Later Mohist Logic, Ethics and Science.* Hong Kong: Chinese University Press. A translation and study of the Mohist Canons (ca. 300 B.C.), propositions that juxtapose ethics, logic, geometry, optics, mechanics, and other subjects. A major contribution.

Historia scientarum. Annual. 1962–1979. Tokyo: Nippon Kagakusi Gakkai. Originally published as *Japanese Studies in the History of science.* Frequently includes studies of China.

Huff, Toby E. 1993. *The Rise of Early Modern Science. Islam, China, and the West.* Cambridge University Press. Argues that science arose only in the West because of its legal concept of corporation, which gave rise to "neutral space and free inquiry, concepts integral to modern science."

Institute of the History of Natural Sciences, Chinese Academy of Sciences. 1983. *Ancient China's Technology and Science.* China Knowledge Series. Beijing: Foreign Languages Press. Short essays by experts on a wide variety of topics, including medicine. Originally published in Chinese for teenagers.

Li Guohao et al., editors. 1982. *Explorations in the History of Science and Technology in China. A Special Number of the "Collections of Essays on Chinese Literature and History."* Shanghai: Chinese Classics Publications House. *Festschrift* for the eightieth birthday of Joseph Needham. Pp. 703–720 are a complete list of Needham's writings to 1980.

Merson, John. 1981. *Culture and Science in China.* Sydney: Australian Broadcasting Commission. Based on two series of radio programs. Splices together excerpts from interviews with Australian, British, and American authorities to make a frequently interesting pseudo-symposium on science and technology and their contexts.

Nakayama, Shigeru; Nathan Sivin, editors. 1973. *Chinese Science. Explorations of an Ancient Tradition.* MIT East Asian Science Series, 2. Cambridge, MA: MIT Press. A collection of essays on many aspects of science and medicine.

Nakayama, Shigeru. 1984. *Academic and Scientific Traditions in China, Japan, and the West,* tr. Jerry Dusenberry. University of Tokyo Press. English version of *Rekishi toshite no gakumon* 歴史としての學問 (Academia as history; Tokyo, 1974). Important comparative study of educational institutions in China, Japan, Europe, and the Middle East.

Needham, Joseph. 1954- . *Science and Civilisation in China* (see under Reference Works) "The single work of scholarship which in our time has raised the banner of human unity most bravely and most triumphantly" (Philip Morrison). Based on a great mass of scattered research in European languages and Chinese, and work in the sources by the author and several Chinese collaborators. Recent volumes are by other authors. Audacious, literate, and arranged for maximal accessiblity, with enormous bibliographies.

Needham, Joseph, et al. 1969. *The Grand Titration. Science and Society in East and West.* London: George Allen & Unwin Ltd. Needham's collected arguments for the primacy of social and economic factors in the conditioning of Chinese scientific achievement. Particularly important is "Science and Society in East and West" (190–217) and his survey of noncyclical time conceptions, "Time and Eastern Man" (218–298). One can follow in this collection of papers, published from 1944 on, the development and modification of the author's views. See the review in *Journal of Asian Studies*, 1971, 30: 870–873.

Needham, Joseph. 1970. *Clerks and Craftsmen in China and the West. Lectures and Addresses on the History of Science and Technology.* Cambridge University Press. Another gathering of contributions published from 1946 on, dealing with a spectrum of themes from the most general aspects of Asian scientific work seen in world perspective to articles on the earliest snow crystal observations. Includes a group of essays on medicine.

Needham, Joseph. 1981. *Science in Traditional China. A Comparative Perspective.* Cambridge, MA: Harvard University Press. Four informal lectures, of which only one is based on previously unpublished research.

Orleans, Leo A., editor. 1980 (published 1981). *Science in Contemporary China.* Stanford University Press. Surveys of the state of Chinese research by leading scientists and engineers, based largely on visits to Chinese institutions but also to some extent on Chinese publications. See esp. the essays on historical background by Sivin and science policy by Suttmeier.

Ronan, Colin A. 1978- . *The Shorter Science and Civilisation in China. An Abridgement of Joseph Needham's Original Text.* 3 vols. to date. Cambridge University Press. A readable but mechanical multi-volume condensation, which does not correct substantive errors, even those corrected by Needham in subsequent volumes of his series. Idiosyncratic short bibliographies. Skimming the big book is likely to be more useful.

Sivin, Nathan, editor. 1977. *Science and Technology in East Asia.* History of Science: Selections from *Isis*, 4. New York: Science History Publications.

Articles published in the international history of science journal over 60 years and still useful.

Yabuuti Kiyosi [=Yabuuchi Kiyoshi]. 1979. The Study of the History of Chinese Science in Kyoto. *Acta Asiatica*, 36: 1–6. On the work over the past half-century of the main Japanese research center in Chinese science.

Science and Society

Huang, Ray; Joseph Needham. 1974. The Nature of Chinese Society: A Technical Interpretation. *East and West*, n.s., 24: 381–401. An attempt to specify material and social factors that impeded the development of capitalism and thus, in the authors' opinion, modern science in China. Cf. Needham 1969. Emphasizes failure to fully develop a money economy.

Nelson, Benjamin. 1974. Sciences and Civilizations, 'East' and 'West.' Joseph Needham and Max Weber. *Boston Studies in the Philosophy of Science*, 11: 445–493. A sociologist's overview of the Scientific Revolution Problem.

Qian, Wen-yuan. 1985. *The Great Inertia. Scientific Stagnation in Traditional China.* London: Croom Helm. A shallow "answer" to the Scientific Revolution Problem uninformed by acquaintance with the primary literature.

Restivo, Sal P. 1979. Joseph Needham and the Comparative Sociology of Chinese and Modern Science. *Research in Sociology of Knowledge, Sciences and Art*, 2: 25–51. Excellent sociological assessment.

Science and Philosophy

Bodde, Derk. 1991. *Chinese Thought, Society, and Science. The Intellectual and Social Background of Science and Technology in Pre-modern China.* Honolulu: University of Hawaii Press. A summing up by a substantial contributor to the history of Chinese philosophy. A major theme is that "written Chinese has . . . hindered more than it has helped the development of scientific ways of thinking in China," but the book does not study the scientific and medical literature, and does not directly link early philosophy and later technical activity or society.

Graham, A. C., translator. 1981. *Chuang-tzu. The Seven Inner Chapters and Other Writings.* London: George Allen & Unwin. By far the best translation to date of a classic influential upon attitudes toward Nature. For documentation see Graham, *Chuang-tzu. Textual Notes to a Partial Translation* (London: School of Oriental and African Studies, University of London, 1982).

Graham, A. C. 1986. *Yin-Yang and the Nature of Correlative Thinking.* Occasional Paper and Monograph Series, 6. Singapore: The Institute of East Asian Philosophies. A philosophic analysis, with attention to pre-Han history, of yin-yang and the Five Phases. See also Sivin, *Traditional Medicine in Contemporary China* (Science, Medicine and Technology in East Asia, 2; Ann Arbor: Center for Chinese Studies, University of Michigan, 1988), 43–94.

Lamont, H. G. 1973–1974. An Early Ninth Century Debate on Heaven: Liu Tsung-yuan's *T'ien shuo* and Liu Yü-hsi's *T'ien lun. Asia Major,* 1973, *18:* 181–208; 1974, *19:* 37–85. Lamont's interpretation is debatable, but the translated documents are important, and deserve to be studied in context.

Major, John S. 1993. *Heaven and Earth in Early Han Thought: Chapters Three, Four, and Five of the Huainanzi.* SUNY Series in Chinese Philosophy and Culture. Albany: State University of New York Press. Important for cosmology and cultural geography.

Yang Hsiung, ca. 4 B.C./1993. *The Canon of Supreme Mystery,* trans. with commentary by Michael Nylan. SUNY Series in Chinese Philosophy and Culture. Albany: State University of New York Press. Translation of a fundamental work on Chinese cosmology, based on the Book of Changes, but a systematic exposé in a form close to that of a prose poem.

Yosida Mitukuni [=Yoshida Mitsukuni]. 1979. The Chinese Concept of Technology: A Historical Approach. *Acta Asiatica,* 36: 49–66. Broad and general reflections on several landmarks of the literature.

The Early Encounter With Europe

D'Elia, Pasquale M. 1960. *Galileo in China.* Cambridge, MA: Harvard University Press. A partisan account of the introduction of astronomy in the early 1600's, but provides a good chronological narrative.

Elman, Benjamin A. 1984. *From Philosophy to Philology. Intellectual and Social Aspects of Change in Late Imperial China* (Harvard East Asian Monographs, 110). Cambridge, MA: Harvard University Council on East Asian Studies. A seminal work on a revolution in the content and social context of thought ca. 1600–1750, with a subtle analysis of the limits of European impact.

Gernet, Jacques. 1985. *China and the Christian Impact. A Conflict of Cultures,* tr. Janet Lloyd. New York: Cambridge University Press. Translation of *Chine et christianisme. Action et réaction* (Bibliothèque des histoires; Paris: La Maison des sciences de l'homme, 1982). For a summary see "Christian and Chinese Visions of the World in the Seventeenth Century," *Chinese Science,* 1980, 4: 1–17. Gernet writes with equal authority on science and religion.

Hashimoto, Keizo. 1988. *Hsü Kuang-ch'i and Astronomical Reform. The Process of the Chinese Acceptance of Western Astronomy 1629–1635.* Osaka: Kansai University Press. A Cambridge dissertation, which among other things documents the European sources of the missionaries' early astronomical translations.

Henderson, John. 1984. *The Development and Decline of Chinese Cosmology.* Neo-Confucian Studies, 11. New York: Columbia University Press. An original look at changes in Chinese philosophy in the seventeenth and early eighteenth centures, with much material that suggests Jesuit influence through astronomy.

Martzloff, Jean-Claude. 1980. La compréhension chinoise des méthodes démonstratives euclidiennes au cours du XVIIe siècle et au début du XVIIIe. In *Actes,* IIe Colloque International de Sinologie. Les rapports entre la Chine et l'Europe au temps des lumières. Centre de Recherches Interdisciplinaires de Chantilly (CERIC), 16–18 septembre 1977. La Chine au temps des lumières, 4, pp. 125–143. Paris: Le Centre. On the character of Chinese responses to Euclid.

Spence, Jonathan D. 1969. *To Change China. Western Advisors in China, 1620–1960.* Boston: Little, Brown. Characteristically reflective, but uncharacteristically based on secondary sources, often unreliable ones.

Spence, Jonathan D. 1983. *The Memory Palace of Matteo Ricci.* New York: Viking Penguin Inc. On the transplantation to China of the classical art of memory. Not important but brilliant and curious. For background see Frances A. Yates' equally brilliant and curious *The Art of Memory* (Harmondsworth: Penguin Books, 1969).

Xi Zezong [=Hsi Tse-tsung] et al. 1973. Heliocentric Theory in China. In Commemoration of the Quincentenary of the Birth of Nicolaus Copernicus. *Scientia Sinica,* 16: 364–376.

Mathematics

Chemla, Karine. 1982. Étude du livre "Reflets des mesures du cercle sur la mer" de Li Ye. Dissertation for doctorat de 3e cycle, University of Paris. 4 vols. Pathbreaking study of mathematical classic *Ts'e yuan hai ching* (1248).

Cullen, Christopher. 1982. An Eighth Century Chinese Table of Tangents. *Chinese Science,* 5: 1–33. On a table of gnomon shadow lengths using third-order finite differences, perhaps the oldest extant tangent table in the world.

Hoe, John. 1977. *Les systèmes d'équations polynômes dans le Siyuan yujian (1303)* (Mémoires de l'Institut des Hautes Études Chinoises, Collège de France, 6). Paris: L'Institut. A dissertation concerned with polynomials in the *Ssu yuan yü chien* (Jade mirror of the four unknowns) of Chu Shih-chieh. Good introductory orientation.

Jami, Catherine. 1990. *Les Méthodes rapides pour la trigonométrie et le rapport précis du cercle (1774). Tradition chinoise et apport occidental en mathématiques* (Mémoires de l'Institut des Hautes Études Chinoises, Collège de France, 32). Paris: L'Institut. Sophisticated analysis of a trigonometric treatise by the Mongol astronomer Minggantu. Important for its conclusions about how traditional methods and assumptions were imposed on techniques borrowed from Europe.

Lam Lay Yong. 1977. *A Critical Study of the Yang hui suan fa. A Thirteenth-Century Chinese Mathematical Treatise.* Singapore University Press. Complete translation of an important text of ca. 1270, with explanations.

Lam Lay Yong; **Ang** Tian Se. 1992. *Fleeting Footsteps. Tracing the Conception of Arithmetic and Algebra in Ancient China.* Singapore: World Scientific. Introduction to ancient Chinese computational methods and translation of *Sun-tzu suan ching,* a textbook of ca. A.D. 400.

Li Yan; Du Shiran. 1987. *Chinese Mathematics. A Concise History*, tr. John N. Crossley & Anthony W.-C. Lun. Oxford Science Publications. Oxford University Press. Uncomprehending translation of Du's competent but positivistic textbook written for Chinese cadres in 1963.

Libbrecht, Ulrich. 1973. *Chinese Mathematics in the Thirteenth Century. The Shu-shu chiu-chang of Ch'in Chiu-shao* (MIT East Asian Science Series, 1). Cambridge, MA: MIT Press. Full of examples and informed discussions of differences in approach between Chinese, Islamic, Indian, and European mathematics. Emphasizes indeterminate equations.

Martzloff, Jean-Claude. 1981. *Recherches sur l'oeuvre mathématique de Mei Wen-ding (1633–1721).* Mémoires de l'Institut des Hautes Études Chinoises, Collège de France, 16. Paris: L'Institut. Dissertation on a great seventeenth-century mathematician's work on arithmetic and elementary geometry.

Martzloff, Jean-Claude. 1988. *Histoire des mathématiques chinoises.* Paris: Masson. Highly competent historical survey, with equal attention to the intellectual and social contexts of mathematics and its technical content.

Mikami, Yoshio. 1913. *The Development of Mathematics in China and Japan.* Leipzig: B. G. Teubner; reprint, New York: Chelsea Publishing Company, n.d. Still worth consulting, particularly for details of operations.

Needham, Joseph, et al. 1959. *Science and Civilisation in China.* Vol. III. *Mathematics and the Sciences of the Heavens and the Earth,* 1–168. Cambridge University Press. A good survey of the literature and classic problems, but in other respects the weakest section of the series, superficial and largely reliant on obsolete Western-language sources.

Vogel, Kurt, translator. 1968. *Neun Bücher arithmetischer Technik.* Ostwalds Klassiker der exakten Wissenschaften, n. s., 4. Braunschweig: View E G. A complete translation of the oldest text of the classical arithmetical tradition, the *Chiu chang suan shu* (late first century A.D.?), with scanty introduction and commentary. Largely based on a Russian translation by Elvira Berezkina.

Wagner, Donald Blackmore. 1978. Liu Hui and Tsu Keng-chih on the Volume of a Sphere. *Chinese Science,* 3: 59–79.

Wagner, Donald Blackmore. 1979. An Early Chinese Derivation of the Volume of a Pyramid: Liu Hui, Third Century A.D. *Historia mathematica,* 6: 164–188. On geometrical proofs, the use of which in China has often been denied.

Wylie, Alexander. 1897/1966. *Chinese Researches.* Reprint, Taipei: Ch'eng-wen Publishing Company. Wylie (1815–1887) was a polymathic missionary who wrote on topics in technology, mathematics, and science that have not been freshly studied since.

Astronomy

Clark, D. M.; F. R. Stephenson. 1977. *Historical Supernovas.* London: Pergamon Press. The best comprehensive attempt to identify supernovas in early astronomical records. Uses Chinese and other materials in a sophisticated way.

Eberhard, Wolfram. 1970. *Sternkunde und Weltbild im alten China. Gesammelte Aufsätze.* Occasional Series, 5. Taipei: Chinese Materials & Research Aids Service Center. A useful collection, all except one article published 1932–1940. Often unreliable for astronomy and for interpretation, but many are the only secondary sources on interesting topics, e.g., aspects of early cosmology, writings of Buddhists on astronomy.

Eagle, C. David; **Liu** Bao-lin. 1992. *Canon of Lunar Eclipses. 1500 B.C. to A.D. 3000.* 3 3.5-inch IBM-compatible computer disks. Richmond, VA: Willmann-Bell, Inc. Highly accurate. Gives tabular information and graphics displays. Similar computer files for solar eclipses and planetary phenomena are available only with books from the same publisher.

Ginzel, F. K. 1906–1914. *Handbuch der mathematischen und technischen Chronologie. Das Zeitrechnungswesen der Völker.* 3 vols. Leipzig: Hinrichs. An introduction to the basic problems of calendar and ephemerides construction and their solutions in various cultures; none of the literature on Chinese astronomy provides full introductory information of this kind. For summary data, see *Explanatory Supplement to the Astronomical Almanac* (Mill Valley, CA: University Science Books, 1992).

Ho Peng Yoke [=Ping-yü]. 1966. *The Astronomical Chapters of the Chin shu. With Amendments, Full Translation and Annotations.* Paris: Mouton & Co. Definitive translation of the Astrological Treatise *(T'ien wen chih* 天文志), which provides abundant data on positional astronomy and the astrological interpretation of observational data. Ho supplements this first translation of its kind with notes and introductory remarks on historiography and observatory practice.

Maeyama, Y[asukatsu]. 1975. On the Astronomical Data of Ancient China (ca. -100 – +200): A Numerical Analysis. *Archives internationales d'histoire des sciences,* 25: 247–276; 1976, 26: 27–58. Through a strikingly original error analysis, redates a fundamental source of star data to roughly 70 B.C.

Maeyama, Y. 1977. The Oldest Star Catalogue of China, Shih Shen's *Hsing ching.* In Maeyama & W. G. Salzer, editors. ΠΡΙΣΜΑΤΑ. *Naturwissenschaftgeschichtliche Studien. Festschrift für Willy Hartner,* 211–245. Wiesbaden: Franz Steiner Verlag GmbH. A broader study that confirms Maeyama's earlier work.

Maspero, Henri. 1938–1939. Les instruments astronomiques des Chinois au temps des Han. *Mélanges chinois et Bouddhiques,* 6: 183–370. Rigorous, accurate, and illuminating. Basic for early astronomy, and among Needham's main sources.

Müller, Paul M. 1975. An Analysis of the Ancient Astronomical Observations with the Implications for Geophysics and Cosmology. Newcastle upon Tyne: The author. Dissertation. Considerable use of East Asian observations; many pertinent references in bibliography.

Nakayama, Shigeru. 1966. Characteristics of Chinese Astrology. *Isis,* 57: 442–454. A concise characterization, discussing the late importation of horoscopic astrology of the Hellenistic type.

Nakayama, Shigeru. 1969. *A History of Japanese Astronomy. Chinese Background and Western Impact.* Cambridge, MA: Harvard University Press. Contains one of the best characterizations to date of Chinese approaches to computational astronomy.

Needham, Joseph, et al. 1959. *Science and Civilisation in China.* Vol. III. *Mathematics and the Sciences of the Heavens and the Earth,* 169–641. Cambridge

University Press. Excellent survey, with much attention to observational astronomy and instruments. Uses little important research by Japanese historians, on which see Shigeru Nakayama, "Japanese Studies in the History of Astronomy," *Japanese Studies in the History of Science,* 1962, 1: 14–22. Chinese astronomy was almost entirely calendrical, that is, oriented toward the production of a complete ephemeris. The reader is misled when Needham dismisses the calendrical problem as trivial, and deals with other aspects of astronomy as though they were autonomous. In most other respects accurate; in case of doubt, check sources.

Schafer, Edward H. 1977. *Pacing the Void. T'ang Approaches to the Stars.* Berkeley: University of California Press. Recreates with brio the sky as seen by medieval poets.

Schlegel, Gustave. 1875. *Uranographie chinoise ou preuves directes que l'astronomie primitive est originaire de la Chine, et qu'elle a été empruntée par les anciens peuples occidentaux a la sphère chinoise.* 2 vols. with star maps in folder. La Haye: Martinus Nijhoff. Despite Schlegel's extravagant views on the antiquity of Chinese astronomy, this remains the basic Western-language reference on Chinese constellations and stars.

Stephenson, F. Richard. 1978. *Applications of Early Astronomical Records.* Monographs on Astronomical Subjects, 4. Bristol: Oxford University Press. A technically sophisticated, but not mathematically difficult, introduction to the topic, using many East Asian observational records.

Stephenson, F. Richard; M. A. Houlden. 1986. *Atlas of Historical Eclipse Maps. East Asia 1500 BC – AD 1900.* Cambridge University Press. Useful introduction, and maps and data on solar eclipses only. For more data on solar and lunar eclipses visible in East Asia see Watanabe Toshio 渡邊敏夫, *Nihon, Chôsen, Chûgoku nisshoku gesshoku hôten* 日本，朝鮮，中國日食寶典 (Canon of solar and lunar eclipses: Japan, Korea, China; Tokyo: Yûzankaku, 1979) or, for computer access, Eagle & Liu 1992.

Swarup, Govind, et al., editors. 1987. *History of Oriental Astronomy.* International Astronomical Union, Colloquium 91, New Delhi, India, November 1985. Cambridge University Press. Short papers on most of the Asian traditions, variable in quality.

Teboul, Michel. 1983. *Les premières théories planétaires chinoises* (Mémoires de l'Institut des Hautes Études Chinoises, Collège de France, 21). Paris: L'Institut. Excellent mathematical analysis of Han planetary techniques and their predecessors, a neglected topic.

Teboul, Michel. 1985. Sur quelques particularités de l'uranographie polaire chinoise. *T'oung Pao,* 71: 1–39. Demonstrates the vagueness of Han nomenclature for stars.

Xu Zhentao. 1989. The Basic Forms of Chinese Sunspot Records. *Chinese Science,* 9: 19–28.

Yabuuchi, Kiyoshi 藪内 清. 1963. Astronomical Tables in China, from the Han to the T'ang Dynasties. In *Chûgoku chûsei kagaku gijutsushi no kenkyû* 中國中世科學技術史の研究 (Studies in the history of medieval Chinese science and technology), ed. Yabuuchi, 445–492. Tokyo: Kadokawa Shoten. A systematic survey of Chinese calendrical treatises and the astronomy they reflect, technically sophisticated. Continued in "Astronomical Tables in China from the Wutai to the Ch'ing Dynasties," *Japanese Studies in the History of Science,* 1963, 2: 94–100, a tabular résumé without discussion.

Yabuuti, Kiyosi. 1979. Researches on the *Chiu-chih li.* Indian Astronomy under the T'ang Dynasty. *Acta Asiatica,* 36: 7–48. Revised translation of an important document for the influence of Indian mathematical astronomy and trigonometric functions in eighth-century China.

See also Major 1993 under Science and Philosophy.

Alchemy and Early Chemical Arts

Eliade, Mircea. 1962. *The Forge and the Crucible.* New York: Harper. Original title *Forgerons et alchimistes* (Paris: Flammarion, 1956). A distinguished general study of the primitive roots of alchemical doctrines. The chapters on Chinese and Indian alchemy are suggestive despite the inadequacy of their secondary sources. For a survey of later developments, see "The Forge and the Crucible: A Postscript," *History of Religions,* 1968, 8: 74–88.

Ho Peng-yoke. 1979. *On the Dating of Taoist Alchemical Texts.* Griffith Asian Papers. Nathan, Queensland: Griffith University. A methodological introduction, with examples. Ho has made important strides in applying dating techniques.

Leicester, Henry M.; Herbert S. Klickstein. 1950. Tenney Lombard Davis and the History of Chemistry. *Chymia,* 5: 1–16. Davis, a professor of chemistry at MIT, edited and published in the 1930's several translations of classical alchemical works done by his graduate students, and wrote seminal studies in the field. An extensive bibliography of these writings is on pp. 6–10. The translations are unreliable.

Li Ch'iao-P'ing. 1948. *The Chemical Arts of Old China.* Easton, PA: Journal of Chemical Education. Little space is devoted to alchemy and little is said that is not said better elsewhere. This is, however, an excellent informal compendium of Chinese chemical technology, explaining traditional methods of making everything from alloys to soy sauce and wine.

Needham, Joseph. The Epic of Gunpowder and Firearms, Developing from Alchemy. In Needham 1981: 27–56 (see under General).

Needham, Joseph, et al. 1974–1983. *Science and Civilisation in China.* Vols. V.2–V.5. Spagyrical Discovery and Invention. Cambridge University Press. A historical survey of alchemy (including the physiological and spiritual disciplines of "internal alchemy"), mainly from the point of view of modern chemistry. See also the comparative study of Chinese, Islamic, and Hellenistic alchemy in V.4, 323–509.

Sivin, Nathan. 1968. *Chinese Alchemy: Preliminary Studies.* Harvard Monographs in the History of Science, 1. Cambridge, MA: Harvard University Press. Mostly discussions and heuristic studies aimed at developing a critical and analytical approach to the history of alchemy. Among the appendixes is a list of published translations of alchemical treatises.

Sivin, Nathan. 1976. Chinese Alchemy and the Manipulation of Time. *Isis,* 67: 513–527. Reprinted in Sivin (ed.), *Science & Technology in East Asia* (New York: Science History Publications, 1976), 108–122.

Sivin, Nathan. 1980. The Theoretical Background of Elixir Alchemy. In Needham et al. 1974–1983: V.4, 210–297. A general reconstruction of the theories Chinese alchemists created to explain their work. For a concise summary see the preceding item, written later but published earlier.

Sung Ying-hsing. 1637/1966. *T'ien-kung k'ai-wu. Chinese Technology in the Seventeenth Century,* trans. E-Tu Zen Sun & Shiou-chuan Sun. University Park: The Pennsylvania State University Press. Translation of a technological encyclopedia containing much information about chemical processes.

Siting (geomancy) and Cartography

Bennett, Steven J. 1978. Patterns of the Sky and Earth: A Chinese Science of Applied Cosmology. *Chinese Science,* 3: 1–26. The best available introduction to geomancy, a misnomer instead of which Bennett proposes "siting."

Feuchtwang, Stephan D. R. 1974. *An Anthropological Analysis of Chinese Geomancy.* Vientiane, Laos: Vithagna. Concerned with both the ideas and the social functions of geomancy. Not a tightly integrated study, nor based on extensive research, but contains useful materials on the siting compass.

Harley, J. Brian; David Woodward. 1994. *The History of Cartography.* Vol. II, Book 2. *Cartography in the Traditional East and Southeast Asian Societies.* University of Chicago Press. Innovative and richly illustrated chapters on China, including celestial mapping and cosmological diagrams.

Science in Twentieth-Century China

Anonymous. 1981. *Thirty Years' Review of China's Science and Technology (1949–1979).* Singapore: World Scientific. Translation of a Chinese yearbook.

Brown, Shannon R. 1977. Foreign Technology and Economic Growth. *Problems of Communism, 26*: 30–40. Broad discussion of technology transfer to China from ca. 1935 on.

Buck, Peter. 1980. *American Science and Modern China 1876–1936.* Cambridge University Press. Based on superficial research on China, and concerned mainly with "the social sources and social consequences of scientific development in the United States." Contains some interesting interpretations, e.g., of the role of Progressivism in forming ideas of Chinese reformers.

Conroy, Richard. 1987. The Disintegration and Reconstruction of the Rural Science and Technology System: Evaluation and Implications. In Ashwani Saith (ed.), *The Reemergence of the Chinese Peasantry. Aspects of Rural Decollectivization* (London: Croom Helm), 137–172. On science policy since the Cultural Revolution.

Gould, Sidney H., editor. 1961. *Sciences in Communist China.* Washington: American Association for the Advancement of Science. Articles by experts; useful among other things to document U.S. ignorance of China during the period.

Orleans, Leo A., editor. 1980. *Science in Contemporary China.* Stanford University Press. Essays on science, engineering, and medicine by prominent U.S. scientists who have taken part in exchange delegations to China since the mid-1970's.

Reardon-Anderson, James. 1991. *The Study of Change. Chemistry in China, 1840–1949.* Studies of the East Asian Institute, Columbia University. New York: Cambridge University Press. An excellent first attempt in a Western language to sum up its topic. Not generally technical; emphasizes institutions, with attention to ideas and language.

Schneider, Laurence A. 1982. The Rockefeller Foundation, the China Foundation, and the Development of Modern Science in China. *Social Science and Medicine, 16*: 1217–1221.

Schneider, Laurence A. 1986. *Lysenkoism in China. Proceedings of the 1956 Qingdao Genetics Symposium.* Special issue of *Chinese Law and Government,* 19. 2. Armonk, NY: M. E. Sharpe. Valuable introduction and summaries of the papers that stopped the persecution of genetics under Soviet Lysenkoist influence.

Sigurdson, Jon. 1980. *Technology and Science in the People's Republic of China: An Introduction.* New York: Pergamon Press. Systematic introduction, based mainly on secondary sources. Useful mainly as documentation of the state of science and technology in 1977–1978, the first stage of the transition between the Cultural Revolution and the modernization of the 1980's.

Simon, Denis Fred; Merle Goldman, editors. 1989. *Science and Technology in Post-Mao China.* Harvard Contemporary China Series, 5. Cambridge, MA: Harvard University Council on East Asian Studies. Wide selection of up-to-date essays by political scientists, historians, and scientists. The best introduction to the topic, but see review in *Minerva,* 1992, 30. 3: 432-439.

Suttmeier, Richard P. 1980. *Science, Technology, and China's Drive for Modernization.* Stanford: Hoover Institution Press. Superior policy study.

Volti, Rudi. 1982. *Technology, Politics, and Society in China.* Boulder, CO: Westview Press. Mainly on agriculture, energy, ground transport, and medicine.

Yamada Keiji. 1971. The Development of Science and Technology in China: 1949–1965. *The Developing Economies,* 9: 502–537. Useful Japanese view.

INDEX

Chinese words, except for book titles, are listed under English translations.